Cambridge Aerospace Series

Editors: Wei Shyy and Vigor Yang

1. J. M. Rolfe and K. J. Staples (eds.): *Flight Simulation*
2. P. Berlin: *The Geostationary Applications Satellite*
3. M. J. T. Smith: *Aircraft Noise*
4. N. X. Vinh: *Flight Mechanics of High-Performance Aircraft*
5. W. A. Mair and D. L. Birdsall: *Aircraft Performance*
6. M. J. Abzug and E. E. Larrabee: *Airplane Stability and Control*
7. M. J. Sidi: *Spacecraft Dynamics and Control*
8. J. D. Anderson: *A History of Aerodynamics*
9. A. M. Cruise, J. A. Bowles, C. V. Goodall, and T. J. Patrick: *Principles of Space Instrument Design*
10. G. A. Khoury (ed.): *Airship Technology*, Second Edition
11. J. P. Fielding: *Introduction to Aircraft Design*
12. J. G. Leishman: *Principles of Helicopter Aerodynamics*, Second Edition
13. J. Katz and A. Plotkin: *Low-Speed Aerodynamics*, Second Edition
14. M. J. Abzug and E. E. Larrabee: *Airplane Stability and Control: A History of the Technologies that Made Aviation Possible*, Second Edition
15. D. H. Hodges and G. A. Pierce: *Introduction to Structural Dynamics and Aeroelasticity*, Second Edition
16. W. Fehse: *Automatic Rendezvous and Docking of Spacecraft*
17. R. D. Flack: *Fundamentals of Jet Propulsion with Applications*
18. E. A. Baskharone: *Principles of Turbomachinery in Air-Breathing Engines*
19. D. D. Knight: *Numerical Methods for High-Speed Flows*
20. C. A. Wagner, T. Hüttl, and P. Sagaut (eds.): *Large-Eddy Simulation for Acoustics*
21. D. D. Joseph, T. Funada, and J. Wang: *Potential Flows of Viscous and Viscoelastic Fluids*
22. W. Shyy, Y. Lian, H. Liu, J. Tang, and D. Viieru: *Aerodynamics of Low Reynolds Number Flyers*
23. J. H. Saleh: *Analyses for Durability and System Design Lifetime*
24. B. K. Donaldson: *Analysis of Aircraft Structures*, Second Edition
25. C. Segal: *The Scramjet Engine: Processes and Characteristics*
26. J. F. Doyle: *Guided Explorations of the Mechanics of Solids and Structures*
27. A. K. Kundu: *Aircraft Design*
28. M. I. Friswell, J. E. T. Penny, S. D. Garvey, and A. W. Lees: *Dynamics of Rotating Machines*
29. B. A. Conway (ed): *Spacecraft Trajectory Optimization*
30. R. J. Adrian and J. Westerweel: *Particle Image Velocimetry*
31. G. A. Flandro, H. M. McMahon, and R. L. Roach: *Basic Aerodynamics*
32. H. Babinsky and J. K. Harvey: *Shock Wave–Boundary-Layer Interactions*
33. C. K. W. Tam: *Computational Aeroacoustics: A Wave Number Approach*
34. A. Filippone: *Advanced Aircraft Flight Performance*
35. I. Chopra and J. Sirohi: *Smart Structures Theory*
36. W. Johnson: *Rotorcraft Aeromechanics vol. 3*
37. W. Shyy, H. Aono, C. K. Kang, and H. Liu: *An Introduction to Flapping Wing Aerodynamics*
38. T. C. Lieuwen and V. Yang: *Gas Turbine Emissions*
39. P. Kabamba and A. Girard: *Fundamentals of Aerospace Navigation and Guidance*
40. R. M. Cummings, W. H. Mason, S. A. Morton, and D. R. McDaniel: *Applied Computational Aerodynamic*
41. P. G. Tucker: *Advanced Computational Fluid and Aerodynamics*
42. Iain D. Boyd and Thomas E. Schwartzentruber: *Nonequilibrium Gas Dynamics and Molecular Simulation*

NONEQUILIBRIUM GAS DYNAMICS AND MOLECULAR SIMULATION

Iain D. Boyd
University of Michigan

Thomas E. Schwartzentruber
University of Minnesota

CAMBRIDGE
UNIVERSITY PRESS

CAMBRIDGE
UNIVERSITY PRESS

University Printing House, Cambridge CB2 8BS, United Kingdom

One Liberty Plaza, 20th Floor, New York, NY 10006, USA

477 Williamstown Road, Port Melbourne, VIC 3207, Australia

4843/24, 2nd Floor, Ansari Road, Daryaganj, Delhi - 110002, India

79 Anson Road, #06-04/06, Singapore 079906

Cambridge University Press is part of the University of Cambridge.

It furthers the University's mission by disseminating knowledge in the pursuit of education, learning, and research at the highest international levels of excellence.

www.cambridge.org
Information on this title: www.cambridge.org/9781107073449
10.1017/9781139683494

First published 2017

A catalogue record for this publication is available from the British Library.

Library of Congress Cataloging in Publication Data
Names: Boyd, Iain D., 1964– author. | Schwartzentruber, Thomas E., 1977– author.
Title: Nonequilibrium gas dynamics and molecular simulation / Iain D. Boyd (University of Michigan), Thomas E. Schwartzentruber (University of Minnesota).
Other titles: Cambridge aerospace series.
Description: Cambridge, United Kingdom ; New York, NY : Cambridge University Press, 2017. | Series: Cambridge aerospace series | Includes bibliographical references and index.
Identifiers: LCCN 2016045940| ISBN 9781107073449 (hardback ; alk. paper) | ISBN 1107073448 (hardback ; alk. paper)
Subjects: LCSH: Gas dynamics – Mathematical models. | Molecular dynamics. | Gas flow – Mathematical models. | Nonequilibrium thermodynamics. | Kinetic theory of gases. | Monte Carlo method.
Classification: LCC QC168 .B63 2017 | DDC 533/.2–dc23
LC record available at https://lccn.loc.gov/2016045940

ISBN 978-1-107-07344-9 Hardback

Contents

Illustrations

Tables

Preface

There are two top-level goals that we aim to address in this book: (1) to provide a description of a gas by considering its most basic constituents, i.e., atoms and molecules, and (2) to introduce readers to computer simulation techniques that are available to analyze a gas at this fundamental level.

The first question we must ask ourselves is, Why should we consider a gas at the molecular level? After all, there are well-established equations and ideas that provide accurate descriptions of gas flow at the macroscopic level that employ variables such as density, flow velocity, temperature, and pressure, and these represent properties that take into account the molecules in an aggregate sense. Certainly, the molecular approach will provide us with a deeper understanding of all gas flows. However, more than that, under certain conditions, the aggregate, or sometimes called fluid, approach is not able to provide a physically accurate picture of the gas. We will find that these conditions arise when there is either not enough time or physical space for a sufficient number of intermolecular collisions to occur to maintain the gas in the well-understood equilibrium state. We refer to such conditions as *nonequilibrium*. To describe nonequilibium flows accurately, we need to study the molecular nature of the gas.

There are a number of important application areas in aerospace engineering where nonequilibrium gas flows arise. In general, we will find that nonequilibrium occurs when the gas flow is at low density and/or involves very small length scales. One important application area for nonequilibrium is the flight of high-speed vehicles at very high altitude in the Earth's atmosphere. Examples include spacecraft returning from orbit, such as the space shuttle, or hypersonic cruise vehicles. These vehicles have a length scale of several meters, and move at very high speed so that the flow field surrounding them involves very high temperatures. However, it is their operation in the low-density environment of near-space that leads to nonequilibrium gas flow phenomena. We focus on low-density, high-temperature air in many of the examples and analyses presented throughout the book.

A second important technology area involving nonequilibrium gas flow is micro- and nanoelectromechanical systems (MEMS/NEMS) that involve fabrication and operation of very small machines based on microfabrication technology. When these devices involve gas flow, the velocities are usually very low and certainly subsonic, and the pressure and temperature are close to atmospheric. In this case, it is the very small length scales, around 1 micron = 10^{-6} m, that may give rise to nonequilibrium gas flow behavior.

A third example of nonequilibrium flow is for small rockets used to maneuver spacecraft when they are in orbit. There are a number of different types of these spacecraft thrusters, but they are generally supersonic, involve relatively low-pressure gas or plasma, and have length scales of a few centimeters. This is a case where it is a combination of low pressure and small length scale that gives rise to nonequilibrium phenomena. The same type of physical situation occurs in related technology areas such as vapor deposition and etching machines employed in the materials processing industry.

Another motivation for considering the molecular properties of gas is that this is required to understand a number of modern optical diagnostics techniques that are used to study gas flows whether they are in a state of equilibrium or not. Examples of such diagnostics include emission spectroscopy and laser-induced fluorescence. These techniques rely on the quantum mechanical energy structure of atoms and molecules to derive basic gas flow information such as density, flow velocity, and temperature.

The book is divided into two parts based on the overall goals, with the first part focusing on fundamental considerations, and the second part dedicated to describing computer simulation methods.

The first section covers three different areas: (1) kinetic theory, (2) quantum mechanics, and (3) statistical mechanics. Important results from these three areas are then brought together to allow analysis of *nonequilibrium processes* in a gas based on molecular level considerations.

Chapter 1 covers kinetic theory, in which the basic idea is to develop techniques to relate the properties and behavior of particles, representing atoms and molecules, to the fluid mechanical aspects of a gas at the macroscopic level. This requires us to provide a basic definition by what is meant by a particle, and how these particles interact with one another through the mechanism of intermolecular collisions. This leads us into a discussion of modeling macroscopic molecular transport processes, such as viscosity and thermal conductivity, that represents one of the first key successes of kinetic theory. We will find that kinetic theory relies on the use of statistical analysis techniques, such as probability density functions, owing to the very large volumes of information involved in tracking the behavior of every single particle in a real gas flow. We will formulate the governing equation of kinetic theory, the Boltzmann equation, in terms of the velocity distribution function. We will find that general solution of the Boltzmann equation is challenging because of its mathematical properties. However, simple solutions are readily available for equilibrium conditions, and these can be further employed for analysis of the properties at a surface in free molecular flows, ones in which there are no intermolecular collisions. We also review methods derived from the Boltzmann equation for analysis of nonequilibrium gas flows.

In Chapter 2, we cover the internal energy structure of atoms and molecules. This involves consideration of the basic ideas of quantum mechanics, where once again a statistical modeling approach is required. However, in this case the need for such an approach is dictated by the Heisenberg Uncertainty Principle related to the wavelike properties that particles possess. Through the introduction of a number of fundamental postulates, we derive the governing equation of quantum mechanics, the Schrödinger equation for a number of different cases. Solution of the Schrödinger Equation gives rise to the quantized energy states that specific atoms and molecules are allowed to occupy. Specifically, we find that there are four different energy modes that different particles may acquire: translational, electronic, rotational, and vibrational, with the last two occurring only in molecules. In this chapter, we also study the actual energy structure of atoms and molecules that occur in high-temperature air as we will need this information in later analyses.

Chapter 3 addresses statistical mechanics in which the aim is to relate particle behavior to macroscopic thermodynamics. This connection is established through the Boltzmann relation, which links the random nature of particle behavior to macroscopic entropy. The random nature is quantified by analysis of how particles can be arranged across the quantized energy states available to them. Once again, a statistical approach is required and this time it is due to the very large number of quantized energy states that we determined in Chapter 2. Two different statistical counting methods are presented, and through analysis we derive partition functions that provide a pathway to classical thermodynamics. We also extend our results to the case of a chemically reacting system for use in later analysis of such phenomena.

Chapter 4 concludes the first part of the book by bringing together ideas from kinetic theory, quantum mechanics, and statistical mechanics in order to analyze finite-rate, nonequilibrium processes. The processes of interest include change in the vibrational energy of a gas and chemical reactions. In each case, these phenomena proceed at the molecular level through intermolecular collisions, and in general require a finite amount of time to reach completion, which is referred to as the *equilibrium state*. We first consider the limiting case where this state is reached instantaneously, and consider equilibrium results for both vibrational energy and chemical composition. We then analyze these same processes at the molecular level, using results from Chapters 1 to 3, to formulate approaches that allow finite-rate, nonequilibrium analysis of vibrational and chemical relaxation.

The second part of the book describes computational simulation approaches for the analysis of nonequilibrium gas phenomena that are based on the fundamental ideas presented in Chapters 1 through 4.

In Chapter 5 the mathematical connection between the Boltzmann equation and the most commonly used forms of the continuum Navier–Stokes equations is developed in the limit of near-equilibrium flow. In the process, a quantitative measure for the accuracy and applicability of the Navier–Stokes equations under nonequilibrium flow conditions is established. This theory reveals how interatomic forces (the model for molecular dynamics calculations) are related to collision cross sections (the model for direct simulation Monte Carlo [DSMC]), and how these cross sections determine the transport properties for viscosity, thermal conductivity, and diffusivity (the models used in continuum computational fluid dynamics calculations). A main goal of Chapter 5 is to establish the collision cross section as a physically meaningful parameter that becomes the key model parameter for the DSMC method.

Chapter 6 describes, in detail, the DSMC method. DSMC is a stochastic particle simulation method that simulates the Boltzmann equation. The applicable flow regimes for the DSMC method are first outlined. Well-established algorithms for calculating collision rates and collision outcomes are presented. The manner in which these collision models determine the gas viscosity, thermal conductivity, and diffusivity are described and example simulations are presented. DSMC models and algorithms for internal energy exchange are described and consistency with continuum models is analyzed. For nonequilibrium flows without chemical reactions, the computational models and algorithms detailed in Chapter 6 enable accurate simulations of the Boltzmann equation for flows ranging from continuum to free molecular.

Finally, Chapter 7 presents DSMC models and algorithms for nonequilibrium reacting flows. High-temperature reacting flows involve significant rotational and vibrational energy excitation and coupling to chemical reactions. The DSMC collision models detailed in Chapter 7 are well established in terms of their physical accuracy and computational efficiency. We present example DSMC simulations for high-temperature reacting air flows, and discuss the most current research and prospects for future DSMC models required for nonequilibrium reacting flows.

This book grew out of two different graduate-level courses taught by the authors. Part I is based on a course on nonequilibrium molecular gas dynamics that is a core graduate class in aerospace engineering at the University of Michigan. It provides the fundamental background needed to understand Part II that is based on an advanced graduate class on computer simulation of gas dynamics in aerospace engineering at the University of Minnesota. In addition to serving as a textbook for such graduate classes, the contents of the book will be useful for researchers in nonequilibrium gas dynamics to understand the basic physical phenomena, as well as how to analyze such flows using computer simulation.

Acknowledgments

Many people have contributed to the development of this book, including all the students who have taken our classes, who found errors and suggested improvements. At Michigan, we would specifically like to thank Eunji Jun for generating a first set of electronic notes, Horatiu Dragnea for typesetting notes, and Erin Farbar. At Minnesota, we would like to thank Ioannis Nompelis for writing the parallel DSMC code used for the simulations in Part II of this text, Paolo Valentini for his research contributions to molecular dynamics modeling of nonequilibrium gases, and Chongling Zhang for his research contributions to particle selection procedures and internal energy transfer modeling as well as for implementing many of the thermochemistry models into the Minnesota DSMC code. Lastly, we thank Kate Boyd for graphic design.

The contents of the book have also benefited tremendously from the opportunities afforded the authors by various funding sources that have allowed us to pursue basic research in nonequilibrium gas dynamics for many years. We would like to thank the Air Force Office of Scientific Research (Aerothermodynamics Program, Space Propulsion Program, Molecular Dynamics Program, Young Investigator Program), the Air Force Research Laboratory (Aerospace Systems Directorate, Space Vehicles Directorate), and the National Air and Space Administration (Ames Research Center, Glenn Research Center, Johnson Space Center).

Part I Theory

1 Kinetic Theory

1.1 Introduction

The primary aim of kinetic theory is to relate molecular level behavior to macroscopic gas dynamics. This is achieved by consideration of the behavior of individual particles, and integrating their collective properties up to the macroscopic level. Consider the simple case of a gas at rest as illustrated in Fig. 1.1. At the macroscopic level, this is an uninteresting situation because all the gas properties, such as density (ρ), pressure (p), and temperature (T), are constants. However, at the molecular level, there is a great deal of activity with particles traveling individually at relatively high speed, and undergoing collisions with other particles. When one considers the behavior of particles at the molecular level, they really only undergo two processes: translational motion in space due to their velocity, and intermolecular collisions with other particles in the gas. While kinetic theory analysis has to consider these two physical phenomena, we will see that it is a complex process. For example, the motions of particles will be divided into considcration of bulk, directed motion, and random, thermal motion. Collisions of particles involve a nonlinear process that includes elastic events where only the particle velocities change, and inelastic processes involving energy exchange with internal modes and even chemical reactions.

1.2 Fundamental Concepts

In this section, we first provide an introduction to some basic concepts and definitions that will be needed to achieve our goal of relating molecular behavior to macroscopic gas dynamics. We then employ these concepts later in the chapter to analyze a number of different gas flow situations.

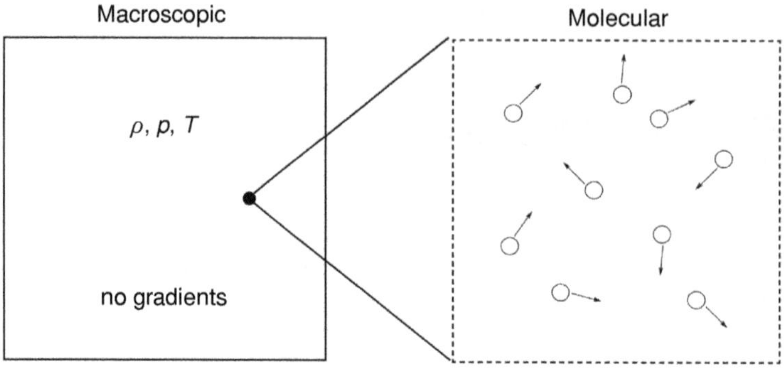

Figure 1.1 Macroscopic and molecular views of a gas at rest.

1.2.1 Particle Model

The *particle* is the fundamental unit in kinetic theory and we will use this term generically to refer to atoms and molecules. Each particle has the following properties:

- Mass (typically around 10^{-26} to 10^{-25} kg)
- Size (typically a few 10^{-10} m)
- Position, velocity, and internal energy

The mass of a particle is simply the sum of the masses of its constituent atoms. Position is the center of mass location of the constituent atoms and velocity is the center-of-mass velocity of those atoms. For molecules, atomic motion relative to the center of mass (i.e., rotation and vibration) contributes to the internal energy of the particle. The sources of internal energy that a particle of a particular chemical species can possess will be treated in detail using quantum mechanics in Chapter 2. In our introductory treatment of kinetic theory, we will ignore the internal energy for now. In addition, to fix ideas, let us focus on a *simple gas*, i.e., one in which all particles are of the same species.

Particle mass is a well-defined quantity, size is not so clear. An atom consists of a nucleus, composed of neutrons and protons, surrounded by orbiting electrons, so how large is it? This is an important question, as particle size determines the nature of intermolecular collisions. In real collisions, particles interact through the field that is formed as a result of the electrostatic Coulomb forces that act between the elementary charges, the protons and electrons, of the interacting bodies. Figure 1.2 shows an example of the potential energy acting between two argon atoms as a function of their distance of separation. The curve illustrates two main points: (1) At large distances of separation, there is a weak attractive force bringing particles closer together; and (2) at small distances of separation, there are strong repulsive forces pushing the particles apart.

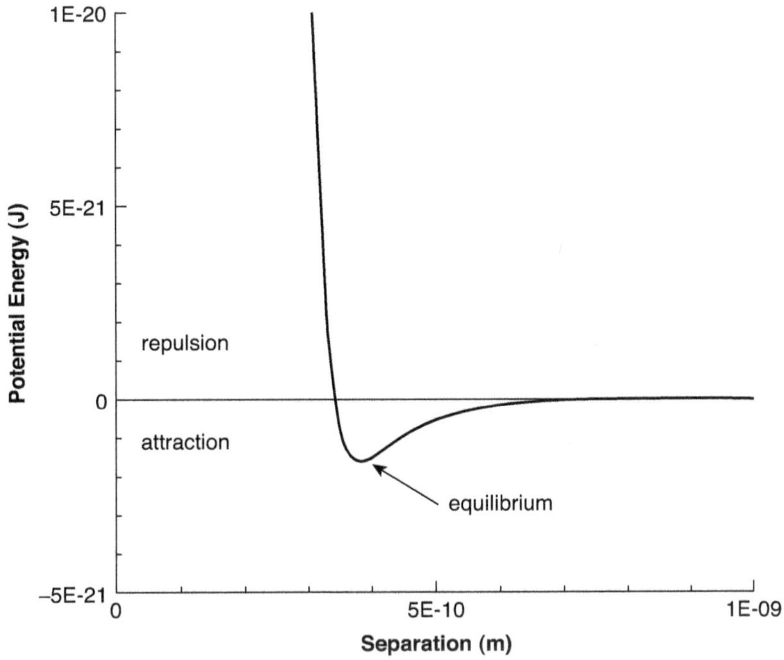

Figure 1.2 Interaction potential for two argon particles.

The weak forces causing *attraction* are important only at very low gas temperature (e.g., less than 100 K) for the relatively simple species that we will focus on, and so this effect can usually be ignored. We therefore concentrate our attention on the repulsive part of the potential field, and, to simplify mathematical analysis, we limit our consideration to two simple models.

(i) Hard sphere (rigid elastic sphere)
This model assumes that each particle has a hard shell, and that a collision occurs only when the surface of one particle is in contact with the surface of another, and so the dynamics resembles that of two billiard balls colliding. Mathematically, this says that the force field between two particles is zero everywhere except at a distance of separation equal to the diameter of one of the spheres. The diameter of the sphere is approximately located at a separation distance in Fig. 1.3, where the potential energy increases rapidly to infinity, at about 3.3×10^{-10} m.

(ii) Inverse power law
It is clear that in the real potential energy field, there is a finite slope as the energy rises, rather than going immediately to infinity, as assumed in the hard sphere model. The next best level of assumption is the inverse power law model that aims to better represent the repulsive part of the real potential using an inverse power law for force:

$$F = \frac{a}{r^{\eta}}$$

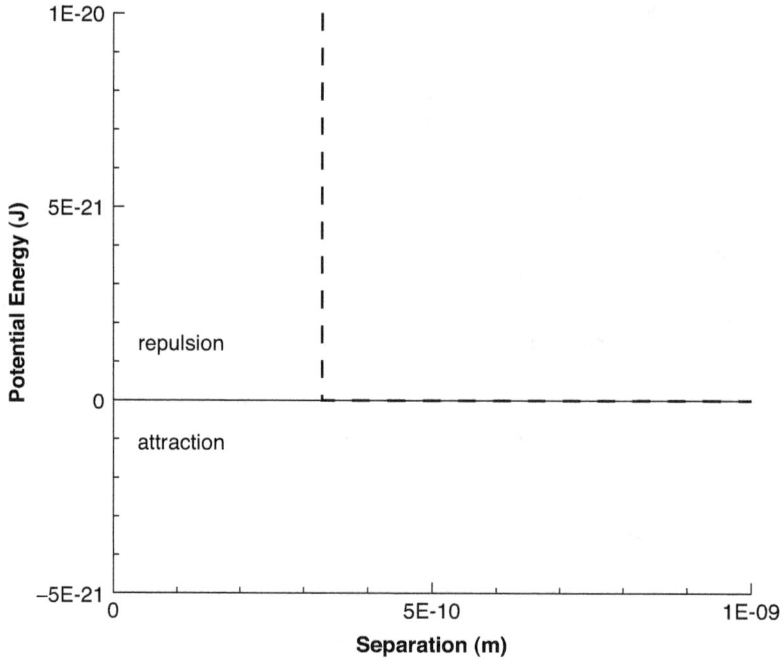

Figure 1.3 Hard-sphere interaction potential for two argon particles.

e.g., for argon, $\eta = 10$, and the associated potential is shown in Fig. 1.4. The parameters a and η can be determined through comparisons with viscosity measurements. In this model, there is no well-defined particle size, the model provides a softer interaction than the hard sphere, and in doing so better reproduces the viscosity temperature dependence of the gas at a macroscopic level. This is formulated rigorously in Chapter 5.

1.2.2 Macroscopic Quantities from Molecular Behavior

To begin to develop relationships between particle behavior and macroscopic gas flow quantities, we start with some simple results based on a collection of particles. As mentioned previously, we assume a small number of basic properties for each particle, i: a mass (m_i), a hard-sphere diameter (d_i), a position $\bar{r}_i = (r_1, r_2, r_3)_i$, and a velocity $\bar{C}_i = (C_1, C_2, C_3)_i$. In the following, we develop simple relations for some of the most fundamental gas flow properties of density, pressure, temperature, and velocity.

(i) *Density*

Consider a small volume containing a total of N particles. The number density is the number of particles per unit volume, and is given by

$$n = \frac{\sum_{i=1}^{N} 1}{\delta V} \tag{1.1}$$

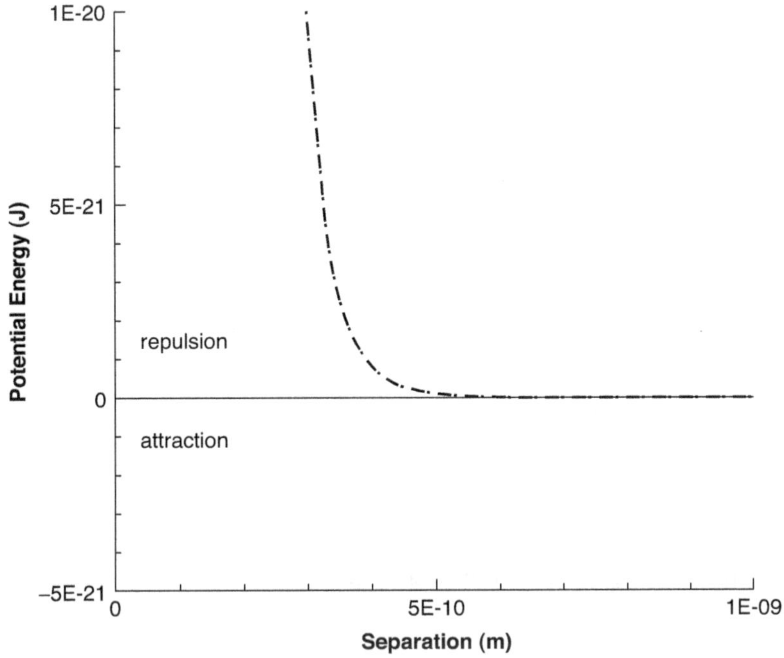

Figure 1.4 Inverse power law interaction potential for two argon particles.

The corresponding *mass density* is given by

$$\rho = \frac{\sum_{i=1}^{N} m_i}{\delta V} \tag{1.2}$$

Note that the results obtained with these expressions are independent of the spatial distribution of the particles within the small volume.

(ii) *Pressure*

Consider a gas in a state of rest, meaning there is no net velocity, inside a cube of volume $V = l^3$. Inside the cubic volume, each particle i will have a unique velocity, $\bar{C}_i$. To derive a result for pressure, we make a number of assumptions:

- The gas is in a state of *thermal equilibrium*: This means that there is no variation of the *number density* and the *velocity distribution function* (VDF) anywhere in the volume. We will discuss the VDF in more depth later; for now it can be considered the probability density function of finding a particle at a particular velocity.
- We will ignore collisions between particles: This is acceptable because we have already assumed equilibrium.
- The interaction of particles with walls is specular, i.e., the sign of the velocity component normal to the wall is reversed and its overall speed is unchanged. Figure 1.5 illustrates the dynamics of a specular wall collision. Also, we implicitly assume that for every wall collision $\bar{C}_i \rightarrow \bar{C}_R$ there is simultaneously somewhere in the system a

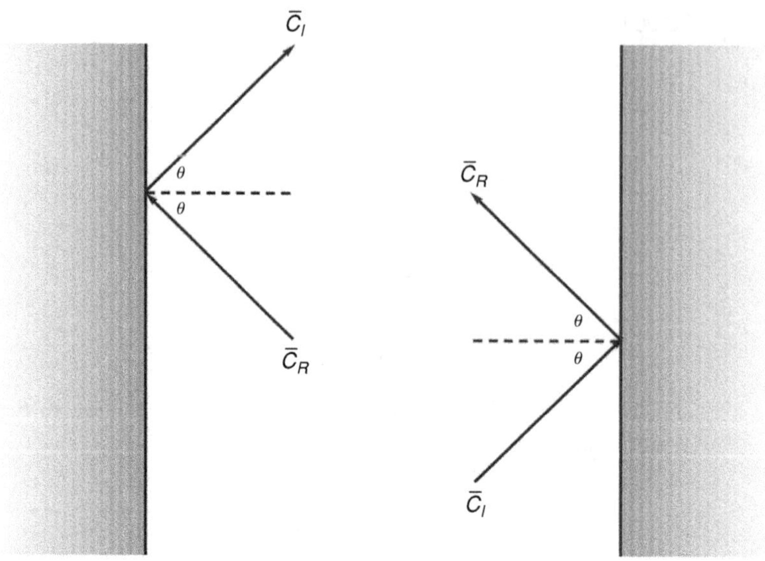

Figure 1.5 A particle undergoing specular reflection from a wall.

corresponding wall collision $\bar{C}_R \to \bar{C}_i$. This assumption is required to maintain the VDF everywhere as constant.

Since there are no intermolecular collisions, and wall collisions only lead to sign changes of velocity components, which means that each particle always has the magnitude of its three velocity components at constant values, then all that changes as a function of time is the sign of those components. Hence, the trajectory of each particle follows that illustrated in Fig. 1.6.

When particle i undergoes a collision with a wall in a r_2–r_3 plane, let us say the plane located at $r_1 = l$, the change of momentum per collision is $2m|C_{1i}|$. Since the particle traverses a distance $2l$ between such collisions, the number of its wall collisions per unit time is $|C_{1i}|/2l$.

The total force exerted by this one particle on this wall

$$= \text{Rate of change of momentum of the particle}$$
$$= (\text{Change of momentum per collision})$$
$$\times (\text{Rate of collisions per unit time})$$
$$= 2m|C_{1i}| \times \frac{|C_{1i}|}{2l} = \frac{m}{l}C_{1i}^2$$

So, the pressure exerted by the particle on this wall

$$= \frac{\text{Force}}{\text{Area}} = \frac{mC_{1i}^2}{ll^2} = \frac{mC_{1i}^2}{V}$$

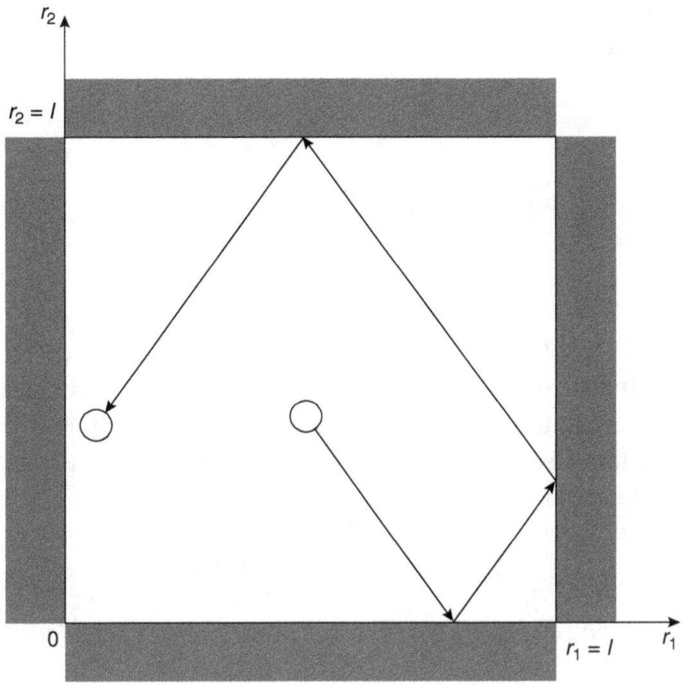

Figure 1.6 Trajectory of a typical particle.

Now, the total pressure exerted on this wall by all the particles in the gas is then

$$p_1 = \frac{1}{V}\sum_{i=1}^{N} m_i C_{1i}^2 \tag{1.3}$$

This is the result for the pressure exerted by the gas in the r_1 direction. Corresponding equations hold for the pressure calculated on the faces in the r_2 and r_3 directions. The average gas pressure is evaluated by taking the sum of these equations for the three coordinate directions and dividing by 3, to obtain

$$p = \frac{1}{3V}\sum_{i=1}^{N} m_i\left(C_{1i}^2 + C_{2i}^2 + C_{3i}^2\right) = \frac{1}{3V}\sum_{i=1}^{N} m_i C_i^2 \tag{1.4}$$

(iii) *Translational Energy and Temperature*
Each particle has *translational energy* due to its kinetic motion:

$$(\epsilon_{\rm tr})_i = \frac{1}{2}m_i C_i^2 = \frac{1}{2}m_i\left(C_1^2 + C_2^2 + C_3^2\right)_i \tag{1.5}$$

and the *total translational energy* of the gas is simply the sum over all particles:

$$E_{\mathrm{tr}} = \sum_{i=1}^{N} (\epsilon_{\mathrm{tr}})_i \tag{1.6}$$

Using Eq. 1.4, we can write

$$E_{\mathrm{tr}} = \frac{3}{2} pV \tag{1.7}$$

Recall the ideal (perfect) gas law: $p = \rho RT = \frac{N'}{V} \hat{R} T$, where $\hat{R}$ is the universal gas constant (8314 J/kg-mol-K), R is the ordinary gas constant, N' is the number of moles in volume V, and T is the translational temperature. Using these results, we may write the total translational energy of the gas as

$$E_{\mathrm{tr}} = \frac{3}{2} N' RT \tag{1.8}$$

A related property is the *average translational energy per particle*:

$$\langle \epsilon_{\mathrm{tr}} \rangle \equiv \frac{E_{\mathrm{tr}}}{N} = \frac{3}{2} \frac{N'}{N} \hat{R} T = \frac{3}{2} \frac{\hat{R}}{\hat{N}} T = \frac{3}{2} kT \tag{1.9}$$

Here, we have introduced a new universal constant:

$$\hat{N} = \frac{N}{N'} \tag{1.10}$$

that is Avogadro's constant, 6.022×10^{26} per kg-mol, which leads to another universal constant widely used in kinetic theory:

$$k \equiv \frac{\hat{R}}{\hat{N}} \tag{1.11}$$

that is Boltzmann's constant, 1.38×10^{-23} J/K. Another related property is the *specific translational energy*

$$e_{\mathrm{tr}} \equiv \frac{E_{\mathrm{tr}}}{\sum_{i=1}^{N} m_i} = \frac{3}{2} \frac{N'}{M} \hat{R} T = \frac{3}{2} RT \tag{1.12}$$

where the total mass is given by

$$M = \sum_{i=1}^{N} m_i \tag{1.13}$$

From these results, we can derive an expression for the *translational temperature* of the gas based on the properties of a collection of particles:

$$T = \frac{2}{3R} \frac{E_{\mathrm{tr}}}{M} = \frac{2}{3R} \frac{\sum \frac{1}{2} m_i C_i^2}{\sum m_i} = \frac{\langle C^2 \rangle}{3R} \tag{1.14}$$

Note: the ordinary gas constant $R = \frac{\hat{R}N'}{M} = \frac{\hat{R}}{M_w}$ where M_w is the molecular weight.

Under our earlier assumption that the particles have no internal structure, the translational energy constitutes the only mode of energy of the gas. In this case, Eq. 1.12 therefore provides the specific energy that is an important variable in thermodynamics, and we will consider further aspects in more detail in Chapter 3. For now, we can evaluate related thermodynamic properties using standard definitions.

The *specific heat at constant volume* for a gas without internal structure:

$$c_v \equiv \left(\frac{\partial e}{\partial T}\right)_V = \frac{de_{tr}}{dT} = \frac{3}{2}R \tag{1.15}$$

The *specific heat at constant pressure* for a thermally perfect gas:

$$c_p = c_v + R = \frac{5}{2}R \tag{1.16}$$

Since our assumed gas model has constant specific heats, it is calorically as well as thermally perfect. The *ratio of specific heats* is defined as

$$\gamma \equiv \frac{c_p}{c_v} = \frac{5}{3} = 1.67 \tag{1.17}$$

This value is confirmed by experimental data measured at room temperature for monatomic gases such as helium, argon, and xenon.

Many common gases, such as N_2, O_2, and NO, are not monatomic, and we will later include additional forms of internal energy to describe fully their associated thermodynamics. However, in terms of their translational motion, kinetic theory does not require us to consider the internal energy modes, and, except for the thermodynamic properties, all of the relations provided in this section apply equally well to atoms and molecules.

(iv) *Velocity and Speed*

The average gas *velocity* vector is simply the mean over all particles:

$$\langle \bar{C} \rangle = \frac{1}{N}\sum_{i=1}^{N} \bar{C}_i = \bar{u} \tag{1.18}$$

and is also called the *flow* or *bulk velocity*. Similarly, the mean square speed is defined by

$$\langle C^2 \rangle = \frac{\sum_{i=1}^{N} m_i C_i^2}{\sum_{i=1}^{N} m_i} = 3RT \tag{1.19}$$

where we have used Eqs. 1.2 and 1.4, and the ideal gas law.

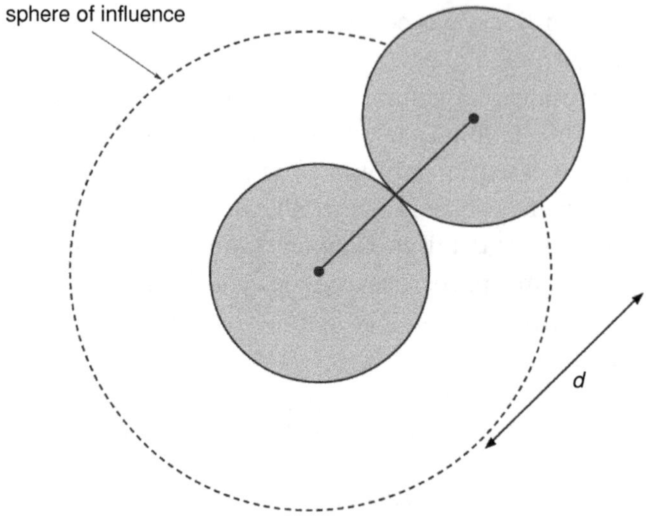

Figure 1.7 Sphere of influence for the collision of two like particles.

Note that these relations are simply averages of particle properties within a small volume element, and make no assumption about how the particle properties are distributed. As will become clear in later chapters, this means that these relations are valid for both equilibrium and nonequilibrium gas states.

1.2.3 Molecular Collisions

In our considerations so far, we have omitted the effects of molecular collisions. Collisions provide the physical mechanism that pushes a gas toward equilibrium. An insufficient number of collisions leads to nonequilibrium. Two important concepts are used to quantify the degree of nonequilibrium in a gas:

(i) *Mean free path*
λ, the average distance traveled by each particle between successive collisions

(ii) *Collision frequency*
Θ, the number of collisions per unit time experienced by each particle, which is related to the mean free time, $\tau = 1/\Theta$, the average time between successive collisions of each particle.

To develop initial results for these important collision quantities, consider a simple gas, of hard-sphere particles with diameter d. For such hard-sphere particles, a collision is defined to occur whenever the distance along the line joining the centers of any two particles is exactly equal to their diameter, d. Put another way, a collision occurs whenever the center of one particle lies on the sphere of influence of another, as illustrated in Fig. 1.7.

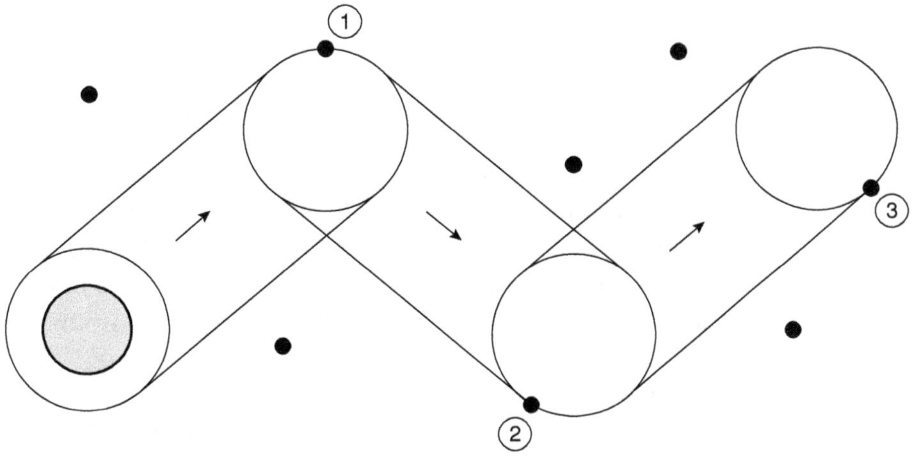

Figure 1.8 Actual path of the particle.

Consider the motion of a single test particle having an average speed that moves in a field of randomly distributed particles that are all at rest, and that have a number density, n, as illustrated in Fig. 1.8. Each time a collision occurs, the test particle has its direction of motion changed, and it then proceeds on a constant direction until its next collision occurs.

To obtain a first result, we simplify the actual path of the test particle to a cylinder (and in doing so, we violate conservation of momentum, but we will correct this later). Along this linear path, shown in Fig 1.9, the volume traced out per unit time by the sphere of influence of the test particle is

$$\pi d^2 \langle C \rangle = \sigma_T \langle C \rangle \tag{1.20}$$

where $\sigma_T \equiv \pi d^2$ is called the *total collision cross section.* The number of particle centers of the particle field encountered within this cylinder per unit time by the test particle is

$$\Theta = n\sigma_T \langle C \rangle \tag{1.21}$$

that, by our collision definition, is also the collision frequency. The distance traveled per unit time by the test particle is $\langle C \rangle$, so the average distance it

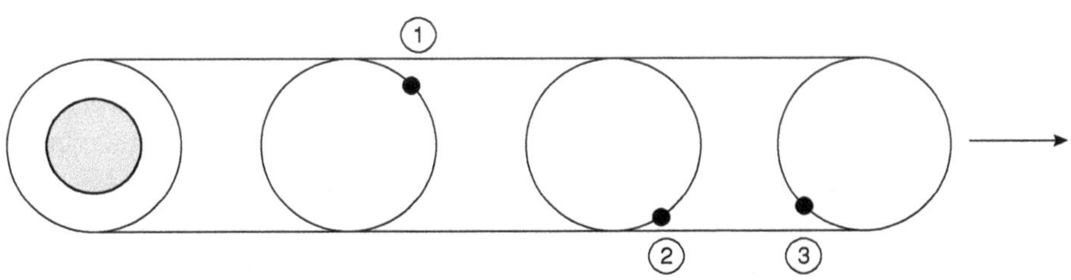

Figure 1.9 Simplified path of the particle.

travels between collisions, its mean free path, is given by

$$\lambda = \frac{\langle C \rangle}{\Theta} = \frac{1}{n\sigma_T} = \frac{m}{\rho\sigma_T} \tag{1.22}$$

We will present a more complete analysis in the next section that includes a number of important aspects missing from this analysis, but in total they only introduce a factor of $\sqrt{(2)}$ in the denominator, so Eq. 1.22 is actually a surprisingly accurate result. At this stage, it is more important to note a number of important conclusions that can be drawn from Eq. 1.22:

(i) λ increases as n decreases, so a rarefied gas has a large mean free path.
(ii) λ increases for small particles, but this is not an important effect for the gases we will consider because there is not a significant variation in d for different chemical species.
(iii) λ does not depend on *temperature*, but this is true only for hard-sphere particles.

An important parameter in characterizing the level of nonequilibrium in a gas flow is the Knudsen number (Kn), which is a dimensionless number defined as the ratio of the mean free path length to a representative physical length scale:

$$\text{Kn} \equiv \frac{\lambda}{L} \tag{1.23}$$

Generally, it is found that nonequilibrium exists for high Knudsen number, e.g., Kn > 0.01. In this situation, there are not enough particle collisions per characteristic length of the flow to establish equilibrium. Therefore, nonequilibrium is found to be important in gas flows at low density and/or small length scale.

1.2.4 Molecular Transport Processes

The phenomena that we have considered so far have involved a gas in *thermodynamic equilibrium*, that is, one in which all macroscopic properties are uniform in space and time. When the gas is out of equilibrium by virtue of a nonuniform spatial distribution of some macroscopic quantity (composition, flow velocity, temperature, etc.), additional phenomena arise as a result of the molecular motion that are called *transport processes*.

When a group of particles move from a region with properties $(\rho, \bar{u}, T)_{\text{A}}$ to a region with properties $(\rho, \bar{u}, T)_{\text{B}}$, they transport some memory of the original conditions with them. This process is illustrated in Fig. 1.10, in which the velocity distributions at two different locations A and B are shown to merge through particle transport in conjunction with intermolecular collisions to produce a new distribution shown as the solid black line.

Table 1.1 Molecular Transport Properties

Property Transported	Result	Macroscopic Cause
Mass	Diffusion	Composition gradient
Momentum	Viscosity	Bulk velocity gradient
Energy	Thermal conductivity	Temperature gradient

The transport of different molecular properties gives rise to different transport processes at the macroscopic level. The names given to each of the main macroscopic transport processes are provided in Table 1.1.

Macroscopic Relations

We start by reviewing a number of physical laws that provide a macroscopic description of the transport phenomena. Subsequently, we will develop a model to describe them at the molecular level based on kinetic theory. We consider the situation illustrated in Fig. 1.11, in which the direction of transport is in the vertical (r_2) direction.

(A) Diffusion

Diffusion concerns the transport of a chemical species A into a gas of species B. The rate of this process is described by *Fick's Law*:

$$\Gamma_A = -D_{AB}\frac{dn_A}{dr_2} \tag{1.24}$$

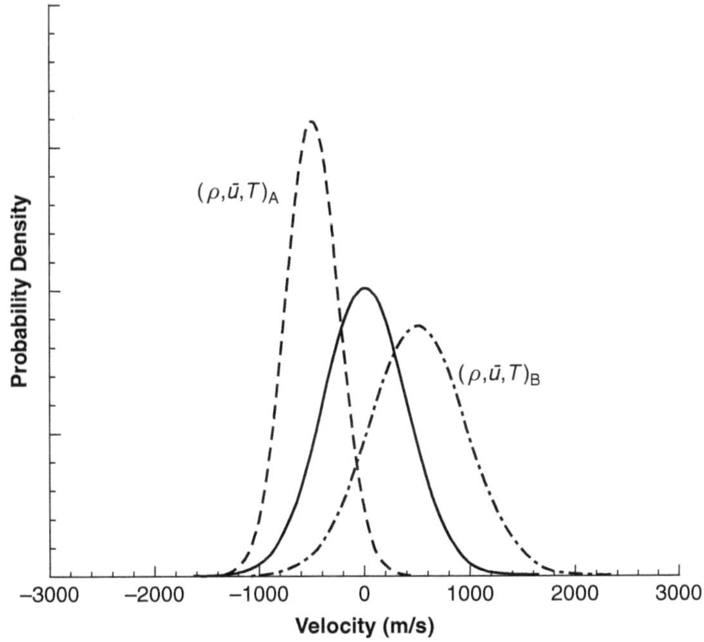

Figure 1.10 Collisional merging of two different VDFs.

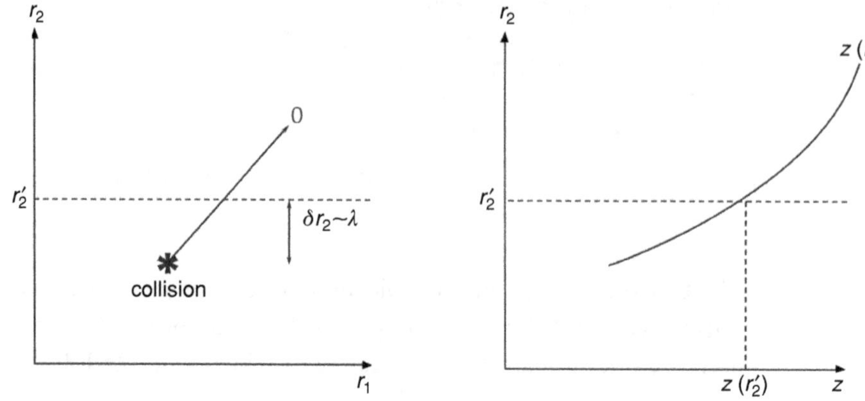

Figure 1.11 Assumed situation for analysis of transport properties.

where Γ_A is the number flux of species A (particles per unit area per unit time), n_A is the number density of species A, and D_{AB} is the diffusion coefficient of species A into species B.

(B) Viscosity

Viscosity involves the transport of momentum of the fluid in the r_1 direction across our surface of constant r_2. A *Newtonian fluid* is said to obey the following relation:

$$\tau = \mu \frac{du_1}{dr_2} \tag{1.25}$$

where τ is the shear stress (flux of momentum per unit area per unit time), u_1 is the macroscopic flow velocity in the r_1 direction, and μ is the viscosity coefficient.

(C) Thermal Conductivity

Thermal conductivity, or heat conduction, involves the transport of energy, and is described by *Fourier's Law*:

$$q = -\kappa \frac{dT}{dr_2} \tag{1.26}$$

where q is the heat flux (energy per unit area per unit time), T is the temperature, and κ is the coefficient of thermal conductivity.

In each case, these relationships are confirmed by laboratory experiments that are needed to determine the transport coefficients, D, μ, and κ. One of the great early successes of kinetic theory was the ability to determine these transport coefficients using a molecular approach. We now outline that analysis.

Consider the general case of a gradient in the r_2 direction of a flow variable, $z(r_2)$ that represents some mean molecular quantity that is measured per particle.

As illustrated in Fig. 1.11, δr_2 is the average distance between the plane $r_2 = r_2'$ and the point where a particle experienced its last collision before crossing that plane, and therefore must be on the order of one mean free path.

The value of z transported by each particle will depend on the last collision and to some extent it will depend on previous collisions. We may write

$$\delta r_2 = \alpha_z \lambda \tag{1.27}$$

where the value of α_z is different for each property z but is always close to 1. In considering the net transport of z across $r_2 = r_2'$ we must include the fluxes in both the positive and negative directions. Thus, the average value of z transported in the positive direction is $z(r_2 - \delta r_2)$, and is $z(r_2 + \delta r_2)$ in the negative direction.

The number flux per unit area and per unit time of particles across r_2' in either direction is proportional to $n \langle C \rangle$, where $\langle C \rangle$ is the average speed of the random molecular motion. Hence, the net flux of z across r_2' in the positive r_2 direction is:

$$\Lambda_z = \eta n \langle C \rangle \left[z(r_2 - \alpha_z \lambda) - z(r_2 + \alpha_z \lambda) \right]_{r_2 = r_2'} \tag{1.28}$$

where η is a constant of proportionality. Now, using a Taylor series expansion to first order, we may write

$$\Lambda_z = -\beta_z n \langle C \rangle \lambda \frac{dz}{dr_2} \tag{1.29}$$

where $\beta_z = 2\eta\alpha_z$ is another constant. Our first-order expansion is accurate only for small λ (and therefore for small Knudsen number).

This relation represents a general model based on molecular considerations for particle transport processes. We now compare Eq. 1.29 to the earlier macroscopic transport expressions, Eqs. 1.24–1.26.

(A) Diffusion

Generally speaking, diffusion concerns a mixture of gases. For simplicity, however, we consider two species with the same properties. To find the flux of species A across $r_2 = r_2'$ we set $z = \frac{n_A}{n}$ and $\Lambda_z = \Gamma_A$ in Eq. 1.29, and so

$$\Gamma_A = -\beta_D n \langle C \rangle \lambda \frac{d(n_A/n)}{dr_2} = -\beta_D \langle C \rangle \lambda \frac{dn_A}{dr_2} \tag{1.30}$$

Comparing to Eq. 1.24, we can deduce that

$$D_{AA} = \beta_D \langle C \rangle \lambda \tag{1.31}$$

with D_{AA} as the coefficient of self-diffusion.

(B) Viscosity

We consider momentum transfer parallel to r_2 and across $r_2 = r_2'$, so that $z = mu_1(r_2)$ and $-\lambda_z = \tau$. Note in fluid mechanics that the sign convention for positive momentum flux is for flux from a fluid flowing vertically above a surface down onto that surface. Thus

$$\tau = \beta_\mu nm \langle C \rangle \lambda \frac{du_1}{dr_2} = \beta_\mu \rho \langle C \rangle \lambda \frac{du_1}{dr_2} \tag{1.32}$$

Comparing with Eq. 1.25 we may deduce that

$$\mu = \beta_\mu \rho \langle C \rangle \lambda \tag{1.33}$$

Now, using Eq. 1.22 for the mean free path:

$$\mu = \beta_\mu \frac{m \langle C \rangle}{\pi d^2} \tag{1.34}$$

Thus, kinetic theory predicts that the viscosity coefficient does not depend on density, and this has been confirmed by experiment.

(C) Thermal Conductivity

Consider flow of total translational energy in the absence of velocity gradients such that $z = E = \frac{3}{2}kT = mc_v T$ and $\Lambda_z = q$ so that

$$q = -\beta_\kappa nm \langle C \rangle \lambda c_v \frac{dT}{dr_2} = -\beta_\kappa \rho \langle C \rangle \lambda c_v \frac{dT}{dr_2} \tag{1.35}$$

Comparing with Eq. 1.26:

$$\kappa = -\beta_\kappa nm \langle C \rangle \lambda c_v = -\beta_\kappa \frac{m \langle C \rangle}{\pi d^2} c_v \tag{1.36}$$

Thus, our kinetic theory approach predicts that the coefficient of thermal conductivity is also independent of density.

Equations 1.30–1.36 provide molecular-based models to account for molecular transport behavior. To make these expressions useful, however, we still need to evaluate the coefficients β_D, β_μ, β_κ. There are several important factors affecting these coefficients that have been omitted from our relatively simple analysis.

(a) As they are transported across r_2', particles should have a distribution of velocities whereas in our model, all particles have precisely the speed $\langle C \rangle$.

(b) Including the effect of previous collisions increases the effective distance δr_2 from which the particles come and thus increases β_z.

(c) Particles with larger translational energy have a higher probability of transporting higher energy and this affects β_κ, as illustrated in Fig. 1.12. Specifically, a particle with a larger velocity component C_2 is more likely to transport a larger translational energy.

Detailed kinetic theory analysis including all of these effects provides the following results for a monatomic, hard-sphere gas (Present 1958):

$$\beta_D = \frac{3\pi}{16} \tag{1.37}$$

$$\beta_\mu = \frac{5\pi}{32} \tag{1.38}$$

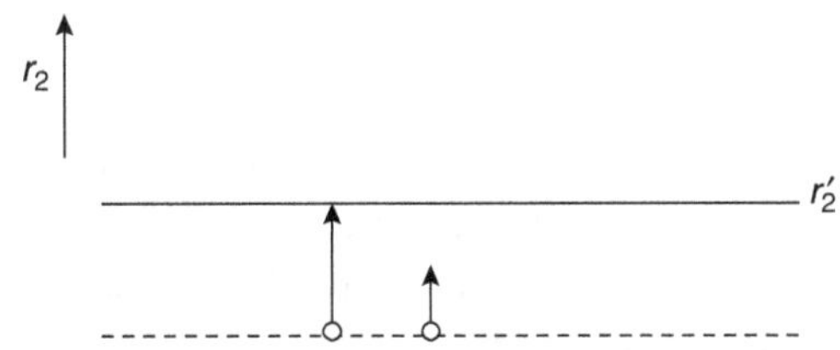

Figure 1.12 Illustration of how the particle velocity component in r_2 affects transport of translational energy.

To find β_κ, eliminate $\rho \langle C \rangle \lambda$ from Eqs. 1.33 and 1.36:

$$\kappa = \frac{\beta_\kappa}{\beta_\mu} \mu c_v \tag{1.39}$$

Now, introduce the Prandtl number:

$$\mathrm{Pr} \equiv \frac{c_p \mu}{\kappa} = \gamma \frac{\mu c_v}{\kappa} \tag{1.40}$$

Thus,

$$\frac{\beta_\kappa}{\beta_\mu} = \frac{\gamma}{\mathrm{Pr}} \tag{1.41}$$

For a monatomic gas, experiments give: $\gamma = 5/3$, $\mathrm{Pr} = 2/3$ for which

$$\kappa = \frac{5}{3}\frac{3}{2} \mu c_v = \frac{5}{2} \mu c_v \tag{1.42}$$

A polyatomic gas has additional forms of internal energy that we will study later. A key point for now is that the transport of nontranslational internal energy must be modeled without the high-energy biasing that applies to translational energy. Thus

$$\kappa = \kappa_{\mathrm{tr}} + \kappa_{\mathrm{int}} = \frac{5}{2} \mu c_{v,\mathrm{tr}} + \mu c_{v,\mathrm{int}} \tag{1.43}$$

$$c_{v,\mathrm{tr}} = \frac{3}{2} R \tag{1.44}$$

$$c_{v,\mathrm{int}} = \frac{R}{\gamma - 1} - c_{v,\mathrm{tr}} \tag{1.45}$$

Completion of this analysis leads eventually to the Eucken formula:

$$\mathrm{Pr} = \frac{4\gamma}{9\gamma - 5} \tag{1.46}$$

Evaluation of Eq. 1.46 for $\gamma = 5/3$ gives $\mathrm{Pr} = 2/3$ as before. For $\gamma = 7/5$, $\mathrm{Pr} = 0.737$, which is very close to the measured value of 0.75 for air at room temperature.

1.3 Kinetic Theory Analysis

In the previous section, we laid the groundwork for the more detailed analysis of kinetic theory presented in this section. Here, we introduce a formal definition of the velocity distribution function (VDF), and use it to develop the fundamental governing equation of kinetic theory: the Boltzmann equation. We will find the equilibrium solution for the Boltzmann equation and some important associated results such as the mean free path and collision frequency. We will also use the equilibrium solution to analyze surface properties for a gas in which there are no intermolecular collisions, referred to as free molecular flow. Finally, we will review some of the techniques available to pursue kinetic theory – based analysis of nonequilibrium gas flow.

1.3.1 Velocity Distribution Function

One cubic meter volume of air at sea level contains more than 10^{25} particles, and each of these particles experiences about 10^{10} collisions every second. As discussed in Section 1.2, each of these particles has position, velocity, and internal energy. These properties continually change in time and space through the mechanism of collisions. Even with modern supercomputers, there is just too much information to hope to track the behavior of every particle using deterministic modeling approaches. Fortunately, it is not necessary either. Therefore, due to the large number of particles in a gas, we need to consider a statistical approach involving distribution functions of velocity and other properties.

The normalized velocity distribution function is the probability density function of finding a particle with a velocity within a small range of velocities. Consider a particle velocity $\bar{C} = (C_1, C_2, C_3)$ and define a small volume in velocity space $d\bar{C} = (dC_1, dC_2, dC_3)$ around this velocity, as shown in Fig. 1.13.

All particles with velocities in the range $(C_1, C_2, C_3) \rightarrow (C_1 + dC_1, C_2 + dC_2, C_3 + dC_3)$ lie within this volume. We define the number of particles with velocity inside $d\bar{C}$ as

$$dN \equiv F(\bar{C})\, d\bar{C} \tag{1.47}$$

where $F(\bar{C})$ is the VDF. For a gas containing N particles, we now introduce a normalized VDF:

$$f(\bar{C}) \equiv \frac{F(\bar{C})}{N}$$

$$dN = Nf(\bar{C})\, d\bar{C} \tag{1.48}$$

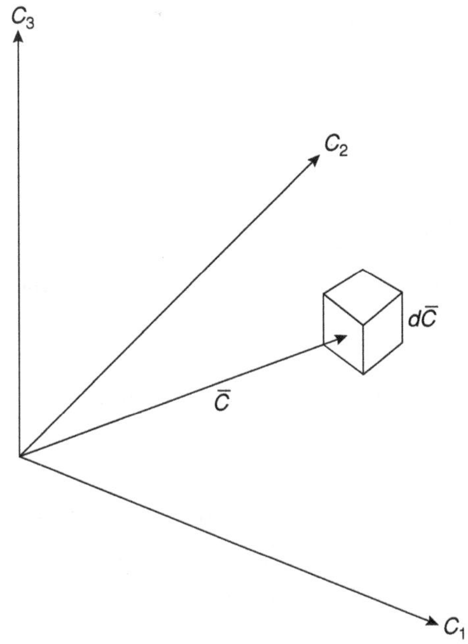

Figure 1.13 Volume element in velocity space.

Since all particles lie somewhere in velocity space, we obtain a *normalization condition*:

$$\int_{-\infty}^{\infty} N f(\bar{C})\, d\bar{C} = N \Rightarrow \int_{-\infty}^{\infty} f(\bar{C})\, d\bar{C} = 1 \tag{1.49}$$

In using a statistical approach, we are often interested in evaluation of average quantities. For a particle property, $Q = Q(\bar{C})$, the *Mean Value Theorem* gives:

$$\langle Q \rangle = \frac{1}{N} \int_{N} Q\, dN = \frac{1}{N} \int_{-\infty}^{\infty} Q(\bar{C}) N f(\bar{C})\, d\bar{C} = \int_{-\infty}^{\infty} Q(\bar{C}) f(\bar{C})\, d\bar{C} \tag{1.50}$$

When the particle property is an integer power of velocity,

$$Q = (\bar{C})^n \tag{1.51}$$

the application of Eq. 1.50 is referred to as taking the nth moment of the VDF. For example:

The zeroth moment ($n = 0$): the normalization condition again:

$$\langle Q \rangle = \int_{-\infty}^{\infty} f(\bar{C})\, d\bar{C} = 1 \tag{1.52}$$

The first moment ($n = 1$): the average particle velocity vector:

$$\langle Q \rangle = \int_{-\infty}^{\infty} \bar{C} f(\bar{C})\, d\bar{C} = \langle \bar{C} \rangle = \bar{u} \tag{1.53}$$

1.3.2 The Boltzmann Equation

To develop the governing equation for a collection of N particles we now introduce the velocity distribution function for phase space that involves the number of particles that lie simultaneously in a region of velocity space ($\bar{C} \to \bar{C} + d\bar{C}$) and physical space ($\bar{r} \to \bar{r} + d\bar{r}$) so that

$$dN = nf(\bar{C})\, d\bar{C}\, d\bar{r} \tag{1.54}$$

where n is the number density. Now, consider an element of phase space $d\bar{C}\, d\bar{r}$ that does not change in shape or size with time. Hence, the rate of change of the number of particles inside $d\bar{C}\, d\bar{r}$ is given by

$$\frac{\partial N}{\partial t} = \frac{\partial}{\partial t}\left(nf(\bar{C})\, d\bar{C}\, d\bar{r}\right) \tag{1.55}$$

Noting in phase space that velocity is constant within $d\bar{r}$ and position is constant within $d\bar{C}$, we now analyze the three physical processes affecting the rate of change of the number of particles in a phase space element.

(1) *Convection across* $d\bar{r}$ due to particle motion at velocity $\bar{C}$, i.e., particles move in and out of the spatial element ($\bar{r} \to \bar{r} + d\bar{r}$). The rate of this process is described by

$$\left(\frac{\partial N}{\partial t}\right)_1 = -\bar{C} \cdot \frac{\partial N}{\partial \bar{r}} = -\bar{C} \cdot \frac{\partial nf}{\partial \bar{r}}\, d\bar{C}\, d\bar{r} \tag{1.56}$$

(2) *Convection across* $d\bar{C}$ due to particle acceleration $\bar{a}$ caused by an external force (such as gravity or electromagnetic fields), i.e., particles move in and out of the velocity element ($\bar{C} \to \bar{C} + d\bar{C}$). The rate of this process is described by

$$\left(\frac{\partial N}{\partial t}\right)_2 = -\bar{a} \cdot \frac{\partial N}{\partial \bar{C}} = -\bar{a} \cdot \frac{\partial nf}{\partial \bar{C}}\, d\bar{C}\, d\bar{r} \tag{1.57}$$

(3) *Intermolecular collisions* that change particle velocities $\bar{C}$ and hence scatter particles in and out of $d\bar{C}$. To analyze the effects of collisions, we will assume a *dilute gas* for which the average spacing between particles, δ, is much larger than the particle size, d. For sea-level air, the ratio δ/d is about 10, and the value will be larger for all gas flows with lower density. The significance for our purposes of the dilute gas assumption is that it allows us to consider only two-body, binary collisions. In nondilute gases, three-body collisions must be accounted for, that complicate significantly the mathematical analysis.

Consider a binary collision between a test particle of velocity $\bar{C}$ and mass m_1 with some other particle with velocity $\bar{Z}$ and mass m_2 that changes the velocities to $\bar{C}'$ and $\bar{Z}'$. To fully analyze the collision dynamics, we note that both linear momentum and energy are conserved:

$$m_1\bar{C} + m_2\bar{Z} = m_1\bar{C}' + m_2\bar{Z}' = (m_1 + m_2)\bar{W} \tag{1.58}$$

$$m_1C^2 + m_2Z^2 = m_1C'^2 + m_2Z'^2 \tag{1.59}$$

where $\bar{W}$ is the center of mass velocity. Introduce the *relative velocity vector*

$$\bar{g} \equiv \bar{C} - \bar{Z} \Rightarrow g = \sqrt{(C_1 - Z_1)^2 + (C_2 - Z_2)^2 + (C_3 - Z_3)^2} \tag{1.60}$$

Note that, even though it is a scalar, the variable g is usually referred to as the *relative velocity*. From these expressions it may be shown that

$$\bar{C} = \bar{W} + \frac{m_2}{m_1 + m_2}\bar{g} \tag{1.61}$$

$$\bar{Z} = \bar{W} - \frac{m_1}{m_1 + m_2}\bar{g} \tag{1.62}$$

Similarly, the post-collision velocities obey

$$\bar{C}' = \bar{W} + \frac{m_2}{m_1 + m_2}\bar{g}' \tag{1.63}$$

$$\bar{Z}' = \bar{W} - \frac{m_1}{m_1 + m_2}\bar{g}' \tag{1.64}$$

From these expressions, the following energy relations may be derived:

$$m_1C^2 + m_2Z^2 = (m_1 + m_2)W^2 + m^*g^2 \tag{1.65}$$

$$m_1C'^2 + m_2Z'^2 = (m_1 + m_2)W'^2 + m^*g'^2 \tag{1.66}$$

where we have introduced the *reduced mass*

$$m^* \equiv \frac{m_1m_2}{m_1 + m_2} \tag{1.67}$$

Comparison of these energy relations with the energy conservation equation shows that the relative velocity is unchanged in the collision, i.e.,

$$g = g' \tag{1.68}$$

The dynamics of a collision can be thought of in three different ways as illustrated in Fig. 1.14: (a) the laboratory frame, (b) the center of mass frame, and (c) in the relative velocity frame. We generally analyze collisions in the relative velocity frame that is reminiscent of the simple mean free path analysis described in Section 1.2, in which we are electing to model the collision of the test particle as a particle moving at velocity $\bar{g}$ in a field of stationary $\bar{Z}$ particles. Recall that in earlier analysis of collisions for hard spheres the sphere of influence had a very well defined collision cross section, $\sigma = \pi d^2$. Real collisions involve solid angle considerations, and two variables are required

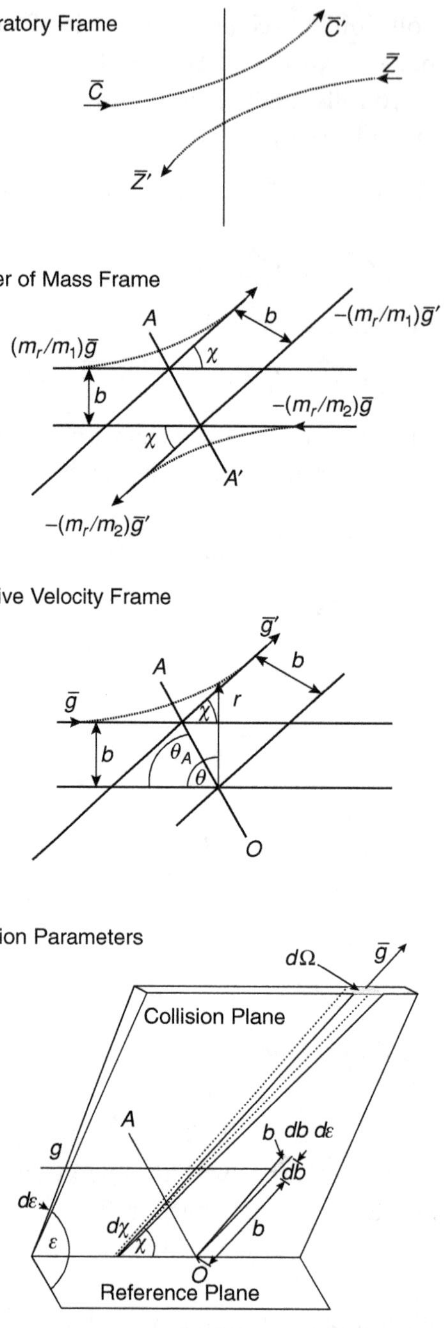

Figure 1.14 Particle scattering in an intermolecular collision.

to uniquely specify the dynamics of a two-body collision, as illustrated in Fig. 1.14. The distance of closest approach of the undisturbed trajectories, b, is often called the impact parameter. The angle ϵ is that between the collision plane and a reference plane. Now consider the plane normal to $\bar{g}$ and containing O, then we define the differential cross section as

$$\sigma d\Omega \equiv b \cdot db \cdot d\epsilon$$

where $d\Omega$ is the unit solid angle about $\bar{g}$ and the total collision cross section is defined as

$$\sigma_T = \int_0^{4\pi} \sigma \, d\Omega \tag{1.69}$$

Using these ideas, we write the volume swept out in space per unit time by each collision as

$$g\sigma \, d\Omega$$

The number of $(\bar{C}, \bar{Z}) \to (\bar{C}', \bar{Z}')$ collisions per unit time experienced by the test particle of velocity $\bar{C}$ is

$$|nf(\bar{Z}) \, d\bar{Z}| \, [g\sigma \, d\Omega] \tag{1.70}$$

The number of $\bar{C}$ particles in the phase space element is

$$nf(\bar{C}) \, d\bar{C} \, d\bar{r} \tag{1.71}$$

Hence, the total rate of such collisions per unit time is

$$n^2 f(\bar{C}) \, f(\bar{Z}) \, g\sigma \, d\Omega \, d\bar{Z} \, d\bar{C} \, d\bar{r} \tag{1.72}$$

Now, using $f(\bar{C}')$, $f(\bar{Z}')$ to represent the post-collision VDFs, the total rate of $(\bar{C}', \bar{Z}') \to (\bar{C}, \bar{Z})$ inverse collisions per unit time is

$$n^2 f(\bar{C}') \, f(\bar{Z}') \, g' \, (\sigma \, d\Omega)' \, d\bar{Z}' \, d\bar{C}' \, d\bar{r} \tag{1.73}$$

Now, the Jacobian for transformation between pre- and post-collision states gives

$$|\sigma \, d\Omega \, d\bar{C} \, d\bar{Z}| = |(\sigma \, d\Omega)' \, d\bar{C}' \, d\bar{Z}'| \tag{1.74}$$

Thus, the rate of increase of particles with velocity $\bar{C}$ due to intermolecular collisions

$$\begin{aligned} &= (\text{rate of collisions generating } \bar{C} \text{ particles}) \\ &- (\text{rate of collisions destroying } \bar{C} \text{ particles}) \\ &= n^2 \left\{ f(\bar{C}') \, f(\bar{Z}') - f(\bar{C}) \, f(\bar{Z}) \right\} g\sigma \, d\Omega \, d\bar{Z} \, d\bar{C} \, d\bar{r} \end{aligned} \tag{1.75}$$

Integration over all of solid angle space and over all particle velocities $\bar{Z}$ gives the final result:

$$\left(\frac{\partial N}{\partial t} \right)_3 = \int_{-\infty}^{\infty} \int_0^{4\pi} n^2 \left\{ f(\bar{C}') \, f(\bar{Z}') - f(\bar{C}) \, f(\bar{Z}) \right\} g\sigma \, d\Omega \, d\bar{Z} \, d\bar{C} \, d\bar{r} \tag{1.76}$$

We now combine Eqs. 1.55 to 1.57, and 1.76, then simplify to obtain the Boltzmann equation

$$\frac{\partial(nf)}{\partial t} + \bar{C} \cdot \frac{\partial(nf)}{\partial \bar{r}} + \bar{a} \cdot \frac{\partial(nf)}{\partial \bar{C}} = \int_{-\infty}^{\infty}\int_{0}^{4\pi} n^2 \left\{ f(\bar{C}') f(\bar{Z}') - f(\bar{C}) f(\bar{Z}) \right\} g\sigma \, d\Omega \, d\bar{Z} \tag{1.77}$$

This is the basic governing equation for all dilute gas dynamics. The right-hand side is called the collision term. The integral-differential nature of the Boltzmann equation makes it difficult to solve analytically, and even computationally it presents a challenge.

Boltzmann Equation Condition for Equilibrium

While the Boltzmann equation is difficult to solve, it can readily be used to find a condition for equilibrium. Recall that for *equilibrium*, there is no variation in the VDF in time or space, which that means that the left-hand side of Eq. 1.77 must be zero. Now, if an entire gas system had zero gradients then it would indeed be in equilibrium but also it would be uninteresting. We can also then introduce the concept of *local equilibrium* that applies over a finite region of space within which there are no gradients. In any case, when the time and space gradients in Eq. 1.77 are zero, then the collision term must also be zero and this requirement is satisfied by the relation

$$f(\bar{C}') f(\bar{Z}') = f(\bar{C}) (\bar{Z}) \tag{1.78}$$

This represents a statement of an important general principle found in all systems governed by finite-rate processes, the *Principle of Detailed Balance*, which states that at equilibrium:

rate of change in the forward direction
= rate of change in the backward direction

Thus, at equilibrium, the rate of collisions removing $\bar{C}$ particles,

$$(\bar{C}, \bar{Z}) \rightarrow (\bar{C}', \bar{Z}') \tag{1.79}$$

is exactly balanced by the rate of the inverse collisions creating $\bar{C}$ particles,

$$(\bar{C}', \bar{Z}') \rightarrow (\bar{C}, \bar{Z}) \tag{1.80}$$

Thus, at equilibrium, there is no change in the VDF due to any of the three physical processes included in the Boltzmann equation.

1.3.3 The H-Theorem of Boltzmann

Similar to the concept of entropy in thermodynamics, the H-Theorem specifies the direction that the collision process must take. Consider the

Boltzmann equation for the following conditions: (1) a simple dilute gas; (2) no external forces, so $\bar{a}$ is zero; and (3) the gas is spatially homogeneous, so that n is a constant and the spatial derivative is zero. Under these conditions, the Boltzmann equation reduces to

$$\frac{\partial f(\bar{C})}{\partial t} = n \int_{-\infty}^{\infty} \int_{0}^{4\pi} \left\{ f(\bar{C}') f(\bar{Z}') - f(\bar{C}) f(\bar{Z}) \right\} g\sigma \, d\Omega \, d\bar{Z} \tag{1.81}$$

The Boltzmann equation describes how the number of particles of velocity $\bar{C}$ changes within an element of phase space. If instead, we want to analyze the change in a particle property $Q = Q(\bar{C})$, we simply multiply both sides by Q to obtain

$$\frac{\partial (f(\bar{C}) Q(\bar{C}))}{\partial t} = n \int_{-\infty}^{\infty} \int_{0}^{4\pi} Q(\bar{C}) \left\{ f(\bar{C}') f(\bar{Z}') - f(\bar{C}) f(\bar{Z}) \right\} g\sigma \, d\Omega \, d\bar{Z} \tag{1.82}$$

If we now integrate over $\bar{C}$ we may write

$$\frac{\partial}{\partial t} \left[\int_{-\infty}^{\infty} f(\bar{C}) Q(\bar{C}) \, d\bar{C} \right] = n \int_{-\infty}^{\infty} \int_{-\infty}^{\infty} \int_{0}^{4\pi} f(\bar{C}) \left\{ f(\bar{C}') f(\bar{Z}') - f(\bar{C}) f(\bar{Z}) \right\} g\sigma \, d\Omega \, d\bar{Z} \, d\bar{C}$$

Setting $Q = \ln|f(\bar{C})|$ and introducing Boltzmann's H-function

$$H \equiv \langle \ln(f) \rangle = \int_{-\infty}^{\infty} f(\bar{C}) \ln|f(\bar{C})| \, d\bar{C} \tag{1.83}$$

where $\langle \, \rangle$ means taking the average over the VDF as considered earlier in Eq. 1.50, we obtain

$$\frac{\partial H}{\partial t} = n \int_{-\infty}^{\infty} \int_{-\infty}^{\infty} \int_{0}^{4\pi} \ln|f(\bar{C})| \left\{ f(\bar{C}') f(\bar{Z}') - f(\bar{C}) f(\bar{Z}) \right\} g\sigma \, d\Omega \, d\bar{Z} \, d\bar{C} \tag{1.84}$$

The right-hand side of this expression is called the *collision integral* $\delta[Q]$ and this has several important properties. First, let us write this quantity out in full:

$$\delta[Q] = n \int_{-\infty}^{\infty} \int_{-\infty}^{\infty} \int_{0}^{4\pi} \ln|f(\bar{C})| \left\{ f(\bar{C}') f(\bar{Z}') - f(\bar{C}) f(\bar{Z}) \right\} g\sigma \, d\Omega \, d\bar{Z} \, d\bar{C} \tag{1.85}$$

Symmetry 1: Since we chose the particle velocities of $\bar{C}$ and $\bar{Z}$ in our collision in an arbitrary manner, we can interchange them $\bar{C} \leftrightarrow \bar{Z}$ without changing the physics to obtain

$$\delta[Q] = n \int_{-\infty}^{\infty} \int_{-\infty}^{\infty} \int_{0}^{4\pi} \ln|f(\bar{Z})| \{f(\bar{C}')f(\bar{Z}') - f(\bar{Z})f(\bar{C})\} g\sigma \, d\Omega \, d\bar{Z} \, d\bar{C} \tag{1.86}$$

Symmetry 2: Similarly, we can interchange the velocities in 1.85 between the forward and backward collisions: $\bar{C} \leftrightarrow \bar{C}'$ and $\bar{Z} \leftrightarrow \bar{Z}'$ to obtain

$$\delta[Q] = n \int_{-\infty}^{\infty} \int_{-\infty}^{\infty} \int_{0}^{4\pi} \ln|f(\bar{C}')| \{f(\bar{C})f(\bar{Z}) - f(\bar{C}')f(\bar{Z}')\} g\sigma \, d\Omega \, d\bar{Z} \, d\bar{C} \tag{1.87}$$

where we have used the earlier results relating the post- and pre-collision properties.

Symmetry 3: Now, in Eq. 1.87, we can interchange $\bar{C}' \leftrightarrow \bar{Z}'$ to write

$$\delta[Q] = n \int_{-\infty}^{\infty} \int_{-\infty}^{\infty} \int_{0}^{4\pi} \ln|f(\bar{Z}')| \{f(\bar{C})f(\bar{Z}) - f(\bar{Z}')f(\bar{C}')\} g\sigma \, d\Omega \, d\bar{Z} \, d\bar{C} \tag{1.88}$$

Now, we combine Eqs. 1.85–1.88 to obtain:

$$\delta[Q] = \frac{n}{4} \int_{-\infty}^{\infty} \int_{-\infty}^{\infty} \int_{0}^{4\pi} \{\ln|f(\bar{C})| + \ln|f(\bar{Z})| - \ln|f(\bar{C}')| - \ln|f(\bar{Z}')|\}$$

$$\times \{f(\bar{C}')f(\bar{Z}') - f(\bar{C})f(\bar{Z})\} g\sigma \, d\Omega \, d\bar{Z} \, d\bar{C} \tag{1.89}$$

This result may be rewritten:

$$\frac{\partial H}{\partial t} = \frac{n}{4} \int_{-\infty}^{\infty} \int_{-\infty}^{\infty} \int_{0}^{4\pi} \ln \left\{ \frac{f(\bar{C})f(\bar{Z})}{f(\bar{C}')f(\bar{Z}')} \right\} \{f(\bar{C}')f(\bar{Z}') - f(\bar{C})f(\bar{Z})\}$$

$$\times g\sigma \, d\Omega \, d\bar{Z} \, d\bar{C} \tag{1.90}$$

Inspection of the right-hand side (RHS) of Eq. 1.90 shows

$$f(\bar{C})f(\bar{Z}) \geq f(\bar{C}')f(\bar{Z}') \Rightarrow \text{RHS} \leq 0$$
$$f(\bar{C})f(\bar{Z}) \leq f(\bar{C}')f(\bar{Z}') \Rightarrow \text{RHS} \leq 0$$

Thus, it follows that the RHS is always negative or equal to zero and so H *never increases*:

$$\frac{\partial H}{\partial t} \leq 0 \tag{1.91}$$

and this is the *Boltzmann H-theorem*. Figure 1.15 illustrates the temporal change in H when a system is initially perturbed at zero time.

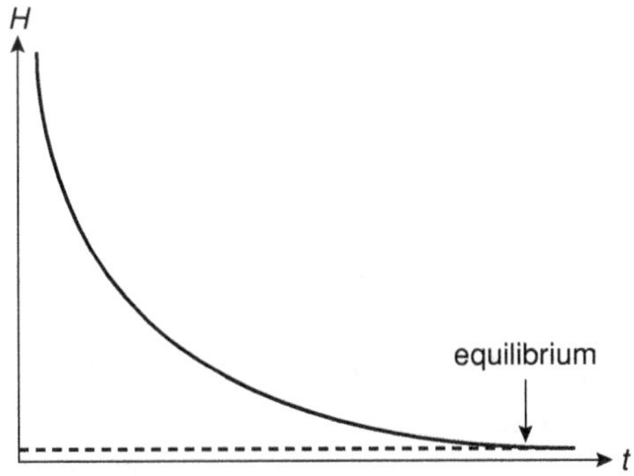

Figure 1.15 Variation of Boltzmann's H-function with time.

In this process, the gas undergoes collisions that continuously reduce the value of H until a plateau is reached. At this point

$$\frac{\partial H}{\partial t} = 0 \Rightarrow \ln\left[\frac{f(\bar{C})\, f(\bar{Z})}{f(\bar{C}')\, f(\bar{Z}')}\right] = 0 \tag{1.92}$$

that is the same as Eq. 1.78, i.e., the condition for equilibrium.

It is also useful to consider the H-theorem in terms of what it tells us *physically*. The fractional change in the VDF over a small time interval is

$$\frac{\partial f / f}{\partial t} = \frac{\partial\,[\ln(f)]}{\partial t} \tag{1.93}$$

So, the average fractional change in the VDF is given by

$$\frac{\partial\,\langle \ln(f) \rangle}{\partial t} = \frac{\partial H}{\partial t} \leq 0 \tag{1.94}$$

Thus, Boltzmann's H-theorem tells us that the average fractional rate of change in the VDF due to collisions can only decrease, and therefore, the effect of collisions is always to move the system closer to equilibrium.

1.3.4 Maxwellian VDF

As stated previously, Eq. 1.92 is equivalent to the Principle of Detailed Balance, and it may be written as

$$\ln|f(\bar{C})| + \ln|f(\bar{Z})| = \ln|f(\bar{C}')| + \ln|f(\bar{Z}')| \tag{1.95}$$

Thus, there is a certain function, $\ln|f(\bar{C})|$, whose sum for the two particles in a collision does not change as a result of the collision. From basic mechanics, we know that mass, momentum, and kinetic energy are properties that are not changed as a result of a collision, and are sometimes called the collision invariants. Thus, we choose the following general form for solution of Eq. 1.95 based on a linear combination of these particle properties:

$$\ln|f(\bar{C})| = a_0 m + m(a_1 C_1 + a_2 C_2 + a_3 C_3) + a_4 \frac{m}{2}\left(C_1^2 + C_2^2 + C_3^2\right) \tag{1.96}$$

where a_i are arbitrary constants. We can rewrite this expression as

$$\ln|f(\bar{C})| = a_4 \frac{m}{2}\left[(C_1 - \alpha_1)^2 + (C_2 - \alpha_2)^2 + (C_3 - \alpha_3)^2\right] + \alpha_4 \tag{1.97}$$

and finally

$$f(\bar{C}) = A \exp\left\{-\beta \frac{m}{2}\left[(C_1 - \alpha_1)^2 + (C_2 - \alpha_2)^2 + (C_3 - \alpha_3)^2\right]\right\} \tag{1.98}$$

where A, β, α_1, α_2, and α_3 are all constants, and the negative sign before β is introduced for later convenience.

Our next goal is to evaluate the constants in Eq. 1.98. First, note that the constants α_j occur only in square terms. Thus, the value of $f(\bar{C})$ depends only on the magnitude of $(C_j - \alpha_j)$, meaning that it is a symmetric function in $(C_j - \alpha_j)$. In turn, this implies that $\langle C_j - \alpha_j \rangle = 0 = \langle C_j \rangle - \langle \alpha_j \rangle$, as α_j is a constant. Thus, $\alpha_j = \langle \alpha_j \rangle = \langle C_j \rangle$, which is the average velocity in one of the three coordinate directions, u_j and so:

$$f(\bar{C}) = A \exp\left\{-\beta \frac{m}{2}\left[(C_1 - u_1)^2 + (C_2 - u_2)^2 + (C_3 - u_3)^2\right]\right\} \tag{1.99}$$

Let us proceed with our analysis for the simple case of a gas at rest for which $u_j = 0$ and so

$$f(\bar{C}) = A \exp\left\{-\beta \frac{m}{2}\left[C_1^2 + C_2^2 + C_3^2\right]\right\} \tag{1.100}$$

We can evaluate the leading constant by making use of the normalization condition:

$$\begin{aligned} 1 &= \int_{-\infty}^{\infty} f(\bar{C})\, d\bar{C} \\ &= A \int_{-\infty}^{\infty} \exp\left(-\frac{m\beta}{2}C_1^2\right) dC_1 \int_{-\infty}^{\infty} \exp\left(-\frac{m\beta}{2}C_2^2\right) dC_2 \int_{-\infty}^{\infty} \exp\left(-\frac{m\beta}{2}C_3^2\right) dC_3 \end{aligned} \tag{1.101}$$

We make use of the following standard integral:

$$\int_{-\infty}^{\infty} \exp\left(-\frac{m\beta}{2}z^2\right) dz = \sqrt{\frac{2\pi}{\beta m}} \tag{1.102}$$

to find that

$$A = \left(\frac{\beta m}{2\pi}\right)^{3/2} \tag{1.103}$$

Thus

$$f(\bar{C}) = \left(\frac{\beta m}{2\pi}\right)^{3/2} \exp\left\{-\beta \frac{m}{2}\left[C_1^2 + C_2^2 + C_3^2\right]\right\} \tag{1.104}$$

We saw in Section 1.2 that

$$\langle C^2 \rangle = \frac{3kT}{m} \tag{1.105}$$

and we use this result, along with taking the second moment of our VDF, to find the remaining constant, β. Therefore,

$$\begin{aligned}
\frac{3kT}{m} &= \int\int\int (C_1^2 + C_2^2 + C_3^2) \left(\frac{\beta m}{2\pi}\right)^{3/2} \\
&\quad \times \exp\left[-\frac{m\beta}{2}(C_1^2 + C_2^2 + C_3^2)\right] dC_1\, dC_2\, dC_3 \\
&= \left(\frac{\beta m}{2\pi}\right)^{3/2} \int_{-\infty}^{\infty} C_1^2 \exp\left[-\frac{m\beta}{2} C_1^2\right] dC_1 \\
&\quad \times \int_{-\infty}^{\infty} \exp\left[-\frac{m\beta}{2} C_2^2\right] dC_2 \int_{-\infty}^{\infty} \exp\left[-\frac{m\beta}{2} C_3^2\right] dC_3 \\
&\quad + \left(\frac{\beta m}{2\pi}\right)^{3/2} \int_{-\infty}^{\infty} \exp\left[-\frac{m\beta}{2} C_1^2\right] dC_1 \int_{-\infty}^{\infty} C_2^2 \\
&\quad \times \exp\left[-\frac{m\beta}{2} C_2^2\right] dC_2 \int_{-\infty}^{\infty} \exp\left[-\frac{m\beta}{2} C_3^2\right] dC_3 \\
&\quad + \left(\frac{\beta m}{2\pi}\right)^{3/2} \int_{-\infty}^{\infty} \exp\left[-\frac{m\beta}{2} C_1^2\right] dC_1 \int_{-\infty}^{\infty} \\
&\quad \times \exp\left[-\frac{m\beta}{2} C_2^2\right] dC_2 \int_{-\infty}^{\infty} C_3^2 \exp\left[-\frac{m\beta}{2} C_3^2\right] dC_3
\end{aligned} \tag{1.106}$$

Now, we make use of the earlier standard integral along with the following result:

$$\int_{-\infty}^{\infty} z^2 \exp\left(-\frac{m\beta}{2} z^2\right) dz = \frac{\pi^{1/2}}{2\left(\frac{\beta m}{2}\right)^{3/2}} \tag{1.107}$$

to obtain

$$\frac{3kT}{m} = 3\frac{\pi^{1/2}}{2\left(\frac{\beta m}{2}\right)^{3/2}} \sqrt{\frac{2\pi}{\beta m}} \sqrt{\frac{2\pi}{\beta m}} \left(\frac{\beta m}{2\pi}\right)^{3/2} = \frac{3}{\beta m} \tag{1.108}$$

Thus,

$$\beta = \frac{1}{kT},\ A = \left(\frac{m}{2\pi kT}\right)^{3/2} \tag{1.109}$$

and this leads to the expression

$$f(\bar{C}) = \left(\frac{m}{2\pi kT}\right)^{3/2} \exp\left[-\frac{m}{2kT}(C_1^2 + C_2^2 + C_3^2)\right] \quad (1.110)$$

Inspection of the right-hand side of this result reveals that it has units of inverse velocity cubed. This is consistent with a need for us to include the size of the velocity element in the expression, as we are interested in evaluating the probability of finding a particle with velocity in the range $\bar{C} \to \bar{C} + d\bar{C}$ that is given by

$$f(\bar{C})\, d\bar{C} = \left(\frac{m}{2\pi kT}\right)^{3/2} \exp\left[-\frac{m}{2kT}(C_1^2 + C_2^2 + C_3^2)\right] d\bar{C} \quad (1.111)$$

Now, let us return to the general case of a gas with a net stream velocity vector, $\bar{u} = (u_1, u_2, u_3)$. Evaluation of parameters A and β is independent of these velocities, and so we obtain the final result for the Maxwellian VDF:

$$f(\bar{C})\, d\bar{C} = \left(\frac{m}{2\pi kT}\right)^{3/2} \exp\left\{-\frac{m}{2kT}\left[(C_1 - u_1)^2 + (C_2 - u_2)^2 + (C_3 - u_3)^2\right]\right\} d\bar{C} \quad (1.112)$$

By inspection of Eq. 1.112, we can deduce the VDF for a single-velocity component:

$$\phi(C_1)\, dC_1 = \left(\frac{m}{2\pi kT}\right)^{1/2} \exp\left\{-\frac{m}{2kT}\left[(C_1 - u_1)^2\right]\right\} dC_1 \quad (1.113)$$

This function is plotted in Fig. 1.16, where we see that it is symmetric about $(C_1 - u_1)$, and has a peak of $1/\sqrt{\pi}$. In addition, for a given chemical species (for which mass is constant):

- A higher temperature gives a wider, flatter VDF.
- A lower temperature gives a narrower, more peaked VDF.

Next we consider the distribution of speed, $C = |\bar{C}|$. We begin with the Maxwellian VDF for a gas at rest, and write it in component form:

$$\begin{aligned} &f(C_1, C_2, C_3)\, dC_1\, dC_2\, dC_3 \\ &= \left(\frac{m}{2\pi kT}\right)^{3/2} \exp\left[-\frac{m}{2kT}(C_1^2 + C_2^2 + C_3^2)\right] dC_1\, dC_2\, dC_3 \end{aligned} \quad (1.114)$$

Now, we make the standard transformation from Cartesian to spherical polar coordinates:

$$f(C, \phi, \theta)\, dC d\phi d\theta = \left(\frac{m}{2\pi kT}\right)^{3/2} C^2 \sin\phi \exp\left[-\frac{mC^2}{2kT}\right] dC d\phi d\theta \quad (1.115)$$

Integrating over ϕ and θ we obtain the Maxwellian speed distribution:

$$\chi(C)\, dC = 4\pi \left(\frac{m}{2\pi kT}\right)^{3/2} C^2 \exp\left[-\frac{mC^2}{2kT}\right] dC \quad (1.116)$$

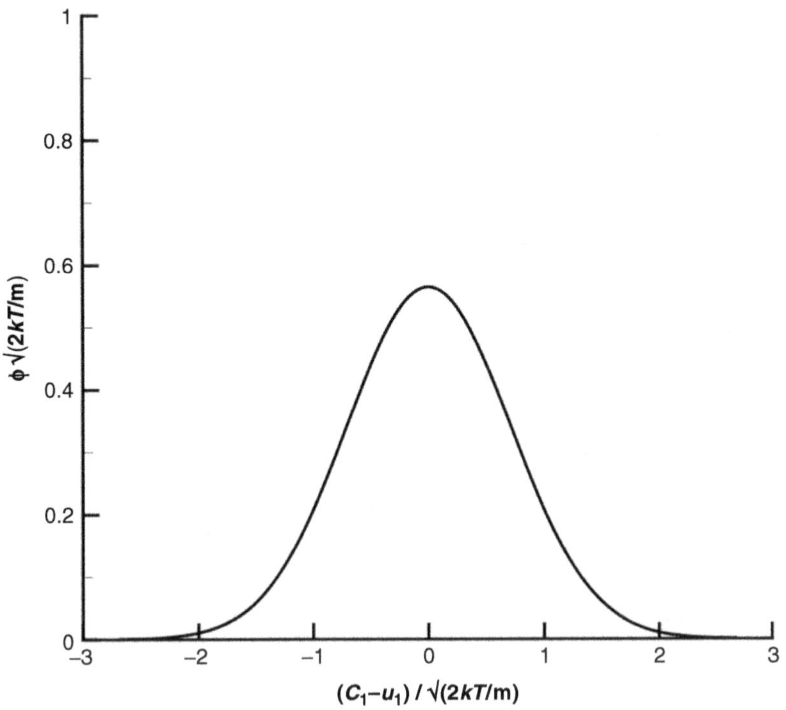

Figure 1.16 Maxwellian VDF in one dimension.

Let's examine some important properties of this distribution.

(1) $\chi(0) = 0$, thus there are no particles exactly at rest.

(2) To find the *most probable speed*, C_{mp}, we find the maximum of χ by differentiation:

$$\chi'(C) = 0$$

$$\Rightarrow 4\pi\left(\frac{m}{2\pi kT}\right)^{3/2}\left\{2C\exp\left(-\frac{mC^2}{2kT}\right) + C^2\left(-\frac{2mC}{2kT}\right)\exp\left(-\frac{mC^2}{2kT}\right)\right\} = 0 \tag{1.117}$$

Hence,

$$C_{\mathrm{mp}} = \sqrt{\frac{2kT}{m}} \tag{1.118}$$

(3) We find the average speed, $\langle C\rangle$, by taking the first moment of the speed distribution (noting that the lower integration limit is zero because we are now considering the speed):

$$\langle C\rangle = \int_0^\infty C\chi(C)\,dC = \sqrt{\frac{8kT}{\pi m}} = \frac{2}{\sqrt{\pi}}C_{\mathrm{mp}} \tag{1.119}$$

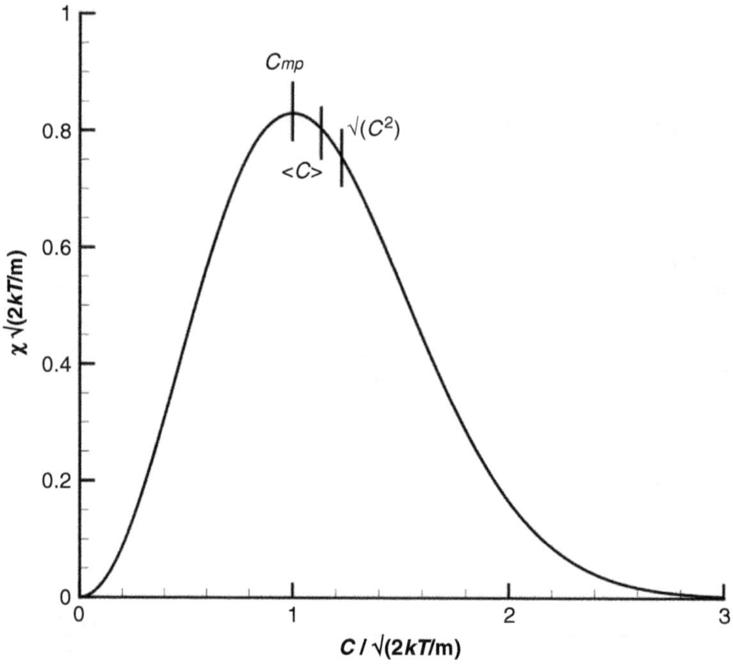

Figure 1.17 Maxwellian speed distribution.

where we used another standard integral:

$$\int_0^\infty z^3 \exp(-az^2)\,dz = \frac{1}{2a^2} \tag{1.120}$$

(4) Finally, we evaluate the root mean square speed: $\sqrt{\langle C^2 \rangle}$

$$\langle C^2 \rangle = \int_0^\infty C\chi(C)\,dC$$

$$= \frac{3kT}{m} \text{ and so } \sqrt{\langle C^2 \rangle} = \sqrt{\frac{3kT}{m}} = \sqrt{\frac{3}{2}}C_{\text{mp}} \approx 1.22C_{\text{mp}} \tag{1.121}$$

where we have used

$$\int_0^\infty z^4 \exp(-az^2)\,dz = \frac{3}{8}\sqrt{\frac{\pi}{a^5}} \tag{1.122}$$

Figure 1.17 shows the Maxwellian speed distribution in normalized form in which the location of the above properties are also indicated. Note that the function is not symmetric: there are more particles traveling faster than the most probable speed than there are traveling slower than it.

1.3.5 Equilibrium Collision Properties

We now use the Maxwellian velocity and speed distributions to develop expressions for the collision rate and mean free path of a gas in thermal equilibrium. Consider the rate of collisions per unit volume of gas between particles of species A at velocity $\bar{C}$ and particles of species B at velocity $\bar{Z}$ in which we transform to relative velocity $\bar{g} \equiv \bar{C} - \bar{Z}$

$$dZ_{AB} = n_A n_B f_A(\bar{C}) f_B(\bar{Z}) g\sigma \, d\Omega \, d\bar{C} \, d\bar{Z} \tag{1.123}$$

Now, insert the Maxwellian VDFs:

$$dZ_{AB} = n_A n_B \exp\left\{-\frac{1}{2kT}\left[m_A C^2 + m_B Z^2\right]\right\} g\sigma \, d\Omega \, dC_1 \, dC_2 \, dC_3 \, dZ_1 \, dZ_2 \, dZ_3 \tag{1.124}$$

To obtain the total collision rate, we must integrate Eq. 1.124 and this is performed through a transformation of velocities that represents a generalization of the earlier considerations in Section 1.3.2. Specifically, we introduce

$$\text{Relative velocity: } \bar{g} \equiv \bar{C} - \bar{Z} \tag{1.125}$$

$$\text{Center of mass velocity: } \bar{W} \equiv \frac{m_A \bar{C} + m_B \bar{Z}}{m_A + m_B} \tag{1.126}$$

From these definitions, we obtain

$$\bar{C} = \bar{W} + \frac{m_B}{m_A + m_B} \bar{g} \tag{1.127}$$

$$\bar{Z} = \bar{W} - \frac{m_A}{m_A + m_B} \bar{g} \tag{1.128}$$

Now, use the *reduced mass*

$$m^*_{AB} \equiv \frac{m_A m_B}{m_A + m_B} \tag{1.129}$$

from which it can be shown that

$$\frac{1}{2} m_A C^2 + \frac{1}{2} m_B Z^2 = \frac{1}{2}(m_A + m_B) W^2 + \frac{1}{2} m^*_{AB} g^2 \tag{1.130}$$

Returning to Eq. 1.124, we must also transform the differential elements, e.g.,

$$dC_1 \, dZ_1 = \left|\frac{\partial(C_1, Z_1)}{\partial(W_1, g_1)}\right| dW_1 \, dg_1 \tag{1.131}$$

where the Jacobian is evaluated using Eqs. 1.127 and 1.128 as

$$\left|\frac{\partial(C_1, Z_1)}{\partial(W_1, g_1)}\right| = \begin{vmatrix} \dfrac{\partial C_1}{\partial W_1} & \dfrac{\partial C_1}{\partial g_1} \\ \dfrac{\partial Z_1}{\partial W_1} & \dfrac{\partial Z_1}{\partial g_1} \end{vmatrix} = \begin{vmatrix} 1 & \dfrac{m_B}{m_A + m_B} \\ 1 & \dfrac{-m_A}{m_A + m_B} \end{vmatrix} = 1 \tag{1.132}$$

We obtain the analogous results in directions 2 and 3. Using these results and inserting a hard-sphere collision cross section, σ_{AB} , Eq. 1.124 becomes

$$dZ_{AB} = n_A n_B \frac{(m_A m_B)^{3/2}}{(2\pi kT)^3} g\sigma_{AB} \exp\left\{-\frac{1}{2kT}\left[(m_A + m_B)W^2 \right.\right.$$
$$\left.\left. + m^*_{AB} g^2\right]\right\} dW_1\, dW_2\, dW_3\, dg_1\, dg_2\, dg_3 \tag{1.133}$$

Now transform both velocities to spherical polar coordinates:

$$dW_1\, dW_2\, dW_3 = W^2 \sin\phi_W\, d\phi_W\, d\theta_W\, dW \tag{1.134}$$

$$dg_1\, dg_2\, dg_3 = g^2 \sin\phi_g\, d\phi_g\, d\theta_g\, dg \tag{1.135}$$

By substituting into Eq. 1.133 and integrating over W and both sets of angles we obtain the final result called the *bimolecular collision rate*:

$$Z_{AB} = n_A n_B \sigma_{AB} \sqrt{\frac{8kT}{\pi m^*_{AB}}} \tag{1.136}$$

where we have used the standard integral:

$$\int_0^\infty z^3 \exp(-az^2)\, dz = \frac{1}{2a^2} \tag{1.137}$$

For a hard-sphere collision between particles of two different species A and B, the collision cross section is evaluated as $\sigma_{AB} = \frac{\pi}{4}(d_A + d_B)^2$. More generally, other collision cross section models have a relative velocity dependence that, when integrated in Eq. 1.133, introduces a different temperature dependence into Eq. 1.136.

A small correction must be applied to Eq. 1.136. In our analysis so far, we say that a particle of species A and velocity $\bar{C}$ collides with a particle of species B and velocity $\bar{Z}$. However, our analysis does not distinguish this from the case of a collision between a particle of species A and velocity $\bar{Z}$ with a particle of species B and velocity $\bar{C}$. For a simple gas, this means we are counting each collision twice. Thus, we introduce a correction as follows:

$$Z_{AB} = \frac{1}{1 + \delta_{AB}} n_A n_B \sigma_{AB} \sqrt{\frac{8kT}{\pi m^*_{AB}}} \tag{1.138}$$

$$\text{where } \delta_{AB} = \begin{cases} 1, & \text{when A} = \text{B} \\ 0, & \text{when A} \neq \text{B} \end{cases} \tag{1.139}$$

Next, we consider the mean free path of species A in a general gas mixture under conditions of thermal equilibrium. Each collision involving a species A particle determines a free path length for species A collisions. When two species A particles collide, two free paths for species A are terminated simultaneously. We need to find the mean value of all free paths for species A. The molecular collision rate, Eq. 1.138, is the rate of ending free paths per unit volume. For A–B collisions, $(1+\delta_{AB})$ free paths for species A are ended per collision. So, the number of free paths ended for species A per unit time for a hard-sphere gas in equilibrium is

$$n_A n_B \sigma_{AB} \sqrt{\frac{8kT}{\pi m^*_{AB}}} \tag{1.140}$$

Thus, the rate of ending free paths per particle of species A through collisions with species B is

$$\Theta_A = \sum_{i=1}^{N_s} \Theta_{Ai} = \sqrt{\frac{8kT}{\pi}} \sum_{i=1}^{N_s} n_i \sigma_{Ai} \sqrt{\frac{1}{m_A} + \frac{1}{m_i}} \tag{1.141}$$

The average distance traveled per unit time by species A is simply the mean speed:

$$\langle C_A \rangle = \sqrt{\frac{8kT}{\pi m_A}} \tag{1.142}$$

So, the mean free path of species A in a gas mixture at equilibrium is given by

$$\lambda_A = \frac{\langle C_A \rangle}{\Theta_A} = \frac{1}{\sum_{i=1}^{N_s} n_i \sigma_{Ai} \sqrt{1 + \frac{m_A}{m_i}}} \tag{1.143}$$

For a simple gas, for which $N_s = 1$, we obtain

$$\lambda = \frac{1}{\sqrt{2} n \sigma} \tag{1.144}$$

which represents the higher fidelity update to the result we obtained in Section 1.2.

1.3.6 Free Molecular Flow onto a Surface

One of the most important applications of the results from equilibrium kinetic theory is in the analysis of surface properties in free molecular, collisionless flow. This approach is typically used, for example, to study spacecraft aerodynamics and heating.

INCIDENT FLUXES

As illustrated in Fig. 1.18, consider a small surface element of area δA in the (r_1, r_2) plane. A free molecular gas stream is incident on the element that is characterized by an equilibrium VDF (Maxwellian) and macroscopic parameters n_i, $\bar{u}_i = (u_1, u_2, u_3)$, T_i.

Therefore, the VDF of the incident stream is given by

$$f_i(\bar{C})\, d\bar{C} = \left(\frac{m}{2\pi k T_i}\right)^{3/2} \exp\left\{-\frac{m}{2kT_i}\left[(\bar{C} - \bar{u}_i)^2\right]\right\} d\bar{C} \tag{1.145}$$

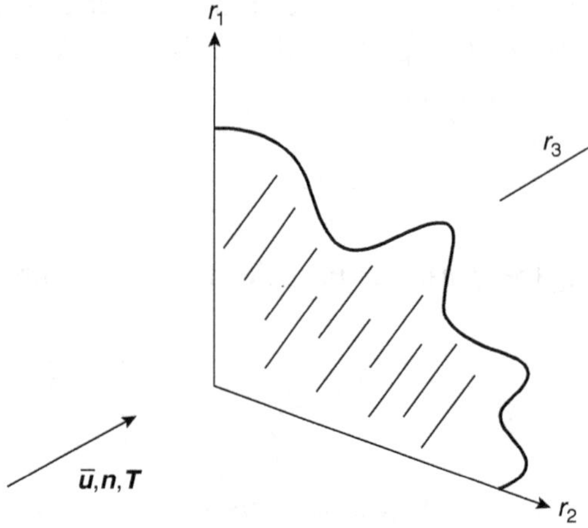

Figure 1.18 Coordinate system for particle fluxes.

The velocity vector of each particle may be written conveniently as $\bar{C} = \bar{u}_i + \bar{C}'$ where $\bar{C}'$ is called the *thermal or random velocity*.

The incident flux of a particle property $Q = Q(\bar{C})$ onto the surface is given by

$$\Gamma_i^Q = \int_{C_1=-\infty}^{\infty} \int_{C_2=-\infty}^{\infty} \int_{C_3=0}^{\infty} n_i C_3 Q(\bar{C}) f_i(\bar{C})\, d\bar{C} \tag{1.146}$$

Note that the lower integration limit in the C_3 direction is zero because backward moving particles will never reach the surface. For the specific case of $Q = Q(C_3)$, we can readily integrate over C_1 and C_2, and express the integral in terms of thermal velocity, to obtain

$$\Gamma_i^Q = n_i \left(\frac{m}{2\pi k T_i}\right)^{1/2} \int_{C_3'=-u_3}^{\infty} (u_3 + C_3') Q(u_3 + C_3') \exp\left(-\frac{mC_3'^2}{2kT_i}\right) dC_3' \tag{1.147}$$

Thus, particles with velocities less than u_3 do not reach the surface. Let us now use Eq. 1.147 to evaluate the incident fluxes of mass, normal momentum, and in an extended form, the energy.

(a) *Incident mass flux*
We set $Q = m$, and name the associated flux quantity the incident mass flux, Φ_i:

$$\Phi_i = mn_i \left(\frac{m}{2\pi k T_i}\right)^{1/2} \int_{-u_3}^{\infty} (u_3 + C_3') \exp\left(-\frac{mC_3'^2}{2kT_i}\right) dC_3' \tag{1.148}$$

Consider the two components of the integral, starting with

$$I_1 = \int_{-u_3}^{\infty} u_3 \exp\left(-\frac{mC_3'^2}{2kT_i}\right) dC_3'$$
$$= \int_0^{\infty} u_3 \exp\left(-\frac{mC_3'^2}{2kT_i}\right) dC_3' + \int_0^{-u_3} u_3 \exp\left(-\frac{mC_3'^2}{2kT_i}\right) dC_3' \tag{1.149}$$

The first term on the right-hand side can be evaluated using a standard integral.

Use the substitution $\frac{mC_3'^2}{2kT_i} = t^2$ in the second term:

$$I_1 = u_3 \frac{1}{2}\left(\frac{2\pi kT_i}{m}\right) + u_3 \int_0^{u_3\left(\frac{m}{2kTi}\right)^{1/2}} \exp\left(-t^2\right)\left(\frac{2kT_i}{m}\right)^{1/2} dt \tag{1.150}$$

Now, we introduce the speed ratio, $s_3 \equiv \frac{u_3}{\sqrt{2kT_i/m}}$, and using the standard error function, erf, we obtain the final result:

$$I_1 = s_3 \frac{\sqrt{\pi}}{2}\frac{2kT_i}{m}\left[1 + erf(s_3)\right] \tag{1.151}$$

Now, we consider the second term of the incident mass flux integral

$$I_2 = \int_{-u_3}^{\infty} C_3' \exp\left(-\frac{mC_3'^2}{2kT_i}\right) dC_3'$$
$$= \int_0^{\infty} C_3' \exp\left(-\frac{mC_3'^2}{2kT_i}\right) dC_3' + \int_{-u_3}^{0} C_3' \exp\left(-\frac{mC_3'^2}{2kT_i}\right) dC_3' \tag{1.152}$$

Using another standard integral for the first term, and using the substitution $t = \frac{mC_3'^2}{2kT_i}$ in the second, we obtain

$$I_2 = \frac{2kT_i}{2m} + \frac{kT_i}{m}\int_{\frac{mu_3^2}{2kT_i}}^{0} \exp(-t)\, dt = \frac{kT_i}{m}\exp\left(-s_3^2\right) \tag{1.153}$$

Thus, the final result for the incident mass flux is

$$\Phi_i = \frac{1}{4} mn_i \sqrt{\frac{8kT_i}{\pi m}}\left[\exp\left(-s_3^2\right) + \sqrt{\pi} s_3 \left\{1 + erf(s_3)\right\}\right] \tag{1.154}$$

Let us consider two important limits:

(1) $s_3 \to 0$: No stream velocity in the r_3 direction, noting that $erf(0) = 0$:

$$\Phi_i = \frac{1}{4} mn_i \sqrt{\frac{8kT_i}{\pi m}} = \frac{1}{4} mn_i \langle C_i \rangle \tag{1.155}$$

Thus, even though there is no net velocity toward the surface, there is a finite amount of mass flux due to the random velocity components in the r_3 direction.

(2) $s_3 \to \infty$: Hypersonic stream in the r_3 direction, noting that $erf(\infty) = 1$:

$$\Phi_i = \frac{1}{4} m n_i \sqrt{\frac{8kT_i}{\pi m}} \sqrt{\pi} \frac{u_3}{\sqrt{\frac{2kT_i}{m}}} 2 = m n_i u_3 = \rho_i u_3 \tag{1.156}$$

In this case, the flux of particles due to the random motion is completely dominated by the flux due to the very high incident velocity.

(b) *Incident normal momentum flux* We set $Q = mC_3$, and name the associated flux quantity the incident pressure, p_i:

$$p_i = m n_i \left(\frac{m}{2\pi kT_i}\right)^{1/2} \int_{-u_3}^{\infty} (u_3 + C_3')^2 \exp\left(-\frac{mC_3'^2}{2kT_i}\right) dC_3'$$

$$= n_i kT_i \left\{ \frac{s_3}{\sqrt{\pi}} \exp\left(-s_3^2\right) + \left(\frac{1}{2} + s_3^2\right)(1 + erf(s_3)) \right\} \tag{1.157}$$

We consider the same two limits as before:

(1) $s_3 \to 0$: $p_i = \frac{1}{2} n_i kT_i$ This result is clearly similar to the ideal gas law except for the factor of one half. This factor arises from the fact that so far we have neglected the momentum exchange that occurs when the particle reflects from the surface.

(2) $s_3 \to \infty$: $p_i = 2 n_i kT_i \left(\frac{u_3^2}{\frac{2kT_i}{m}}\right) = m n_i u_3^2 = \rho_i u_3^2$

This result corresponds to the Newtonian pressure model often used in hypersonic vehicle analysis (Bertin 1994).

(c) *Incident energy flux*

We set $Q = \frac{1}{2} mC^2 = \frac{1}{2} m(C_1^2 + C_2^2 + C_3^2)$ and name the associated flux quantity the incident heat flux, $\dot{q}_i$. This expression for Q is not just a function of C_3 and so we must return to the full, original form of the incident flux equation, Eq. 1.146, to obtain

$$\dot{q}_i = \frac{1}{2} m n_i \left(\frac{m}{2\pi kT_i}\right)^{3/2}$$

$$\cdot \left\{ \int_{-\infty}^{\infty} C_1^2 \exp\left(-\frac{mC_1'^2}{2kT_i}\right) dC_1' \int_{-\infty}^{\infty} \exp\left(-\frac{mC_2'^2}{2kT_i}\right) dC_2' \int_{0}^{\infty} \exp\left(-\frac{mC_3'^2}{2kT_i}\right) dC_3' \right.$$

$$+ \int_{-\infty}^{\infty} \exp\left(-\frac{mC_1'^2}{2kT_i}\right) dC_1' \int_{-\infty}^{\infty} C_2^2 \exp\left(-\frac{mC_2'^2}{2kT_i}\right) dC_2' \int_{0}^{\infty} \exp\left(-\frac{mC_3'^2}{2kT_i}\right) dC_3'$$

$$\left. + \int_{-\infty}^{\infty} \exp\left(-\frac{mC_1'^2}{2kT_i}\right) dC_1' \int_{-\infty}^{\infty} \exp\left(-\frac{mC_2'^2}{2kT_i}\right) dC_2' \int_{0}^{\infty} C_3^2 \exp\left(-\frac{mC_3'^2}{2kT_i}\right) dC_3' \right\} \tag{1.158}$$

All of these integrals can be evaluated analytically and eventually lead to the final result of

$$\dot{q}_i = \frac{\Phi_i}{m}\left(\frac{1}{2}mu^2 + \frac{5}{2}kT_i\right) - \frac{1}{8}n_i\langle C_i\rangle kT_i \exp\left(-s_3^2\right) \tag{1.159}$$

Once again, we consider the results in the two limits:

(1) $s_3 \to 0$: $\dot{q}_i = \frac{1}{4}n_i\langle C_i\rangle\left\{\frac{1}{2}mu^2 + 2kT_i\right\}$ Note that we allow for finite kinetic energy of the free stream even if $u_3 = 0$. If the gas is completely at rest ($u_1 = u_2 = 0$) there is still finite heat transfer due to the random velocity components.

(2) $s_3 \to \infty$: $\dot{q}_i = \frac{n_i m u_3}{m}\left(\frac{1}{2}mu^2\right) = \frac{1}{2}\rho_i u_3 u^2$

NET FLUXES

To determine the net fluxes of mass, momentum, and energy, we must first evaluate the reflected fluxes. The reflected fluxes are obtained under the assumption that they originate from an imagined reservoir of gas in thermal equilibrium that lies below the real surface. The virtual gas is assigned a nondrifting Maxwellian VDF characterized by macroscopic parameters n_r, $\bar{u}_r = 0$, T_r, where the subscript r refers to the reflected gas.

We can make use of the same approach as we employed for the incident fluxes to write the corresponding general flux equation for the reflected flux of property Q as follows:

$$\Gamma_r^Q = \int_{C_1=-\infty}^{\infty}\int_{C_2=-\infty}^{\infty}\int_{C_3=-\infty}^{0} n_r C_3 Q(\bar{C}) f_r(\bar{C})\, d\bar{C} \tag{1.160}$$

Now, we can immediately write down the corresponding reflected flux properties since we have already evaluated the limiting case of zero velocity. Thus:

(a) *Reflected mass flux*

Use Eq. 1.154 with $(s_3)_r = 0$:

$$\Phi_r = \frac{1}{4}mn_r\sqrt{\frac{8kT_r}{\pi m}} \tag{1.161}$$

In the case where there is no chemistry on the surface, the net mass flux to the surface is zero, so

$$\Phi_r = \Phi_i \tag{1.162}$$

By equating Eqs. 1.154 and 1.161, we obtain the following expression for the number density of the fictional reservoir gas:

$$n_r = n_i\left(\frac{T_i}{T_r}\right)^{1/2}\left\{\exp\left(-s_3^2\right) + \sqrt{\pi}s_3[1 + erf(s_3)]\right\} \tag{1.163}$$

(b) *Normal momentum flux*

We now use Eq. 1.157 with $(s_3)_r = 0$ to find the reflected pressure:

$$p_r = \frac{1}{2} n_r k T_r = \frac{1}{2} n_i k (T_i T_r)^{1/2} \left\{ \exp\left(-s_3^2\right) + \sqrt{\pi} s_3 [1 + erf(s_3)] \right\} \quad (1.164)$$

The net normal momentum flux gives the overall surface pressure

$$p = p_i - (-p_r) = \text{Eq. } 1.157\text{–}1.164 \quad (1.165)$$

Let us consider a couple of specific, limiting cases:

(1) $s_3 \to 0$ and $T_i = T_r$: $p = n_i k T_i$ Thus, we obtain a result consistent with the perfect gas law when we account for both the incident and reflected momentum fluxes.

(2) $s_3 \to \infty$: $p = \rho_i u_3^2$

(c) *Energy flux*

Finally, we use Eq. 1.159 with $(s_3)_r = 0$:

$$\dot{q}_r = 2kT_r \frac{\Phi_r}{m} = 2kT_r \frac{\Phi_i}{m} \quad (1.166)$$

The net energy flux is then found as

$$\dot{q} = \dot{q}_i - \dot{q}_r = \text{Eq. } 1.159\text{–}1.166 \quad (1.167)$$

In experiments characterizing particle interactions with surfaces, it is found that reflected particles are not completely thermalized to the wall temperature, so that $T_r \neq T_w$. This phenomenon is referred to as *incomplete accommodation* and is often characterized by an accommodation coefficient, α, that represents the fraction of collisions that are *diffuse*. In diffuse reflection, a particle scatters from a surface with velocity components determined by the Maxwellian VDF at the surface temperature. Hence, $(1 - \alpha)$ represents the fraction of collisions that are specular. In specular reflection, the only change to a particle's velocity is that the component normal to the surface direction is reversed in sign. This results in zero shear stress and zero heat transfer. For engineering surfaces, typically the accommodation coefficient lies in the range of $0.8 \leq \alpha \leq 1.0$.

Introducing the accommodation coefficient into our analysis gives the following results:

$$p = \alpha(p_i + p_r) + (1 - \alpha)(2p_i) = (2 - \alpha)p_i + \alpha p_r \quad (1.168)$$

$$\dot{q} = \alpha(\dot{q}_i - \dot{q}_r) + (1 - \alpha)(0) = \alpha(\dot{q}_i - \dot{q}_r) \quad (1.169)$$

Example 1.1 We use our expressions above to calculate the profiles of surface pressure and heat transfer rate for the front surface of the *Sputnik* spacecraft (a sphere with a diameter of 58.5 cm) orbiting Earth at an altitude of 220 km.

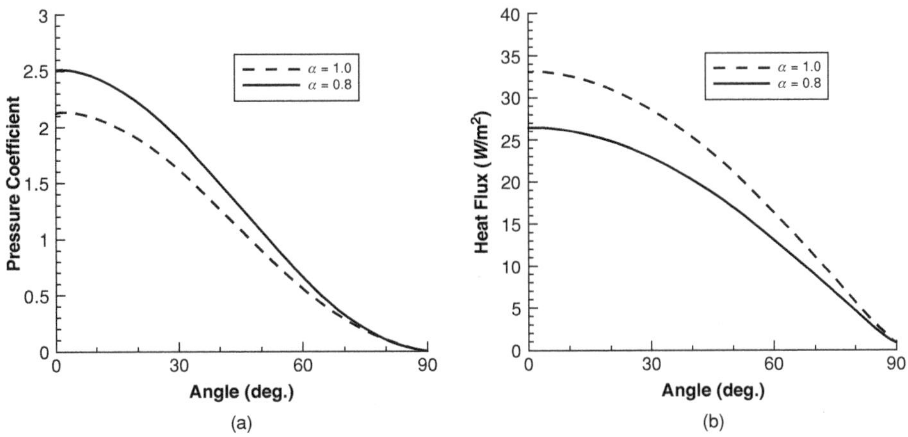

Figure 1.19 Free molecular analysis of the surface properties of the *Sputnik* spacecraft: (a) pressure coefficient; (b) heat flux.

Incident flow conditions: $p_i = 5.01 \times 10^{-5}$ Pa, $T_i = 899$ K, $\rho_i = 1.37 \times 10^{-10}$ kg/m^3, $U_i = 7,780$ m/s

Surface conditions: $T_r = 500$ K, $\alpha = 0.8$ and 1.0

Nondimensional parameters:

(i) Assuming a hard sphere diameter $d = 4 \times 10^{-10}$ m to evaluate the mean free path, and using the *Sputnik* diameter as the characteristic length scale:
- Knudsen number, Kn = 600, so the flow is free molecular.

(ii) Speed ratio, $s_3 = 9$, so we are in the hypersonic limit.

Plots are provided in Fig. 1.19 of pressure coefficient and heat flux as a function of angle around the sphere, with zero being the stagnation point. These results show that

- The coefficient of pressure is increased at lower accommodation coefficient.
- The heat flux is reduced at lower accommodation coefficient.

AERODYNAMIC FORCES

Following a similar approach for the incident and reflected fluxes of tangential momentum, the net result for shear stress is

$$\tau = \alpha p_i s_t \left\{ \frac{\exp\left(-s_3^2\right)}{\sqrt{\pi}} + s_3[1 + erf(s_3)] \right\} \tag{1.170}$$

where $s_t = \frac{(u_1^2+u_2^2)^{1/2}}{(2kT_i/m)^{1/2}}$. Using the definitions shown in Fig. 1.20, we can now employ our results for p and τ to evaluate the aerodynamic forces of lift and

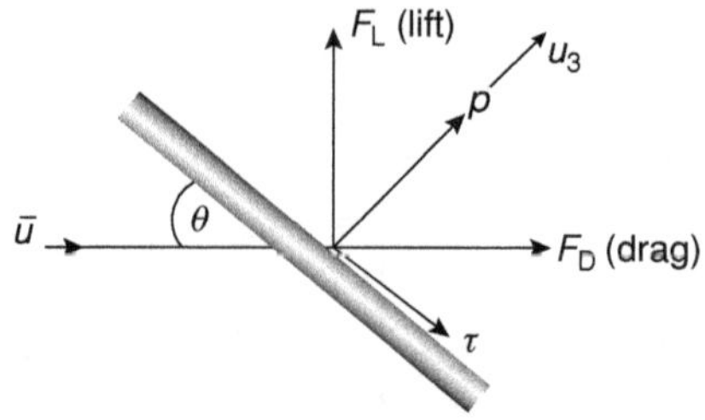

Figure 1.20 Aerodynamic forces on a surface element.

drag on a straight surface element of area δA:

$$\delta F_L = (p\cos(\theta) - \tau \sin(\theta))\delta A \tag{1.171}$$

$$\delta F_D = (p\sin(\theta) + \tau \cos(\theta))\delta A \tag{1.172}$$

The total aerodynamic forces of a complex shape may then be obtained by integrating these force contributions over all surface elements:

$$F_L = \int_S \delta F_L \,,\, F_D = \int_S \delta F_D \tag{1.173}$$

The aerodynamic coefficients are defined in the usual way:

$$C_L = \frac{2F_L}{\rho_i u_i^2 A}\,,\, C_D = \frac{2F_D}{\rho_i u_i^2 A} \tag{1.174}$$

For a flat plate of area A, and assuming fully diffuse reflection ($\alpha = 1$), we obtain the following results for the lift and drag coefficients:

$$C_L = \cos(\theta)\left[\frac{erf(s_3)}{s^2} + \sqrt{\pi}\frac{\sin(\theta)}{s_w}\right] \tag{1.175}$$

$$C_D = \frac{2}{\sqrt{\pi}}\left[\frac{\exp\left(-s_3^2\right)}{s} + \sqrt{\pi}\sin(\theta)\left(1 + \frac{1}{2s^2}\right)erf(s_3) + \pi\frac{\sin^2(\theta)}{s_w}\right] \tag{1.176}$$

where $s_w = \sqrt{\frac{mu^2}{2kT_w}}$.

In the hypersonic limit (taking each of s, s_w, and $s_3 \gg 1$), we obtain:

$$C_L \to \frac{\sin(\theta)}{\cos(\theta)} s\sqrt{\pi\frac{T_w}{T_i}} \Rightarrow F_L \propto s \tag{1.177}$$

$$C_D \to 2\sin(\theta) \Rightarrow F_D \propto s^2 \tag{1.178}$$

1.3.7 Kinetic-Based Analysis of Nonequilibrium Flow

So far, we have analyzed an equilibrium gas and free molecular flow. The former is characterized by very high collision rates and very small Knudsen number, whereas the latter is characterized by a lack of collisions and very high Knudsen number. In between these two limiting cases lies the regime where there are collisions, but not enough to maintain Maxwellian VDFs. The analysis of this regime requires solution of the Boltzmann equation in some form or other. Direct solution of the Boltzmann equation is challenging both analytically and computationally. In this section, we outline an approach that allows the derivation of sets of partial differential equations (PDEs) that can be used to model gas flows under conditions of equilibrium and near nonequilibrium. The second half of this book is dedicated to describing the direct simulation Monte Carlo (DSMC) computational method for analysis of strongly nonequilibrium gas flows.

Maxwell derived the Equation of Change by taking moments of the Boltzmann equation to produce

$$\frac{\partial}{\partial t}(n\langle Q\rangle) + \frac{\partial}{\partial \bar{r}}(n\langle \bar{C}Q\rangle) = [\dot{Q}] \tag{1.179}$$

where $Q(\bar{C})$ is a particle property, and $[\dot{Q}]$ is the rate of change of property Q due to intermolecular collisions. Various sets of PDEs can be formed by substituting different forms of Q into Eq. 1.179. We consider the two most widely used sets of equations in the following.

EULER EQUATIONS

We begin by making the assumption that the gas flow is locally in a state of thermal equilibrium everywhere. Thus, the local VDFs are Maxwellian and are given by local flow conditions of velocity and temperature. In addition, to maintain the VDFs in equilibrium, the local collision rate must be infinitely large, so that the local mean free path, λ, and therefore the local Knudsen number, are zero. In turn, this implies that the molecular transport coefficients are also zero because they are proportional to λ.

We now form a set of PDEs by inserting various forms of Q into Eq. 1.179. For the Euler equations, we use forms of Q that are kept constant during each collision, so that $[\dot{Q}] = 0$ in each case. Specifically, we use the particle mass, momentum, and energy, i.e., the collision invariants.

First, we use particle mass, $Q = m$, to obtain

$$\frac{\partial \rho}{\partial t} + \frac{\partial(\rho\langle \bar{C}\rangle)}{\partial \bar{r}} = 0 \tag{1.180}$$

We write the particle velocity using the random component as $\bar{C} = \bar{u} + \bar{C}'$ and note for a Maxwellian VDF that $\langle \bar{C}'\rangle = 0$ because of its symmetric nature. Thus

$$\frac{\partial \rho}{\partial t} + \frac{\partial(\rho \bar{u})}{\partial \bar{r}} = 0 \tag{1.181}$$

Next, we take particle momentum, $m\bar{C} = m(\bar{u} + \bar{C}')$, and this leads to

$$\frac{\partial\left(\rho\langle \bar{u} + \bar{C}'\rangle\right)}{\partial t} + \frac{\partial\left(\rho\langle(\bar{u} + \bar{C}')(\bar{u} + \bar{C}')\rangle\right)}{\partial \bar{r}} = 0 \tag{1.182}$$

Now: $\langle(\bar{u}+\bar{C}')(\bar{u}+\bar{C}')\rangle = (\bar{u}\bar{u} + 2\bar{u}\langle\bar{C}'\rangle + \langle\bar{C}'\bar{C}'\rangle)$ and thus

$$\frac{\partial(\rho\bar{u})}{\partial t} + \frac{\partial\left(\rho\bar{u}\bar{u} + \rho\langle\bar{C}'\bar{C}'\rangle\right)}{\partial\bar{r}} = 0 \tag{1.183}$$

For a Maxwellian VDF, $\rho\langle\bar{C}'\bar{C}'\rangle = p$, so we finally obtain

$$\frac{\partial(\rho\bar{u})}{\partial t} + \rho\bar{u}\frac{\partial\bar{u}}{\partial\bar{r}} + \frac{\partial p}{\partial\bar{r}} = 0 \tag{1.184}$$

Finally, we consider particle kinetic energy $Q = \frac{1}{2}mC^2$, that leads to

$$\frac{\partial\left(\frac{1}{2}\rho\langle C^2\rangle\right)}{\partial t} + \frac{\partial\left(\frac{1}{2}\rho\langle\bar{C}C^2\rangle\right)}{\partial\bar{r}} = 0 \tag{1.185}$$

For a Maxwellian VDF, $\langle C^2\rangle = 3p/\rho$, $\langle\bar{C}'C^2\rangle = 0$ so we finally obtain

$$\frac{3}{2}\frac{\partial p}{\partial t} + \frac{3}{2}\frac{\partial(p\bar{u})}{\partial\bar{r}} + p\frac{\partial\bar{u}}{\partial\bar{r}} = 0 \tag{1.186}$$

Thus, we have generated a set of five PDEs and these are often referred to as the Euler equations in macroscopic gas dynamics.

Let us consider at the molecular level what happens when a gas described by the Euler equations encounters a solid surface. Specifically, we need to consider boundary conditions for mass, momentum, and energy.

(1) Mass: We use the same approach as we considered in the free molecular analysis of Section 1.3.6, namely, that in the absence of surface chemistry, the incident and reflected mass fluxes must be equal: $\Phi_i = \Phi_r$.
(2) Momentum: Each particle undergoes *specular reflection* as this results in zero shear stress, that is consistent with there being no transport processes in this gas. At the same time, there will be finite pressure at the surface.
(3) Energy: Specular reflection also means that each collision of a particle with the surface is elastic, so that there is no heat transfer, and hence the surface interaction is described in thermodynamic terms as being *adiabatic*.

NAVIER–STOKES EQUATIONS

To derive higher fidelity gas dynamics equations from Maxwell's Equation of Change, it is necessary to consider a VDF that is not the equilibrium Maxwellian form and particle properties Q that are not collision invariants. Note in Eq. 1.179 that the time derivative of $\langle Q\rangle$ depends on the divergence of the next higher moment, $\langle\bar{C}Q\rangle$. This represents a closure problem when $[\dot{Q}] \neq 0$ that has been addressed in two different ways that give essentially identical final results and are described in more detail by (Gombosi 1994).

(1) *Chapman–Enskog approach*: Here, a specific form of the VDF is assumed that represents a small perturbation from the equilibrium Maxwellian:

$$f_{\mathrm{CE}}(\bar{C})\,d\bar{C} = \Gamma(\bar{C})f_{\mathrm{Maxwellian}}(\bar{C})\,d\bar{C} \tag{1.187}$$

where $\Gamma(\bar{C}) = \Gamma(\bar{C}, \tau, q)$ and is of $O(1)$.

(2) *Grad's method*: Specific relations are assumed between the second and fourth moments that provide closure.

Using $Q = m, m\bar{C}, \frac{1}{2}mC^2$ as we did to derive the five Euler equations, and further setting $Q = mC_iC_j$ and $Q = mC_iC_jC_k$, $i, j, k = 1, 3$ we obtain a set of 20 PDEs involving shear stress and heat flux tensors. Replacing the heat flux tensor by a vector reduces the set to 13 PDEs. Finally, by assuming that variations in shear stress and heat flux are small (and so their derivatives can be neglected), we obtain a set of five PDEs that are called the Navier–Stokes equations.

This set of equations is valid for Kn $< 10^{-2}$, allows for small perturbations from equilibrium in the VDF, and includes molecular transport processes. Once again, let us consider the boundary conditions for interaction of the gas with a solid surface.

(1) Mass: Same as for the Euler equations.
(2) Momentum: As described in Kennard (1938), Maxwell's slip velocity tangent to the surface is given by

$$u_t = \frac{2-\alpha_M}{\alpha_M}\frac{2\lambda_w}{3}\frac{\partial u}{\partial r_n} \tag{1.188}$$

where λ_w is the mean free path at the wall, and α_M is the *tangential momentum accommodation coefficient*, which is the fraction of particles reflected diffusely. When the local wall Knudsen number, $\lambda_w/\Delta r_n$, is small then $u_t \to 0$, and we recover the familiar no slip velocity condition. Alternatively, at finite wall Knudsen number, $u_t > 0$ so there will be velocity slip.
(3) Energy: Kennard (1938) also develops Smoluchowsky's temperature jump condition:

$$T_w - T_s = \frac{2-\alpha_T}{\alpha_T}\frac{2\lambda_w}{3}\frac{\partial T}{\partial r_n} \tag{1.189}$$

where T_w is the gas temperature at the wall, T_s is the surface wall temperature, and α_T is the *thermal accommodation coefficient* that is the fraction of particles whose energy is fully accommodated to the surface temperature. When the local wall Knudsen number, $\lambda_w/\Delta r_n$ is small then $T_w \to T_s$, and we recover the familiar isothermal wall condition. Alternatively, at finite wall Knudsen number, $T_w > T_s$ so there will be a temperature jump.

For analysis of equilibrium and near-equilibrium gas flows, the Euler and Navier–Stokes equations can be solved numerically using computational fluid dynamics (CFD) techniques. The next higher set of PDEs in the hierarchy is called the Burnett equations that involves using $Q = mC_iC_jC_kC_l$. This equation set is very difficult to solve numerically and requires higher order boundary conditions.

1.3.8 Free Molecular Flow Analysis

Finally, we describe a mathematical procedure for solution of a specific free molecular flow problem. Consider the Boltzmann equation without any external force and with the collision integral set to zero, i.e., free molecular or collisionless flow:

$$\frac{\partial(nf)}{\partial t} + \bar{C}\cdot\frac{\partial(nf)}{\partial\bar{r}} \tag{1.190}$$

For unsteady flow, this is an initial value problem. Consider a case in which

$$n(\bar{r}, t=0)f(\bar{C}, \bar{r}, t=0) = n_i(\bar{r})f_i(\bar{C}, \bar{r}) \tag{1.191}$$

Equation 1.190 has the form of the Liouville equation that has the important property that the solution (nf) is constant along characteristics of the equation:

$$n(\bar{r}, t)f(\bar{C}, \bar{r}, t) = n_i(\bar{r})f_i(\bar{C}, \bar{r}-\bar{C}t) \tag{1.192}$$

To find the average value of a particle property $Q(\bar{C})$, we use

$$n(\bar{r}, t)\langle Q(\bar{r}, t)\rangle = \int_{-\infty}^{\infty} Q(\bar{C})n_i(\bar{r})f_i(\bar{C}, \bar{r}-\bar{C}t)\,d\bar{C} \tag{1.193}$$

Now, using the transformation

$$\bar{r}' = \bar{r} - \bar{C}t \Rightarrow r_1' = r_1 - C_1 t \Rightarrow dr_1' = -dC_1 t \text{ and so}$$

$$d\bar{C} = \left|-\frac{1}{t^3}\right| d\bar{r}'$$

Thus,

$$n(\bar{r}, t)\langle Q(\bar{r}, t)\rangle = \frac{1}{t^3}\int Q\bar{C})n_i(\bar{r})f_i\left(\frac{\bar{r}-\bar{r}'}{t}, \bar{r}'\right)\,d\bar{r}' \tag{1.194}$$

where the integration limits are the physical domain of the gas at zero time.

To illustrate the solution approach, consider a homogeneous equilibrium gas separated from vacuum by a thin wall at $r_1 = 0$. At $t = 0$, the wall is removed, and the gas expands freely into the vacuum. Provided that $r_1 \ll \lambda$ we will have collisionless flow and Eq. 1.194 is valid. Let us consider a one-dimensional expansion for which

$$n(r_1, t)\langle Q(r_1, t)\rangle = \frac{1}{t^3}\int Q(\bar{C})n_i(r_1)f_i\left(\frac{r_1 - r_1'}{t}, r_1'\right)\,dr_1' \tag{1.195}$$

where $r_1' = r_1 - C_1 t$. Recall the one-dimensional Maxwellian VDF from Eq. 1.40 for zero initial velocity:

$$\phi(C_1)\,dC_1 = \frac{\beta}{\sqrt{\pi}}\exp\left\{-\beta^2 C_1^2\right\}dC_1 \tag{1.196}$$

where we introduce $\beta = \sqrt{\frac{m}{2kT}}$ for convenience. For given initial density and temperature:

$$n(r_1,t)\langle Q(r_1,t)\rangle = \frac{1}{t}\int_{-\infty}^{0} \exp\left(-\frac{\beta_i^2(r_1-r_1')^2}{t^2}\right) d\left(\frac{\beta_i r_1'}{t}\right) \tag{1.197}$$

Using this expression, let us develop solutions for the first two moments of the distribution.

(1) Density: Set $Q = 1$ (the zeroth-order moment of the VDF)

$$\frac{n(r_1,t)}{n_i(r_1)} = \frac{1}{\sqrt{\pi}}\int_{-\infty}^{0} \exp\left(-\frac{\beta_i^2(r_1-r_1')^2}{t^2}\right) d\left(\frac{\beta_i r_1'}{t}\right) \tag{1.198}$$

Multiply each side by 2 and let

$$\frac{\beta_i(r_1-r_1')}{t} = Z \Rightarrow d\left[-\frac{\beta_i r_1'}{t}\right] = dZ \tag{1.199}$$

to obtain

$$\frac{2n(r_1,t)}{n_i(r_1)} = \frac{2}{\sqrt{\pi}}\int_{\frac{\beta_i r_1}{t}}^{\infty} \exp(-Z^2)\,dZ = 1 - \frac{2}{\sqrt{\pi}}\int_{0}^{\frac{\beta_i r_1}{t}} \exp(-Z^2)\,dZ \tag{1.200}$$

Introducing the error function:

$$erf(a) \equiv \frac{2}{\sqrt{\pi}}\int_{0}^{a} \exp(-x^2)\,dx \tag{1.201}$$

and the complementary error function $erf\,c(a) = 1 - erf(a)$, we obtain the solution

$$\frac{n(r_1,t)}{n_i(r_1)} = \frac{1}{2} erf\,c\left(\frac{\beta_i r_1}{t}\right) \tag{1.202}$$

(2) Velocity: Set $Q = C_1$, and use standard integrals:

$$n(r_1,t)\langle C_1(r_1,t)\rangle = \frac{n1(r_1)}{2\sqrt{\pi}\beta_i}\exp\left[-\left(\frac{\beta_i r_1}{t}\right)^2\right] \tag{1.203}$$

so that

$$\langle C_1(r_1,t)\rangle = \frac{1}{\sqrt{\pi}\beta_i}\frac{\exp\left[-\left(\frac{\beta_i r_1}{t}\right)^2\right]}{erf\,c\left(\frac{\beta_i r_1}{t}\right)} \tag{1.204}$$

As an illustrative example, the plots in Fig. 1.21 show solutions for $-\lambda_i \leq r_1 \leq \lambda_i$ at $t = (0.1, 0.2, 0.3)\ \tau_i = \lambda_i/\langle C\rangle$.

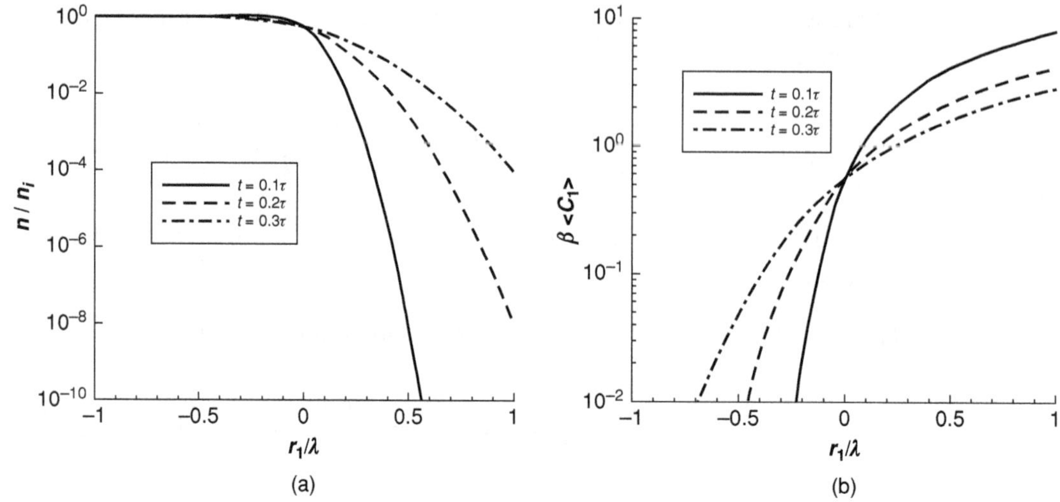

Figure 1.21 Collisionless jet expansion: profiles of (a) number density and (b) velocity.

Note: Since $erf(0) = 0$ and $erfc\,(0) = 1$, then at $r_1 = 0$ for all time:

$$\frac{n}{n_i} = \frac{1}{2}, \langle C_1 \beta_i \rangle = \frac{1}{\sqrt{\pi}} \tag{1.205}$$

From these plots, we can see for our expanding gas system that the density falls in the container and increases in the jet because of particle transport. At the same time, the velocity rises in the container, due to depletion of the population with negative values of C_1, and falls in the jet, due to an increase of the population with small, positive values of C_1.

1.4 Summary

In this chapter, we introduced the basic ideas of kinetic theory starting from the basic properties of an individual particle and building up to the Boltzmann equation that describes the evolution of the velocity distribution function in time and space. The equilibrium solution of the Boltzmann equation allowed us to study the most important collision properties of a gas. These results were used to develop results for the basic flux properties onto a surface under free molecular conditions, allowing analysis of aerodynamic forces and heat transfer to a vehicle. The connections between kinetic theory and the familiar macroscopic equations of fluid mechanics were established. Flow field analysis of an unsteady, expanding jet was analyzed under free molecular conditions. Many of the most important ideas from kinetic theory are employed in Chapter 4 to study finite-rate processes, and are seen widely in the second part of the book in providing the fundamental basis of molecular simulation methods.

1.5 Problems

1.1 The Knudsen number (Kn) is defined as the ratio of the mean free path to the characteristic length of the flow. It is generally accepted that free molecular flow (i.e., collisionless flow) occurs for $\mathrm{Kn} > 1$, and continuum flow occurs for $\mathrm{Kn} < 0.01$. In between lies the transition regime. Use the following simple model for the variation of the atmospheric density with altitude to determine the altitude range for which the NASA Crew Exploration Vehicle (CEV) capsule may be regarded as being in transition flow:

$$\rho/\rho_0 = \exp(-\alpha h)$$

where $\rho_0 = 1.225\ \mathrm{kg/m^3}, \alpha = 1.4 \times 10^{-4}$ per meter, h is altitude in meters. (Note: Average air particle has mass 4.8×10^{-26} kg and diameter 4×10^{-10} m. The diameter of the CEV is 5 m and this may be taken as the characteristic length of its flight.)

1.2 Consider an equilibrium mixture of two species A and B with diameters d_A and d_B. Find simple expressions for the collision frequency of one A-particle with all other particles, and for the mean free path of A-particles.

1.3 Use the standard definitions of Knudsen number ($\mathrm{Kn} = \lambda/L$), Reynolds number ($\mathrm{Re} = \rho u L/\mu$), Mach number ($\mathrm{Ma} = u$/sound-speed), and viscosity coefficient (μ) of a hard-sphere gas to relate Kn as a function only of Re, Ma, and γ. What does this relation tell us about conditions for nonequilibrium? [Note: $\langle C \rangle = \sqrt{8RT/\pi}$, sound-speed $= \sqrt{\gamma RT}$]

1.4 Consider a monatomic gas with temperature and velocity gradients dT/dr_2 and du_1/dr_2. Using the methods outlined in this chapter, show the flux of molecular energy in the r_2-direction is given by

$$\Lambda = -\kappa \frac{dT}{dr_2} - \frac{5}{2} u_1 \mu \frac{du_1}{dr_2} = q - \frac{5}{2} u_1 \tau$$

where κ is the coefficient of thermal conductivity, μ is the viscosity coefficient, q is the heat flux, and τ is the shear stress.

1.5 Consider air at an altitude of 5 km for which $T = 256$ K, $p = 54{,}000$ $\mathrm{N/m^2}$, $\rho = 0.736\ \mathrm{kg/m^3}$; $\mu = 1.33 \times 10^{-5}$ kg/m/s. The average molecular weight of air is 28.9 kg/kg-mol. Using these values, provide estimates of the following quantities in SI units:
(a) Molecular speed
(b) Molecular mass
(c) Molecular diameter
(d) Number density
(e) Mean free path
(f) Collision frequency

1.6 A simple gas of mass m in equilibrium with a number density n and temperature T is enclosed in a container. The gas escapes into a vacuum through a small circular hole of area A in the container. If the wall thickness is infinitely small, and the hole size is smaller than a mean free path:

(a) Show that the number of molecules escaping from the hole per unit time per unit area is $n\langle C\rangle/4$.

(b) Determine the normalized velocity distribution for the particles escaping through the hole.

(c) Show that the mean kinetic energy of escaping particles is 4/3 greater than that for the particles in the container. Explain this difference.

1.7 Consider a simple gas of particles with mass m in a state of thermal equilibrium at temperature T.

(a) Obtain an expression for the probability of particles having kinetic energy $\epsilon = 1/2mC^2$ in the range ϵ to $\epsilon + d\epsilon$.

(b) Show that the most probable value of ϵ is $kT/2$.

(c) Show that $\langle\epsilon\rangle$ = 3kT/2 (Hint: You may need to use the gamma function, noting that $\Gamma(1/2) = \sqrt{\pi}$ and $\Gamma(x+1) = x\Gamma(x)$ where

$$\Gamma(x) = \int_0^\infty t^{x-1} \exp(-t)\, dt$$

(d) Using the error function

$$erf(x) = \frac{2}{\sqrt{\pi}} \int_0^x \exp(-z^2)\, dz$$

show that the fraction of particles with kinetic energy greater than a specified value ϵ is

$$1 + \frac{2}{\sqrt{\pi}} \sqrt{\frac{\epsilon}{kT}} \exp\left(-\frac{\epsilon}{kT}\right) - erf\left(\sqrt{\frac{\epsilon}{kT}}\right)$$

1.8 For each part, generate a single plot of the following distribution functions under equilibrium conditions and comment on any differences:

(a) Distribution of speed with mean velocity = 0 m/s

i. For xenon at T = 300 K

ii. For argon at T = 300 K

(b) Distribution for velocity in one direction

i. For argon at T = 300 K with mean velocity = 0 m/s

ii. For argon at T = 3000 K with mean velocity = 0 m/s

iii. For argon at T = 300 K with mean velocity = 1000 m/s

Molecular weights: xenon = 131.3 kg/kg-mol, argon = 40 kg/kg-mol.

1.9 Spheres of radius 1 nm (= 10^{-9} m) are used as part of a flow visualization diagnostic in air at STP (pressure of 1 atm, and temperature of 288 K) that has a horizontal flow velocity of 10 m/s. Assume each sphere is at rest, has a temperature of 288 K, the air molecules reflect diffusely from the spheres, and are hard spheres of diameter 4×10^{-10} m. Do the following:

(a) Determine the regime of the air flow around a single sphere.

(b) Using results from this Chapter, obtain expressions for the surface pressure and heat transfer rate on a general surface element of the sphere.

(c) Plot the angular variation of pressure and heat transfer rate for the upper half of the sphere. Comment on your results.

2 Quantum Mechanics

2.1 Introduction

In kinetic theory, we analyzed particle motions and collisions without concerning ourselves about the details of their internal structure. In this chapter, we address internal energy structure without analysis of collisions. Our discussion concerns *quantum physics*. Based on postulates, we will develop the Schrödinger equation. We will solve this equation to determine the allowable, quantized energy levels for a particular chemical species. Finally, we will briefly review the naming conventions employed for identification of the energy levels.

2.2 Quantum Mechanics

The most basic concept of quantum mechanics is wave–particle duality. The dual nature of radiation was formulated by de Broglie, who combined two separate ideas:

(1) Einstein proposed that an energy field radiating at frequency ν (or angular frequency ω) can be modeled as a collection of photons (particles) each of energy:

$$\epsilon = h\nu = \hbar\omega \tag{2.1}$$

where h is Planck's constant $= 6.626 \times 10^{-34}$ J · s, and $\hbar \equiv h/2\pi$.

(2) Compton proposed that an energy field radiating at wavelength $\lambda = c/\nu$ (or wave number $\bar{k}$, where c is the speed of light) can be modeled as a collection of photons each having momentum:

$$p = \frac{h}{\lambda} = \hbar\bar{k} \tag{2.2}$$

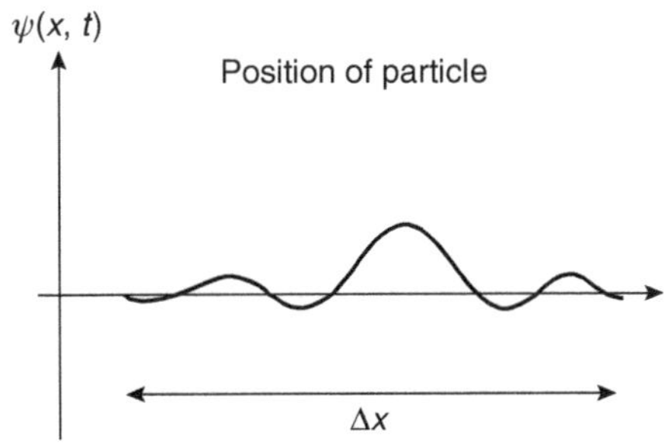

Figure 2.1 Wave packet representation of particle behavior.

De Broglie turned these ideas around by proposing that a particle of energy ϵ and momentum p can be modeled as a wave with the following properties:

- frequency (ν)

$$\nu = \frac{\epsilon}{h} \text{ or } \omega = \frac{\epsilon}{\hbar} \tag{2.3}$$

- wavelength (λ)

$$\lambda = \frac{h}{p} \text{ or } \bar{k} = \frac{p}{\hbar} \tag{2.4}$$

These simple equations involve a difficult physical question: How can a particle in a gas behave like a wave?

Before answering this question, let us first develop a general mathematical framework. The waves associated with particles are called *pilot waves*. The wavelike properties are localized around the physical position of a particle as illustrated in Fig. 2.1 and lie within a finite region Δx close to the particle. Each pilot wave is modeled as a *harmonic oscillator* with the general formulation

$$\exp[i(\bar{k}x - \omega t)] \tag{2.5}$$

To describe a real wave associated with a particle, a number of pilot waves are superimposed using Fourier transformations to form a *wave packet*:

$$\psi(x, t) = \int_{\bar{k}}^{\bar{k}+\Delta\bar{k}} a(\bar{k}) \exp[i(\bar{k}x - \omega t)]\, d\bar{k} \tag{2.6}$$

which represents the amplitude of the wave packet as a function of time and space.

The function $a(\bar{k})$ is a weighting factor for the contribution of the pilot wave of wave number $\bar{k}$ to the wave packet. The wave packet travels with group velocity that is consistent with the particle velocity and is given by

$$V_g = \frac{d\omega}{d\bar{k}} = \frac{d\epsilon}{dp} \tag{2.7}$$

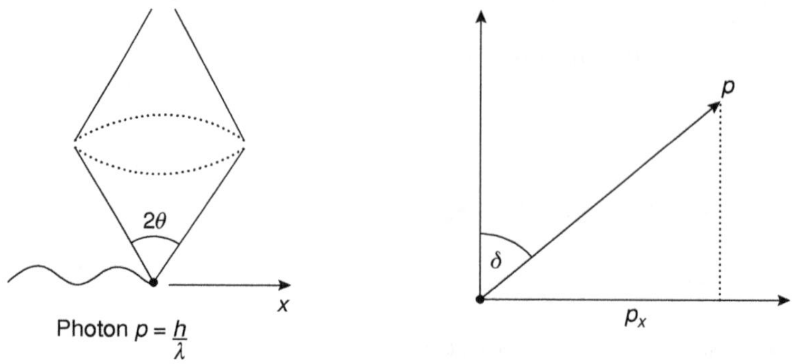

Figure 2.2 Thought experiment for measurement of particle properties.

2.2.1 Heisenberg Uncertainty Principle

At the macroscopic level, we assume that we can exactly determine gas properties such as density and temperature. The measurement of such quantities uses large sample sizes of particles, e.g., 10^{20} in a cubic centimeter of air at sea level, and any uncertainties at the molecular level are averaged out. This is not the situation for measuring the properties of a single particle. Uncertainties in determining such properties give rise to the wavelike behavior of particles, and are quantified by the *Heisenberg Uncertainty Principle*.

As a thought experiment, consider the motion of a single particle in the x-direction for which we want to measure its position and momentum, simultaneously. We observe the particle using a microscope subtending an angle 2θ at the particle location, as shown in Fig. 2.2. To "see" the particle, it must be illuminated by a photon that has momentum

$$p = \frac{h}{\lambda} \tag{2.8}$$

To be collected by the microscope, the photon must collide with the particle and be scattered into the angle 2θ. In this process, the momenta of both the particle and the photon will be changed. Thus, to detect the particle, its properties must be changed.

After scattering, the x-component of momentum of the photon is given by

$$p_x = p \sin\delta = \frac{h}{\lambda}\sin\delta \tag{2.9}$$

To be observed by the microscope

$$|\delta| \le \theta \Rightarrow -\frac{h}{\lambda}\sin\theta \le p_x \le \frac{h}{\lambda}\sin\theta \tag{2.10}$$

The uncertainty in p_x is therefore given by

$$|\Delta p_x| \approx \frac{h}{\lambda}\sin\theta \tag{2.11}$$

From optics, the resolving power of a microscope for light made up of photons of wavelength λ is

$$\Delta x \approx \frac{\lambda}{\sin\theta} \tag{2.12}$$

which is the accuracy of detecting the position of the particle. Therefore, the total uncertainty of determining particle momentum and position simultaneously is

$$\Delta x \cdot \Delta p_x \approx h = 6.6 \times 10^{-34}\ \mathrm{J \cdot s} \tag{2.13}$$

This is the mathematical form of the *Heisenberg Uncertainty Principle.* The magnitude of the uncertainty is negligible at the macroscopic level. The significance of the result is that it indicates that a *statistical approach* is needed to analyze the dual nature of particles.

Correspondence Principle

Proposed by Bohr, the Correspondence Principle states that *any quantum mechanical theory must satisfy the laws of classical physics in a regime where classical physics is valid.* Bohr defined the valid regime for classical physics as systems with large quantum numbers, and more generally we consider it to be a regime in which quantum effects are not important. The significance of the Correspondence Principle is that it allows us to use ideas from classical physics to construct concepts for quantum mechanics.

2.2.2 The Schrödinger Equation

The Schrödinger equation is the fundamental governing equation of quantum mechanics and is derived by invoking a number of fundamental postulates. Solution of the Schrödinger equation for a particular system yields the wave function, $\psi(\bar{r}, t)$. Further postulates provide mathematical operators that yield particle properties when applied to $\psi(\bar{r}, t)$. Note that these properties are known only within the limits imposed by the Heisenberg Uncertainty Principle.

Postulates

The first postulates concern the association of waves with particles. The waves are not related to real physical processes, but rather to the uncertainties associated with determination of particle properties.

Postulate 1

A wave function $\psi(\bar{r}, t)$ exists that can describe the properties of each physical system.

Postulate 2 (The Born Postulate)

The probability of locating a particle at time t in a volume $d\bar{r}$ at position $\bar{r}$ is

$$P(\bar{r}, t)\, d\bar{r} = \psi^*(\bar{r}, r)\psi(\bar{r}, t)\, d\bar{r} \tag{2.14}$$

where ψ^* is the complex conjugate of ψ. Thus

$$P(\bar{r}, t) = \psi^* \psi = |\psi^2| \tag{2.15}$$

is the probability density, and $\psi(\bar{r}, t)$ is the probability amplitude. Since the particle must be located somewhere in physical space, we can introduce a normalization condition:

$$\int_{-\infty}^{\infty} \psi^* \psi \, d\bar{r} = 1 \tag{2.16}$$

Wave functions of real systems are made to satisfy this condition by writing

$$\psi_n = \frac{\psi}{C^{1/2}}, \psi_n^* = \frac{\psi^*}{C^{1/2}} \tag{2.17}$$

where C is a constant independent of time.

Postulate 3

The dynamical variables of position, momentum, and energy of a particle are given by the following linear operators that act on the wave function:

$$\bar{r}_{\text{op}} = \bar{r} \tag{2.18}$$

$$\bar{p}_{\text{op}} = -i\hbar\nabla \tag{2.19}$$

$$\bar{\epsilon}_{\text{op}} = i\hbar\frac{\partial}{\partial t} \tag{2.20}$$

In general, for each particle quantity, $B(\bar{r}, \bar{p})$, there exists an operator $B_{\text{op}} = B(\bar{r}, -i\hbar\nabla)$. In this way, we can directly relate classical mechanics to quantum mechanics.

Expectation (Mean) Values The Heisenberg Uncertainty Principle says that particle properties cannot be determined exactly. We can, however, evaluate the expectation or mean value for a given wave function ψ. If the probability that a particle lies between r and $r + dr$ is $P(r, t)\,dr$ then the expectation (mean) particle position is:

$$\langle r \rangle = \int_{-\infty}^{\infty} rP(r, t)\,dr \tag{2.21}$$

Now, using the second (Born) postulate:

$$\langle r \rangle = \int_{-\infty}^{\infty} \psi(r, t) r \psi(r, t)\,dr \tag{2.22}$$

where the normalized wave functions are used. This idea is generalized as follows.

Postulate 4 The expectation value of any dynamic variable $B(\bar{r}, \bar{p})$ associated with operator $B_{op}(\bar{r}, -i\hbar\nabla)$ is

$$\langle B \rangle = \int_{-\infty}^{\infty} \psi^*(\bar{r}, t) B_{op} \psi(\bar{r}, t) \, d\bar{r} \tag{2.23}$$

For example, consider momentum in the x-direction:

$$\langle p_x(x, t) \rangle = \int_{-\infty}^{\infty} \psi^*(x, t) \left(-i\hbar \frac{\partial}{\partial x} \right) \psi(x, t) \, dx = -i\hbar \int_{-\infty}^{\infty} \psi^*(x, t) \frac{\partial \psi(x, t)}{\partial x} \, dx \tag{2.24}$$

Some of the problems provided at the end of the chapter require the reader to prove that these postulated operators and definitions do indeed provide exact correspondence between wave and particle descriptions of physical properties.

The Schrödinger Equation

From classical physics, the total energy of a particle is given by the sum of its kinetic and potential energies (the Hamiltonian):

$$\epsilon = \frac{p^2}{2m} + V(\bar{r}, t) \tag{2.25}$$

where $V(\bar{r}, t)$ is the potential field in which the particle moves. For our purposes, the potential field will be that established from electrostatic forces generated between the protons and electrons of real atoms and molecules. Using the Correspondence Principle, Schrödinger replaced the classical variables in Eq. 2.25 with the corresponding quantum mechanical values using the operators defined in Postulate 3. Thus

$$-\frac{\hbar^2}{2m} \nabla^2 + V(\bar{r}, t) = i\hbar \frac{\partial}{\partial t} \tag{2.26}$$

We now apply this operator identity to the wave function, $\psi(\bar{r}, t)$:

$$-\frac{\hbar^2}{2m} \nabla^2 \psi(\bar{r}, t) + V(\bar{r}, t) \psi(\bar{r}, t) = i\hbar \frac{\partial \psi(\bar{r}, t)}{\partial t} \tag{2.27}$$

and this is the Schrödinger equation. Clearly, this is a second-order partial differential equation that has in general complex solutions. If we now employ separation of variables in the space and time domains using $\psi(\bar{r}, t) \equiv \psi(\bar{r}) T(t)$ and focus on the spatial component, we obtain the time-independent *Schrödinger equation*:

$$-\frac{\hbar^2}{2m} \nabla^2 \psi(\bar{r}) + V(\bar{r}, t) \psi(\bar{r}) = \epsilon \psi(\bar{r}) \tag{2.28}$$

Solution of Eq. 2.28 represents an eigenvalue problem. Physical solutions are obtained only for certain discrete (or quantized) values of energy ϵ and are determined by specification of auxiliary conditions. Postulate 2 is one such condition, and others are stated as follows.

Postulate 5

The amplitude function $\psi(\bar{r})$ obtained by solving the time-independent Schrödinger equation, and its first derivative, $\nabla\psi(\bar{r})$, must be finite, continuous and single valued. This requires the waves associated with particles to have reasonable physical behavior.

In summary, for the potential functions $V(\bar{r})$ of real particles, acceptable solutions of the time-independent Schrödinger equation occur only for discrete, quantum, eigenvalues:

$$\epsilon_0, \epsilon_1, \epsilon_2, \ldots, \epsilon_n, \ldots$$

where subscript n is called the quantum number. The corresponding eigenfunctions are the associated wave functions:

$$\psi_0, \psi_1, \psi_2, \ldots, \psi_n, \ldots$$

Degeneracy The *energy level* is the amount of energy measured in Joules. In practical problems, it often occurs that several linearly independent eigenfunctions may be associated with the same *energy level.* This occurs when more than one quantum number is required to describe the system. Each set of these multiple quantum numbers is defined as an energy state. Thus, there may be several degenerate *energy states* at the same *energy level.*

For example, a diatomic molecule has quantized rotational and vibrational energies described by quantum numbers J and v, respectively. Cases arise where different combinations of J and v give the same total energy level.

2.2.3 Solutions of the Schrödinger Equation

We now consider solutions of the time-independent Schrödinger equation for the different forms of particle energy.

TRANSLATIONAL ENERGY

Consider a particle in motion inside a cube of length L in the absence of any field forces ($V(\bar{r}) = 0$ in Eq. 2.28). Thus

$$-\frac{\hbar^2}{2m}\left(\frac{\partial^2\psi}{\partial x^2} + \frac{\partial^2\psi}{\partial y^2} + \frac{\partial^2\psi}{\partial z^2}\right) = \epsilon\psi \tag{2.29}$$

In terms of boundary conditions, the particle cannot exist on or outside the cube walls, thus

$$\psi(0, y, z) = \psi(x, 0, z) = \psi(x, y, 0) = 0 \tag{2.30}$$

$$\psi(L, y, z) = \psi(x, L, z) = \psi(x, y, L) = 0 \tag{2.31}$$

Equation 2.29 is solved using separation of variables:

$$\psi(x, y, z) = \psi_1(x)\psi_2(y)\psi_3(z) \tag{2.32}$$

resulting in three partial differential equations:

$$\begin{aligned} \frac{\partial^2 \psi_1}{\partial x^2} + \frac{2m}{\hbar^2}\epsilon_1\psi_1 = 0 \\ \frac{\partial^2 \psi_2}{\partial y^2} + \frac{2m}{\hbar^2}\epsilon_2\psi_2 = 0 \\ \frac{\partial^2 \psi_3}{\partial z^2} + \frac{2m}{\hbar^2}\epsilon_3\psi_3 = 0 \end{aligned} \tag{2.33}$$

where ϵ_1, ϵ_2, and ϵ_3 are the translational energies of the particle in the x, y, and z directions, respectively, such that the total translational energy is

$$\epsilon = \epsilon_1 + \epsilon_2 + \epsilon_3 \tag{2.34}$$

By using the boundary conditions, Eq. 2.31, it may be shown that the eigenvalues in the x direction are

$$\epsilon_1 = \frac{h^2}{8m}\frac{n_1^2}{L^2}, \; n_1 = 1, 2, 3, \ldots \tag{2.35}$$

where the integer n_1 is the quantum number for translational motion in the x direction. Further, using the normalization condition (Postulate 2), the corresponding eigenfunctions are

$$\psi_1 = \left(\frac{2}{L}\right)^{1/2} \sin\left[\left(\frac{2m\epsilon_1}{\hbar^2}\right)^{1/2} x\right] \tag{2.36}$$

Combining with the analogous results in y and z gives the overall results:

$$\epsilon = \frac{h^2}{8mL^2}(n_1^2 + n_2^2 + n_3^2) \tag{2.37}$$

$$\psi = \left(\frac{8}{L^3}\right)^{1/2} \sin\left(\frac{n_1\pi x}{L}\right) \sin\left(\frac{n_2\pi y}{L}\right) \sin\left(\frac{n_3\pi z}{L}\right) \tag{2.38}$$

where the integers n_1, n_2, n_3 are the translational quantum numbers in the x, y, and z directions, and are all greater than zero. Thus, we obtain the surprising result that translational energy of a particle is quantized, and it can take on only certain clearly specified values. It is also rather nonintuitive that the translational energy levels should depend on the length scale, L. This is related to the number of discrete waves that can be accommodated within a finite region of space. If we consider the ground state translational energy in any one direction, for which the translational quantum number is 1, then the wave function will be exactly the first half of a sine wave with zeros at 0 and L. Thus, this basic wave is directly scaled with L and this property applies to all of the quantized energy levels. Owing to the very small quantum spacing of the translational energy mode, this question really becomes important only

Table 2.1 First Few Quantized Translational Energy Levels

ϵ	n_1	n_2	n_3	g
$\frac{3h^2}{8mL^2}$	1	1	1	1
	2	1	1	
$\frac{6h^2}{8mL^2}$	1	2	1	3
	1	1	2	
	2	2	1	
$\frac{9h^2}{8mL^2}$	2	1	2	3
	1	2	2	

at very small length scales, but it is important to note that this fundamental quantum property can be put to good use in constraining physical processes in nanoscale technologies.

Table 2.1 lists the lowest translational energy levels obtained with different sets of the translational quantum numbers, and the final column indicates the degeneracy, g.

Degeneracy occurs at many levels of translational energy and will increase at higher energy levels. Although translational energy is quantized, the separation between adjacent energy levels is extremely small. Consider the particular case of a molecule of nitrogen (N_2) in a 1 cm^3 cube, for which

$$\Delta\epsilon_t = \frac{h^2}{8\,mL^2} = 10^{-38}\,\text{J} \tag{2.39}$$

By comparison, using a result from classical kinetic theory, the average translational energy at room temperature is

$$\langle\epsilon_{\text{tr}}\rangle = \frac{3}{2}kT = 6\times 10^{-21}\,\text{J} \tag{2.40}$$

Thus, there will be an enormous number (approximately 10^{17}) quantized levels that lie below the average translational energy. So, in analyzing the spectrum of translational energy, the spacing between levels is so small that the energy distribution can be approximated by a continuous (nonquantized) function. This is the approach we used in kinetic theory.

2.2.4 Two-Particle System

To derive the quantum energy states for the rotational, vibrational, and electronic modes, we need to consider a general, two-particle system. Let one of the particles have position (x_1, y_1, z_1) and momentum p_1, and the other particle has position (x_2, y_2, z_2) and momentum p_2. The Hamiltonian for the combined system is

$$\frac{p_1^2}{2m_1} + \frac{p_2^2}{2m_2} + V = \epsilon \tag{2.41}$$

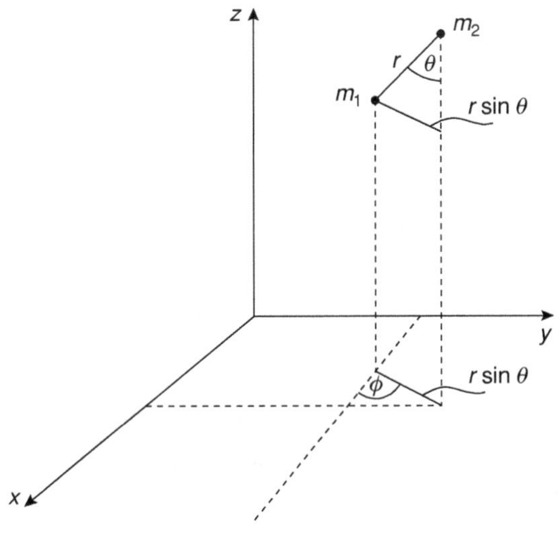

Figure 2.3 Two-particle coordinate system.

Following a similar approach as we used earlier, we can develop the associated time-independent Schrödinger equation:

$$\frac{\hbar^2}{2m_1}\nabla_1^2\psi + \frac{\hbar^2}{2m_2}\nabla_2^2\psi + (\epsilon - V)\psi = 0 \tag{2.42}$$

where $\psi = \psi(x_1, y_1, z_1, x_2, y_2, z_2)$, and $V = V(x_1, y_1, z_1, x_2, y_2, z_2)$.

We now introduce two spatial transformations as illustrated in Fig. 2.3:
(1) Center of mass coordinates:

$$X = \frac{m_1 x_1 + m_2 x_2}{m_1 + m_2}$$
$$Y = \frac{m_1 y_1 + m_2 y_2}{m_1 + m_2}$$
$$Z = \frac{m_1 z_1 + m_2 z_2}{m_1 + m_2}$$

(2) Relative coordinates:

$$x = x_2 - x_1 = r\sin\theta\cos\phi$$
$$y = y_2 - y_1 = r\sin\theta\sin\phi$$
$$z = z_2 - z_1 = r\cos\theta$$

Next, we assume that the overall potential can be separated as follows:

$$V(X, Y, Z, x, y, x) = V_e(X, Y, Z) + V_i(x, y, z) \tag{2.43}$$

where V_e is the external field acting on the combined system at its center of mass, and V_i is the internal field controlling the interaction between the two particles.

Next, we use separation of variables on the wave function:

$$\psi(X, Y, Z, x, y, z) = \psi_e(X, Y, Z)\psi_i(x, y, z) \tag{2.44}$$

to obtain

$$\frac{\hbar^2}{2m_t}\left(\frac{\partial^2\psi_e}{\partial X^2} + \frac{\partial^2\psi_e}{\partial Y^2} + \frac{\partial^2\psi_e}{\partial Z^2}\right) + (\epsilon_e - V_e)\psi_e = 0 \tag{2.45}$$

$$\frac{\hbar^2}{2\mu}\left(\frac{\partial^2\psi_i}{\partial x^2} + \frac{\partial^2\psi_i}{\partial y^2} + \frac{\partial^2\psi_i}{\partial z^2}\right) + (\epsilon_i - V_i)\psi_i = 0 \tag{2.46}$$

where the total mass, $m_t = m_1 + m_2$ and the reduced mass, $\mu = \frac{m_1 m_2}{m_1+m_2}$.

The total energy of the system is $\epsilon = \epsilon_i + \epsilon_e$ in which ϵ_i is the energy due to *internal motion*, and ϵ_e is the energy due to *center of mass motion*.

Now, we transform the internal wave function into spherical polar coordinates and separate variables:

$$\psi_i(x, y, z) = R(r)\Phi(\phi)\Theta(\theta) \tag{2.47}$$

$$r \in [0, \infty], \phi \in [0, 2\pi], \theta \in [0, \pi]$$

For the component Φ, the following ordinary differential equation is obtained:

$$\frac{d^2\Phi}{d\phi^2} + \beta\Phi = 0 \tag{2.48}$$

that has the following complex solution:

$$\Phi(\phi) = \exp(i\beta^{1/2}\phi) = \cos(\beta^{1/2}\phi) + i\sin(\beta^{1/2}\phi) \tag{2.49}$$

Postulate 5 says that Φ must be single valued, thus

$$\Phi(\phi) = \Phi(\phi + 2\pi) \tag{2.50}$$

and this condition requires that $\beta^{1/2} = m_l = 0, \pm 1, \pm 2, \ldots$

Next, for the component Θ, the following ordinary differential equation is obtained:

$$\frac{1}{\sin\theta}\frac{d}{d\theta}\left(\sin\theta\frac{d\Theta}{d\theta}\right) + \left(\alpha - \frac{m_l^2}{\sin\theta}\right)\Theta = 0 \tag{2.51}$$

that only has physical solutions when

$$\alpha = l(l+1) \tag{2.52}$$

where

$$l = i + |m_l|, i = 0, 1, 2, \ldots \tag{2.53}$$

and l is called the *orbital angular momentum quantum number*.

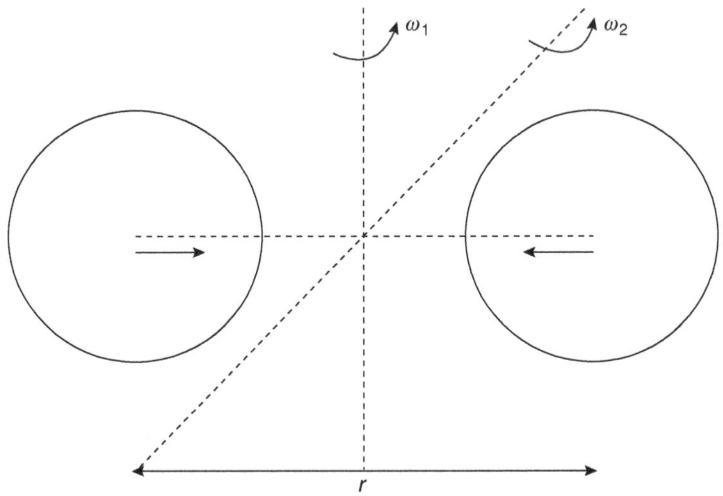

Figure 2.4 Rotations and vibrations of a diatomic particle.

Applying the quantum mechanical angular momentum operator for these wave functions gives

$$L^2 = l(l+1)\hbar^2 \tag{2.54}$$

where L is the orbital angular momentum and its square is shown by this relation to be quantized. Further analysis also indicates that

$$L_z = m_l \hbar \tag{2.55}$$

where L_z is the component of orbital angular momentum along the inter-particle axis, that is also found to be quantized, and m_l is called the *magnetic quantum number*.

2.2.5 Rotational and Vibrational Energy

We are now in a position to consider the internal motions of diatomic molecules such as N_2, O_2, and NO. In such molecules, ψ_i governs the rotations and vibrations of the two atoms as they interact with each other internally, and ψ_e governs the external translational motion of their combined center of mass. We consider the internal energies to be independent of the translational motion. In our first consideration of these energy modes, we assume that rotations and vibrations are also independent of one another.

Rotational Energy

Let us introduce the *rigid rotor* model for rotation in which the two atoms are separated by a fixed distance, r_e, and so have the geometry of a dumb-bell as illustrated in Fig. 2.4. From classical mechanics, the rotational energy is given by

$$\epsilon_{\text{rot}} = \frac{1}{2} I \omega^2 \tag{2.56}$$

where $I = \mu r^2$ is the moment of inertia of our geometry, μ is the reduced mass, and ω is the angular velocity. Similarly, the classical mechanics expression for angular momentum is $L = I\omega$ so that

$$\epsilon_{\text{rot}} = \frac{L^2}{2I} = \frac{l(l+1)\hbar^2}{2\mu r_e^2} \tag{2.57}$$

For the rigid rotor, we replace l by J and write the rotational energy as

$$\epsilon_{\text{rot}} = \frac{\hbar^2}{2\mu r_e^2} J(J+1) \tag{2.58}$$

where $J = 0, 1, 2 \ldots$ is the rotational quantum number.

Equation 2.58 shows that the rotational energy is quantized, and this arises directly from the fact that the square of angular momentum is quantized. Also, note that the minimum rotational energy is zero when $J = 0$, unlike the minimum translational energy that is very small, but finite. Returning back to the solution of the two-particle Schrödinger equation, we can see that the degeneracy for the rigid rotor is $g_J = 2J + 1$. This result can be proved by example from Eq. 2.53. Finally, it should also be clear that the spacing between adjacent rotational levels, $\Delta\epsilon_{\text{rot}}$, increases with J.

Vibrational Energy

For the vibrations of the two atoms along their chemical bond in a diatomic molecule, we assume that the potential that dictates their interaction represents a *harmonic oscillator*, i.e.,

$$V(r) = \frac{k(r - r_e)^2}{2} \tag{2.59}$$

where the spring constant, $k = 4\pi^2\mu\nu^2$ and ν is the oscillating frequency.

By analysis of the purely radial dependence of the two-particle amplitude function, $R(r)$, that employs Eq. 2.59 as the potential, and after significant manipulation we obtain

$$\frac{d^2H}{dy^2} - 2y\frac{dH}{dy} + (\lambda - 1)H = 0 \tag{2.60}$$

where

$$H(y) = yR(y)\exp\left(\frac{y^2}{2}\right), y = \left(\frac{2\pi\nu\mu}{\hbar}\right)^{1/2}(r - r_e), \lambda = \frac{2\epsilon_v}{h\nu} \tag{2.61}$$

and ϵ_v is the vibrational energy. Equation 2.60 is the Hermite equation, for which the only solutions are

$$(\lambda - 1) = 2v \text{ with } v = 0, 1, 2, \ldots \tag{2.62}$$

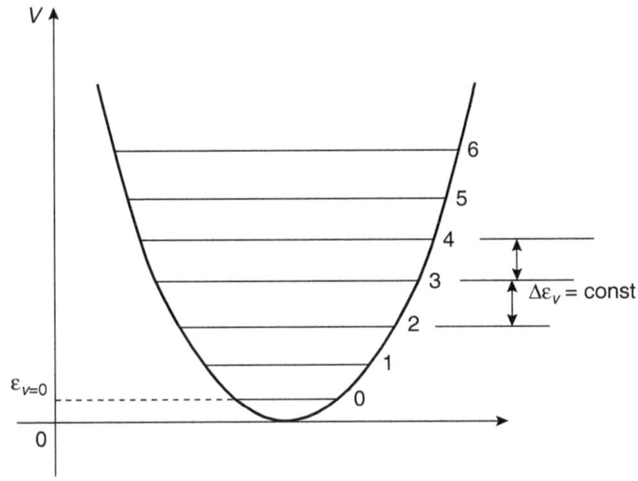

Figure 2.5 Quantized vibrational energy levels for the harmonic oscillator model.

Thus, we obtain the following expression for the vibrational energy:

$$\epsilon_{\text{vib}} = \lambda \frac{h\nu}{2} = \left(\upsilon + \frac{1}{2}\right) h\nu \tag{2.63}$$

where $\upsilon = 0, 1, 2, \ldots$ is the vibrational quantum number.

Equation 2.63 shows that the vibrational energy is quantized, that the minimum vibrational energy is nonzero when $\upsilon = 0$, that the degeneracy, $g_\upsilon = 1$, for all levels, and the spacing between adjacent vibrational levels, $\Delta\epsilon_{\text{vib}}$, is constant. These properties are illustrated along with the harmonic oscillator potential in Fig. 2.5.

Using the rigid rotor and harmonic oscillator models, let us consider the energy spacing of the rotational and vibrational modes for the N_2 molecule:

$$\Delta\epsilon_{\text{rot}} = \frac{\hbar^2}{2\mu r_e^2} = 4.04 \times 10^{-23}\ \text{J} \equiv k\theta_{\text{rot}} \tag{2.64}$$

$$r_e = 1.09 \times 10^{-10}\ \text{m} \tag{2.65}$$

$$\Delta\epsilon_{\text{vib}} = h\nu = 4.68 \times 10^{-20}\ \text{J} \equiv k\theta_{\text{vib}} \tag{2.66}$$

$$\nu = 7.06 \times 10^{13}\ \text{Hz} \tag{2.67}$$

where we introduce θ_{rot} and θ_{vib} as the characteristic temperatures for rotational and vibration, respectively, that provide a convenient way to characterize the quantum spacing of different energy modes for different molecules. For this particular example, we can see that $\Delta\epsilon_{\text{rot}}$ is three orders of magnitude smaller than $\Delta\epsilon_{\text{vib}}$. For most diatomic molecules, it is found that $\Delta\epsilon_{\text{rot}}$ is small enough that quantum effects can be ignored, although one important exception is hydrogen, H_2. It is also generally the case for most diatomic molecules that the vibrational energy spacing, $\Delta\epsilon_{\text{vib}}$, is large enough that quantum effects cannot be neglected.

2.2.6 Electronic Energy

A particle can take on different electronic energies when the electrons that surround its nucleus occupy different orbits. In general, the ground electronic state represents the electron configuration that is most stable. Any transfer of an electron to a higher energy orbit represents an excited electronic state. Here, for simplicity, we consider the special theoretical case of the *hydrogenic particle* that consists of a nucleus of Z positive charges with a single electron orbiting in a Coulomb field for which the electrostatic potential is

$$V(r) = -\frac{Ze^2}{4\pi\epsilon_0 r} \tag{2.68}$$

where ϵ_0 is the permittivity of free space. We use this potential in the purely radial form of the two-body Schrödinger equation to eventually obtain the associated electronic energy:

$$\epsilon_{\text{el}} = -\frac{Ze^2\mu}{32\pi^2\epsilon_0^2\hbar^2}\frac{1}{n^2}, \; n = 1, 2, 3, \ldots \tag{2.69}$$

where n is called the *principle quantum number*. Note that the convention for electronic energy is that ϵ_{el} is zero when an electron is completely removed from the atom. This process requires input of energy, and thus all electronic energy levels are negative. Electronic energy is a form of potential energy, so it is really *changes* in electronic energy that are important. Three further quantum numbers are required to completely specify an electronic state:

(a) The *orbital angular momentum quantum number*, $l = 0, 1, 2, \ldots, n-1$
(b) The *magnetic quantum number*, $m_l = 0, \pm 1, \pm 2, \ldots, \pm l$
(c) The spin quantum number, $m_s = \pm\frac{1}{2}$

The first two of these quantum numbers are derived directly from our considerations of the two-particle Schrödinger equation, and the third accounts for the fact that the orbiting electrons can spin in two different directions. While the electronic energy level is uniquely specified by n, combinations of the other three quantum numbers give rise to degenerate states. For each value of l there are $2l + 1$ values of m_l and two different values of m_s. Hence, the total degeneracy is

$$g_n = \sum_{l=0}^{n-1} 2(2l+1) = 2n^2 \tag{2.70}$$

Finally, note that the total internal energy of a particle including all possible forms of energy is given by $\epsilon_i = \epsilon_{\text{rot}} + \epsilon_{\text{vib}} + \epsilon_{\text{el}}$.

2.3 Atomic Structure

Let us now consider in detail the internal energy structure of real atoms. The *hydrogenic particle* has only one electron orbiting the nucleus. A real atom has many electrons orbiting, and each of these electrons has its own kinetic energy. The interaction between the multiple electrons and the positive charges in the nucleus creates a complicated electrostatic potential that gives rise to a range of angular momentum coupling effects. The solution of the Schrödinger equation including all of these effects is impossible in a general form. We simplify the problem by assuming that each electron moves in a spherically symmetric electrostatic field generated by all the other charged particles. We retain the same set of quantum numbers to describe each electron (n, l, m_l, m_s).

2.3.1 Electron Classification

First, we require a naming system for the electrons orbiting the nucleus of an atom. All electrons with the same value of n, the principle quantum number, occupy the same orbital shell with the following symbolic designation:

n	1	2	3	4	5	...
Symbol	K	L	M	N	O	...

For each value of n, the possible values of l are $0, 1, 2, \ldots n - 1$. All electrons with the same values of n and l lie in the same orbital subshell with the following symbolic designation:

1	0	1	2	3	4	...
Symbol	s	p	d	f	g	...

These letters were chosen based on the characteristics of the spectroscopic lines associated with each type of state, i.e., "s" indicates "sharp," "p" indicates "principal", "d" indicates "diffuse," and "f" indicates "fundamental."

In classifying an electron, the convention employed is to use the numerical value of n and the symbol for l. For example, a $3p$ electron (or subshell) has $n = 3$ and $l = 1$. In the absence of a magnetic field, the values of m_l and m_s represent degeneracies.

The *Pauli Exclusion Principle* states that no two electrons in the same atom may have the same set of quantum numbers: (n, l, m_l, m_s). This allows the construction of a table of electron classification as shown in Table 2.2 that is related to the periodic table.

2.3.2 Angular Momentum

Electronic states are characterized by their electron configuration and their angular momentum properties. In a multielectron atom, each electron i has

Table 2.2 Table of Electron Classification

	K	L		M		
n	1	2		3		
l	0	0	1	0	1	2
	s	s	p	s	p	d
m_l	0	0	−1, 0, 1	0	−1, 0, 1	−2, −1, 0, 1, 2
m_s	$\pm\frac{1}{2}$	$\pm\frac{1}{2}$	$\pm\frac{1}{2}, \pm\frac{1}{2}, \pm\frac{1}{2}$	$\pm\frac{1}{2}$	$\pm\frac{1}{2}, \pm\frac{1}{2}, \pm\frac{1}{2}$	$\pm\frac{1}{2}, \pm\frac{1}{2}, \pm\frac{1}{2}, \pm\frac{1}{2}, \pm\frac{1}{2}$
No. # of states ($2n^2$)	2	8		18		

orbital angular momentum vector, $\bar{l}_i$, and spin angular momentum vector, $\bar{s}_i$. Magnetic fields associated with $\bar{l}_i$ and $\bar{s}_i$ give rise to different forms of angular momentum coupling:

(a) j–j coupling ($\bar{l}_i$–$\bar{l}_j$ coupling, and $\bar{s}_i$–$\bar{s}_j$ coupling) that is relatively weak, and
(b) L–S coupling ($\bar{l}_i$–$\bar{s}_i$ coupling) that is usually dominant.

To account for this coupling, we assume a vector model that results in an orbital angular momentum vector: $\bar{L} = \sum_i \bar{l}_i$ and $|l_i| = \sqrt{l_i(l_i+1)}\hbar$. Similarly, the resulting spin angular momentum vector is given by $\bar{S} = \sum_i \bar{s}_i$ and $|s_i| = \sqrt{s_i(s_i+1)}\hbar$. Finally, the resultant total angular momentum vector: $\bar{J} = \bar{L} + \bar{S}$. Within these assumptions, it may be shown that $\bar{L}$, $\bar{S}$, $\bar{J}$ are quantized as follows:

$$|\bar{L}| = \sqrt{L(L+1)}\hbar \tag{2.71}$$

$$|\bar{S}| = \sqrt{S(S+1)}\hbar \tag{2.72}$$

$$|\bar{J}| = \sqrt{J(J+1)}\hbar \tag{2.73}$$

where L is the orbital angular momentum quantum number, S is the spin angular momentum quantum number, and J is the total angular momentum quantum number.

A detailed quantum mechanical analysis provides rules for the permissible values of these quantum numbers as follows:

L: An integer (0, 1, 2,…)
S: A half-integer (0, 1/2, 1, 3/2, 2,…)
J: Determined by L and S as follows:

$$J = (L+S), (L+S-1), (L+S-2), \ldots, |L-S| \tag{2.74}$$

Equation 2.74 can be generalized for the possible outcomes P from two quantum numbers P_1, P_2:

$$P = (P_1 + P_2), (P_1 + P_2 - 1), \ldots, |P_1 - P_2|$$
$$P = \sqrt{P(P+1)}\hbar$$

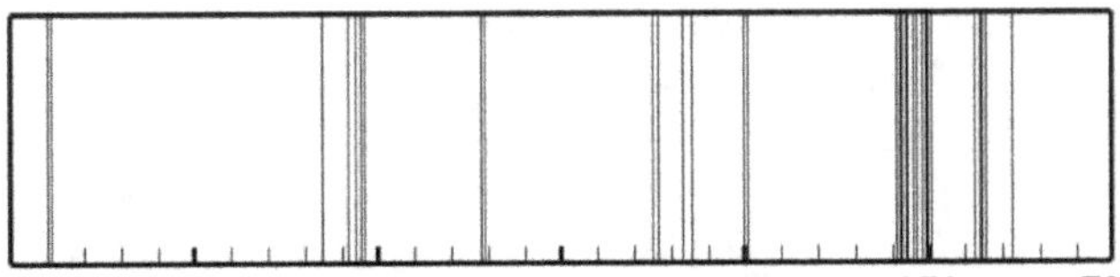

Figure 2.6 Portion of the nitrogen atom energy spectrum showing fine structure.

As an example, consider a two-electron atom, e.g., He, for which $l_1 = 3$; $l_2 = 2$; $s_1 = \frac{1}{2}$; $s_2 = \frac{1}{2}$. Let us find all values of J for the specific case of $L = 4$, $S = 1$.

Using Eq. 2.74 for $l_1 = 3$; $l_2 = 2 \Rightarrow L = 5, \mathbf{4}, 3, 2, 1$
Similarly, for $s_1 = \frac{1}{2}$; $s_2 = \frac{1}{2} \Rightarrow S = \mathbf{1}, 0$
Finally, for $L = 4$, $S = 1$, $J = 5, 4$, and 3, so there are three different values.

2.3.3 Spectroscopic Term Classification

There is a special nomenclature used to designate the angular momentum states of an atom. Based on L–S coupling, the general form of the scheme is

$$^{2S+1}L_J$$

in which numerical values of S and J are employed while L is represented symbolically as follows:

L	0	1	2	3	4	...
Symbol	S	P	D	F	G	...

For example: $^2S_{\frac{1}{2}}$ indicates that $L = 0$, $S = 1/2$, and in this case J can equal only 1/2. Many additional examples are provided for air atoms in Table 2.3.

Another important spectroscopic property is the *multiplicity* that is the total number of permissible values of J given L and S (either $2L + 1$ or $2S + 1$). The physical manifestation of multiplicity is fine structure observed in the spectrum of an energy state as illustrated for atomic nitrogen in Fig. 2.6. Radiation is emitted at wavelength λ corresponding to energy

$$\epsilon = \frac{hc}{\lambda} \tag{2.75}$$

and small departures from this energy are observed due to the angular momentum coupling discussed previously. Some specific examples of the relationship between spectroscopic term classification and multiplicity are as follows:

All S terms ($L = 0$) are singlet.
All P terms ($L = 1$) are singlet, or doublet, or triplet.
All D terms ($L = 2$) are singlet, or doublet, or triplet, or quartet, or quintet.

Table 2.3 Lowest Lying Electronic States of Air Atoms

Atom	Configuration	Term	Multiplicity	Energy (eV)	Degeneracy
N	$2s^2\,2p^3$	$^4S_{3/2}$	1	0	4
N	$2s^2\,2p^3$	$^2D_{5/2,3/2}$	2	2.39	10
N	$2s^2\,2p^3$	$^2P_{3/2,1/2}$	2	3.58	6
N	$2s^2\,2p^3\,3s$	$^4P_{5/2,3/2,1/2}$	3	10.33	12
N	$2s^2\,2p^3\,3s$	$^2P_{3/2,1/2}$	2	10.68	6
N	$2s\,2p^4$	$^4P_{5/2,3/2,1/2}$	3	10.93	12
N	$2s^2\,2p^2\,3p$	$^2S_{1/2}$	1	11.60	2
N	$2s^2\,2p^2\,3p$	$^4D_{7/2,5/2,3/2,1/2}$	4	11.75	20
N	$2s^2\,2p^2\,3p$	$^4P_{5/2,3/2,1/2}$	3	11.84	12
N	$2s^2\,2p^2\,3p$	$^4S_{3/2}$	1	12.00	4
N	$2s^2\,2p^2\,3p$	$^2D_{5/2,3/2}$	2	12.00	10
N	$2s^2\,2p^2\,3p$	$^2P_{3/2,1/2}$	2	12.12	6
N	$2s^2\,2p^2\,3s$	$^2D_{5/2,3/2}$	2	12.36	10
O	$2s^2\,2p^4$	$^3P2$	1	0	5
O	$2s^2\,2p^4$	$^3P1$	1	0.0196	3
O	$2s^2\,2p^4$	$^3P0$	1	0.0281	1
O	$2s^2\,2p^4$	$^1D2$	1	1.97	5
O	$2s^2\,2p^4$	$^1S0$	1	4.20	1
O	$2s^2\,2p^3\,3s$	$^5S2$	1	9.16	5
O	$2s^2\,2p^3\,3s$	$^3S1$	1	9.54	3

In summary, the electronic state of an atom is completely described using the term classification and the subshell configuration.

Note that the degeneracy for each component of a multiplet is given by $g_J = 2J + 1$ and the total degeneracy is simply the summation $g = \sum_J g_J$.

2.3.4 Excited States

So far, we have considered the case in which the electrons of the atom all occupy the lowest possible subshell: This is called the *ground electronic state*. An atom becomes electronically excited when one of the electrons moves to a higher energy subshell position and this occurs via intermolecular collision with other particles or through absorption of radiation.

Each atom has a large number of excited state configurations. Table 2.3 provides information for some of the lowest electronic states of the two primary atoms found in air. Note that the energy levels are measured data obtained from spectra. They are not easily calculated from first principles.

Consider the nitrogen atom, N. In its ground state, the $1s$ and $2s$ orbital shells are filled, and the $2p$ shell is half filled. Electronic excitation first occurs

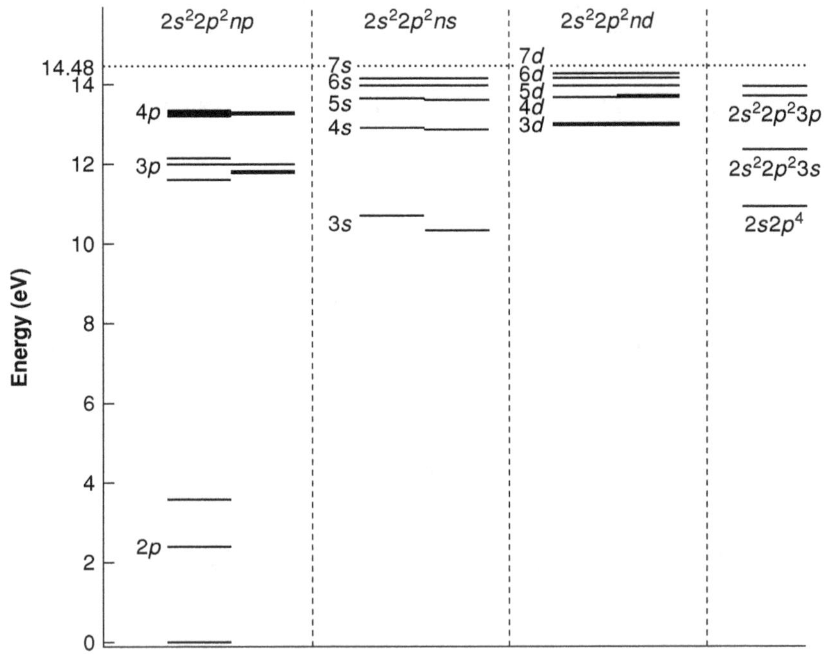

Figure 2.7 Energy level diagram for atomic nitrogen.

when one of the electrons from the $2p$ subshell occupies a higher energy state within the same orbit. These excited states are illustrated on an energy level diagram in Fig. 2.7 for the nitrogen atom. The energy scale uses the electron volt, eV (1 eV = 1.6×10^{-19} J).

For nitrogen, the ground and first two excited states all occur within the $2p$ subshell:

$$^4S,\ ^2D_{5/2,3/2},\ ^2P_{3/2,1/2}$$

The third excited state occurs when an electron moves into the $3s$ subshell and is a triplet:

$$^4P_{5/2,3/2,1/2}$$

When the electron is completely removed at an energy of 14.48 eV, the nitrogen atom is ionized (indicated as NII or N^+) and this ion particle has its own internal energy structure and is considered a different chemical species.

2.4 Structure of Diatomic Molecules

A diatomic molecule (AB) is formed when two atoms merge together in a stable state, in which the electronic configuration of the individual atoms is completely changed. In addition to translational and electronic energy, molecules also possess rotational and vibrational energy. In this section, we consider the permissible energy states for real diatomic molecules.

2.4.1 Born–Oppenheimer Approximation

Consider a diatomic molecule AB consisting of N electrons. The time-independent Schrödinger equation for this system is

$$\frac{1}{m_A}\nabla_A^2\psi + \frac{1}{m_B}\nabla_B^2\psi + \frac{1}{m_e}\sum_{i=1}^{N}\nabla_i^2\psi + \frac{8\pi^2}{h^2}(\epsilon - V)\psi = 0 \tag{2.76}$$

where ∇_j^2 indicates that the derivative is taken with respect to the coordinates of particle j, ψ is the overall amplitude function, and ϵ is the total internal energy excluding the external translation motion of the center of mass.

The potential V consists of three components:

(1) Interactions between the nuclei A and B
(2) Interactions between the N electrons
(3) Interactions between AB and the electrons

We assume that the overall potential field is a simple summation of these three parts:

$$V = V_{A-B} + V_{e-e} + V_{AB-ee} \tag{2.77}$$

The *Born–Oppenheimer* approximation simplifies solution of Eq. 2.76 by assuming that the nuclear and electronic motions are independent. That is, the overall wave function is given by the product of the electronic and nuclear wave functions:

$$\psi = \psi_e\psi_n \tag{2.78}$$

where $\psi_e = \psi_e(x, y, z, \ldots, x_N, y_N, z_N)$ and $\psi_n(x_A, y_A, z_A, x_B, y_B, z_B)$.

Substitution into Eq. 2.76 results in the following two Schrödinger equations:

$$\frac{1}{m_e}\sum_{j=1}^{N}\nabla_j^2\psi_e + \frac{8\pi^2}{h^2}(\epsilon^{el} - V)\psi_e = 0 \tag{2.79}$$

where ϵ^{el} is the electronic energy and V is the overall potential, and

$$\frac{1}{m_A}\nabla_A^2\psi_n + \frac{1}{m_B}\nabla_B^2\psi_n + \frac{8\pi^2}{h^2}(\epsilon_m - \epsilon_n^{el})\psi_n = 0 \tag{2.80}$$

where ϵ_n^{el} is the electronic energy solution from Eq. 2.79 with principal quantum number n and ϵ_m is the total energy of the particle.

As illustrated in Fig. 2.5, the overall potential V in Eq. 2.79 depends on the interatomic spacing r_{AB}. Owing to molecular vibration, r_{AB} changes continually. So, for each value of r_{AB}, we must solve Eq. 2.79 for the associated potential, $V(r_{AB})$, to obtain the corresponding energy $\epsilon_n^{el}(r_{AB})$.

Thus, because of the continual variation of r_{AB}, the electronic states of molecules consist of several continuously varying functions, each one representing a different principle quantum number, n. Figure 2.8 provides a

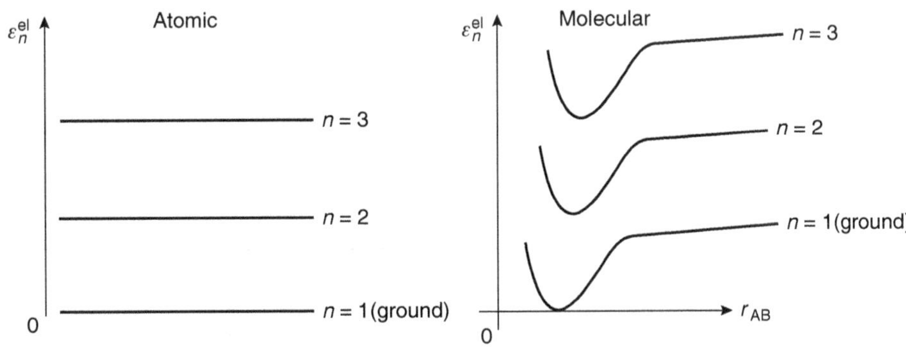

Figure 2.8 Comparison of atomic and molecular electronic states.

notional comparison of this important difference between the electronic states of atoms and molecules.

After solving Eq. 2.79 for a given value of r_{AB}, the total energy of the molecule ϵ_m is found from Eq. 2.80 in which the electronic energy ϵ_n^{el} acts as the potential function. For each value of r_{AB}, there are a number of discrete solutions for ϵ_m corresponding to each electronic principle quantum number, n. We write these solutions as $\epsilon_{n,m}$.

Consider $\epsilon_n^{el}(r_{AB})$ which is the potential energy function and is different for each value of n. The ground state ($n = 1$) is illustrated in Fig. 2.9.
To interpret the potential, consider the associated force field:

$$F_{AB} = -\frac{d\epsilon_1^{el}(r_{AB})}{dr_{AB}} \tag{2.81}$$

We may deduce several key points. First, at large r_{AB}, ϵ_1^{el} = constant, and so $F_{AB} = 0$. Thus, at large distances of separation, there is no force acting between nuclei A and B, they behave as separate atoms, and molecule AB is

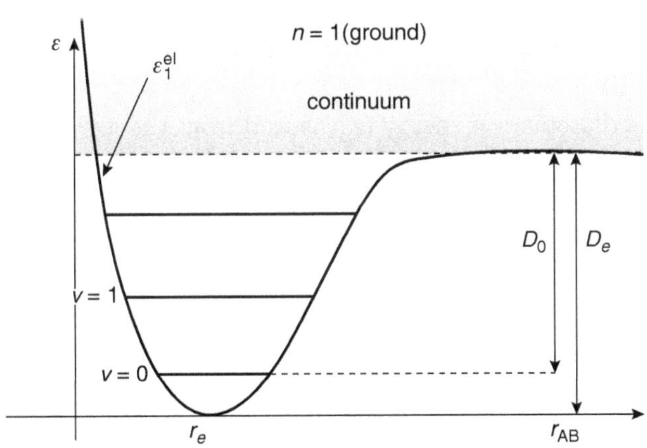

Figure 2.9 Schematic of the ground electronic potential function.

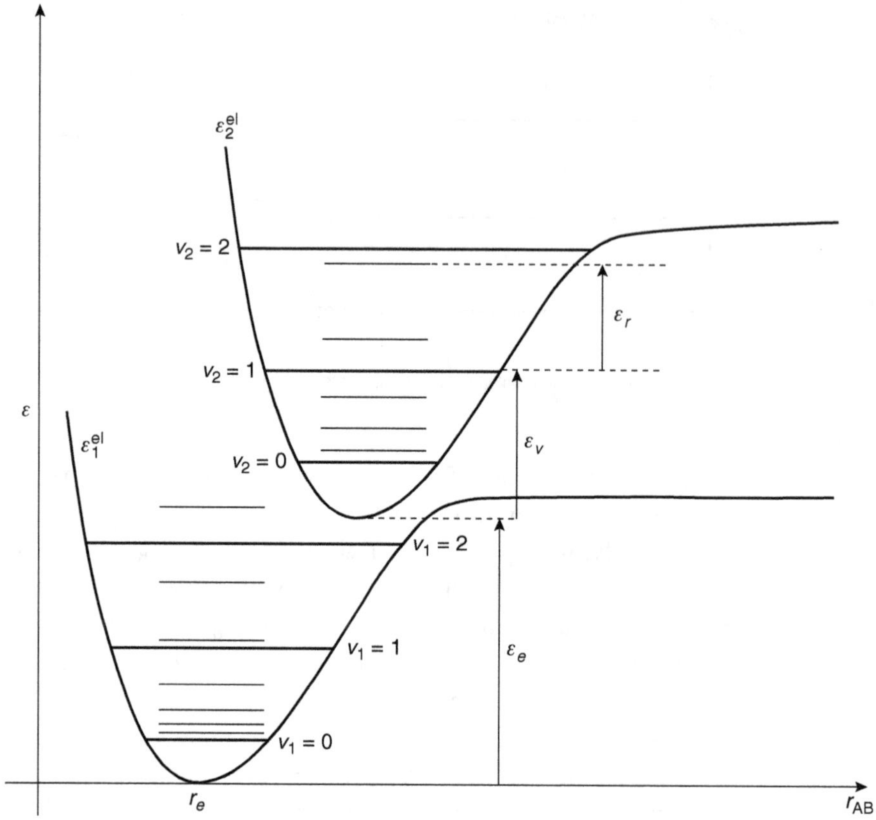

Figure 2.10 Illustration of ground and first electronically excited molecular states.

said to have dissociated. Second, at $r_{\mathrm{AB}} > r_e$, the internuclear force is attractive (negative), and at $r_{\mathrm{AB}} > r_e$, the force is repulsive (positive). Hence, the atoms oscillate back and forward about the equilibrium point, r_e.

It should also be noted that there is no stable configuration at energies above the asymptotic limit that is labeled as the continuum of atomic states. This limit is defined in two different ways:

(1) D_e: The electronic binding energy (measured from the minimum of the potential)
(2) D_0: The dissociation energy (measured from the lowest vibration energy level)

The ground and first electronically excited molecular states are illustrated in Fig. 2.10.

In each vibrational level, there is a series of rotational energy states. Thus, the total internal energy is given by $\epsilon_m = \epsilon_{\mathrm{el}} + \epsilon_{\mathrm{vib}} + \epsilon_{\mathrm{rot}}$. Finally, note in terms of the energy spacing of the various modes for air molecules, that

$$\Delta\epsilon_{\mathrm{rot}}(\approx 10^{-3}\ \mathrm{eV}) \ll \Delta\epsilon_{\mathrm{vib}}(\approx 0.1\ \mathrm{eV}) \ll \Delta\epsilon_{\mathrm{el}}(\approx 1\ \mathrm{eV}) \tag{2.82}$$

2.4.2 Rotational and Vibrational Energy

Consider determination of ϵ_{rot} and ϵ_{vib} for a given electronic state. For the rigid rotor and harmonic oscillator models:

$$\epsilon_{\text{rot}} = k\theta_{\text{rot}}J(J+1)\ (J = 0, 1, 2, ..)$$

$$\epsilon_{\text{vib}} = k\theta_{\text{vib}}\left(v + \frac{1}{2}\right)\ (v = 0, 1, 2, ..)$$

In spectroscopy, these energy levels are expressed in wave-number units of cm^{-1} by dividing by hc and by 100:

$$F(J) \equiv \frac{\epsilon_{\text{rot}}}{100hc} = B_e J(J+1) \tag{2.83}$$

$$G(v) \equiv \frac{\epsilon_{\text{vib}}}{100hc} = \omega_e\left(v + \frac{1}{2}\right) \tag{2.84}$$

These variables are called the *rotational* and *vibrational term values*, respectively, in which B_e is the rotational constant, and ω_e is the vibrational wave number. These parameters have different values for each electronic state of each molecule.

There are two important shortcomings of the above model:

(1) In spectroscopy measurements, the vibrational levels are found to not be evenly spaced as predicted by the harmonic oscillator potential. A better model is provided by the Morse potential:

$$V(r_{\text{AB}}) = D_e\{1 - \exp[-\beta(r_{\text{AB}} - r_e)]^2\} \tag{2.85}$$

where β is a constant that is different for each potential. This model provides vibrational energy spacing that decreases monotonically from the ground state to the dissociation limit.

(2) In our model, there is no coupling between the rotational and vibrational energies whereas in practice, it is found that ϵ_{rot} and ϵ_{vib} do affect each other. Only the effect of vibration on rotation is important physically because a molecule typically rotates about 1000 times for each full vibrational cycle. Specifically, variation in r_{AB} through vibration changes angular velocity and therefore ϵ_{rot}.

To account for these effects, we introduce the *nonrigid rotor/anharmonic oscillator (NRR/AHO)* model as follows:

$$G(v) + F_v(J) = \omega_e\left(v + \frac{1}{2}\right) - \omega_e x_e\left(v + \frac{1}{2}\right)^2 + \omega_e y_e\left(v + \frac{1}{2}\right)^3 + \cdots$$
$$+ B_v J(J+1) - D_v J^2(J+1)^2 + \cdots \tag{2.86}$$

$$\begin{aligned} B_v &= B_e - \alpha_e\left(v + \frac{1}{2}\right) + \cdots \\ D_v &= D_e + \beta_e\left(v + \frac{1}{2}\right) + \cdots \end{aligned} \tag{2.87}$$

Table 2.4 Ground State Molecular Constants for the NRR/AHO Model

Species	B_e (cm^{-1})	α_e (cm^{-1})	ω_e (cm^{-1})	$\omega_e x_e$ (cm^{-1})	$\omega_e y_e$ (cm^{-1})	θ_{rot} (K)	θ_{vib} (K)	r_e (10^{-10} m)	D_0 (eV)
N_2	1.998	0.0179	2357.6	14.06	0.00751	2.87	3390	1.088	9.76
O_2	1.445	0.0158	1580.2	12.07	0.05460	2.08	2280	1.207	5.12
NO	1.704	0.0178	1903.6	13.97	−0.0012	2.45	2740	1.151	6.48

The quantities $x_e, y_e, B_e, D_e, \alpha_e, \beta_e$ are molecular constants that vary with species and electronic state (note: D_e here is not the electronic binding energy discussed earlier).

From Eq. 2.86, the *vibrational term* is:

$$G(v) = \omega_e\left(v + \frac{1}{2}\right) - \omega_e x_e\left(v + \frac{1}{2}\right)^2 + \omega_e y_e\left(v + \frac{1}{2}\right)^3 \cdots \tag{2.88}$$

This expression provides a higher order approximation to the vibrational energy levels than the harmonic oscillator. The vibrational energy is now coupled directly to rotational energy using the *rotational term*:

$$F_v(J) = B_v J(J+1) - D_v J^2(J+1)^2 + \cdots \tag{2.89}$$

Since $\omega_{el} \gg \omega_{el}x_{el}$ and $B_{el} \gg \alpha_{el}$, the expression $G(v) + F_{vib}(J)$ reduces to the sum of the rigid rotor plus harmonic oscillator as a first-order approximation at small quantum numbers. Table 2.4 provides values of the key parameters required by the NRR/AHO model for the ground electronic state of the three primary air molecules. It may be shown that

$$D_e = \frac{4B_e^3}{\omega_e^2} \tag{2.90}$$

and so D_e is not listed in the table. Also, note the following useful expressions for the *characteristic temperatures of rotation and vibration*:

$$\theta_{rot} = \frac{B_e 100hc}{k} \tag{2.91}$$

$$\theta_{vib} = \frac{\omega_e 100hc}{k} \tag{2.92}$$

The NRR/AHO model may be used to construct an energy level diagram. In each vibrational energy level, there is a unique set of allowable ro-vibrational levels. As illustrated in Fig. 2.11, the many overlapping levels make molecular spectra difficult to interpret.

2.4.3 Electronic States

Similar to atoms, we must describe the effects of orbital and spin angular momenta on the electrons surrounding a molecule. It is found that the orbital

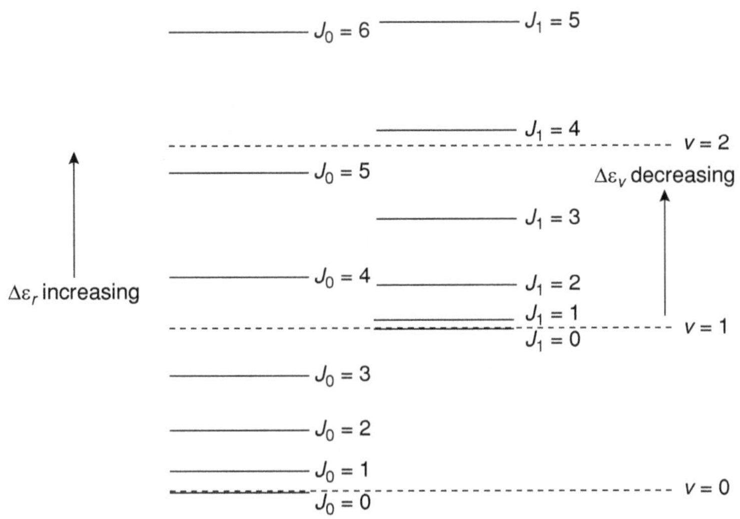

Figure 2.11 Schematic diagram of ro-vibrational levels.

angular momentum vector $\bar{L}$ in a diatomic molecule is quantized only along its internuclear axis owing to axial symmetry of the potential field. This component is written

$$L_{\mathrm{AB}} = \pm\Lambda\hbar \tag{2.93}$$

where $\Lambda = 0, 1, 2, \ldots$ is the *electron orbital quantum number*. The $\pm$ indicates that the component may exert itself in either direction along AB. The quantum number is designated symbolically as follows:

Λ	0	1	2	3	4	...
Symbol	Σ	Π	Δ	Φ	Γ	...

The spin angular momentum is also quantized along AB:

$$S_{\mathrm{AB}} = \Sigma\hbar \tag{2.94}$$

where the *spin quantum number* can assume integral or half-integral values:

$$\Sigma = 0, \pm 1, \pm 2, \ldots, \pm S \text{ or } \Sigma = \pm\frac{1}{2}, \pm\frac{3}{2}, \ldots, \pm S \tag{2.95}$$

and S is the overall spin quantum number. Molecular electronic states are written as

$$^{2S+1}(\text{symbol for } \Lambda) \tag{2.96}$$

For example: $^1\Sigma \Rightarrow S = 0, \Lambda = 0$

The quantity (2S + 1) is the *multiplicity* that represents the total number of possible components along AB resulting from orbit-spin coupling interactions.

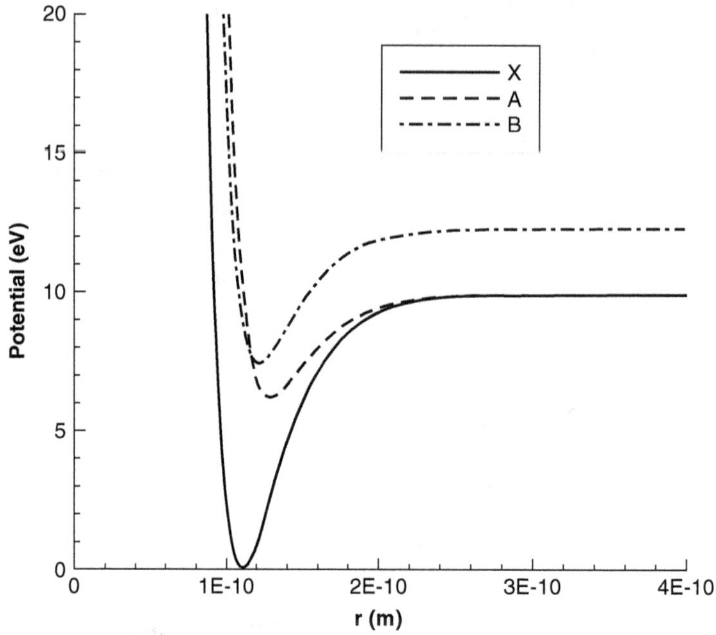

Figure 2.12 Potential energy diagram for the three lowest lying electronic states of molecular nitrogen.

A further feature of classifying molecular electronic states concerns *symmetries* of the electronic amplitude function ψ_e obtained from the Schrödinger equation. Specifically

(1) If ψ_e is unchanged after inversion of the electron coordinates about the molecular center, then a right subscript g (from *gerade*, German for equal) is used, e.g., ${}^1\Sigma_g$ Otherwise, u (*ungerade*) is employed if the sign of ψ_e is changed. This consideration applies only to homonuclear diatomic molecules.

For $\Lambda = 0$ states only, i.e., Σ states, a right superscript of "+" is used if there is no change in the sign of ψ_e after reflection of the electron coordinates in a plane through the internuclear axis AB. A right superscript of "−" is used if there is a change.

In terms of degeneracy, from Eq. 2.93, for $\Lambda > 0$, there is a degeneracy of two owing to the $\pm$ sign. In addition, from Eq. 2.95, there is a further degeneracy of $2S + 1$. Thus

$$g_{\mathrm{el}} = \begin{cases} 2(2S+1) & (\Lambda > 0) \\ 2S+1 & (\Lambda = 0) \end{cases} \tag{2.97}$$

Note, for the vibrational and rotational energies, we still have $g_{\mathrm{vib}} = 1, g_J = (2J+1)$.

Using the Morse potential, Eq. 2.76, the potentials of the three lowest lying electronic states of N_2 (X, A, and B) are shown in Fig. 2.12. The information that can be gathered from their spectroscopic designations are as follows.

Ground state: $X^1\Sigma_g^+$

$$\begin{aligned} E &= \quad 0\ \text{cm}^{-1} \\ X &\Rightarrow \quad \text{ground state (always)} \\ 1 &\Rightarrow \quad S = 0 \\ \Sigma &\Rightarrow \quad \Lambda = 0 \\ g &\Rightarrow \quad \text{gerade} \\ + &\Rightarrow \quad \text{no change} \\ g_{\text{el}} &= \quad 1;\ \text{multiplicity} = 1 \end{aligned} \tag{2.98}$$

First excited state: $A^3\Sigma_u^+$

$$\begin{aligned} E &= \quad 5.0 \times 10^4\ \text{cm}^{-1} = 6.2\,\text{eV} \\ 3 &\Rightarrow \quad S = 1 \\ \Sigma &\Rightarrow \quad \Lambda = 0 \\ u &\Rightarrow \quad \text{ungerade} \\ + &\Rightarrow \quad \text{no change} \\ g_{\text{el}} &= \quad 3;\ \text{multiplicity} = 3 \end{aligned} \tag{2.99}$$

Second excited state: $B^3\Pi_g$

$$\begin{aligned} E &= \quad 5.9 \times 10^4\ \text{cm}^{-1} = 7.3\,\text{eV} \\ 3 &\Rightarrow \quad S = 1 \\ \Pi &\Rightarrow \quad \Lambda = 1 \\ g &\Rightarrow \quad \text{gerade} \\ g_{\text{el}} &= \quad 6;\ \text{multiplicity} = 3 \end{aligned} \tag{2.100}$$

2.5 Summary

In this chapter, the basic ideas of quantum mechanics were introduced and used to formulate the Schrödinger equation. Solutions to the Schrödinger equation yielded quantized energy states for all four energy modes relevant to a gas, namely the translational, rotational, vibrational, and electronic modes. The systems used to uniquely label atomic and molecular energy states were described, and information provided for the particles of interest in air. The information provided in this chapter will be used in Chapter 3 to describe the thermodynamic state of a gas, and will be seen again in the molecular simulation of relaxation and chemical processes.

2.6 Problems

2.1 Prove that the linear operators given in Postulate 3 for momentum and energy provide the results proposed in wave-particle duality when applied to a single pilot wave (you may assume this wave exists over some finite region of space):

$$\psi(x,t) = \exp[i(kx - \omega t)]$$

2.2 The quantized distribution for a property J of a particle is given by

$$f(J)\,dJ \propto (2J+1)\exp[-0.007 \times J \times (J+1)]$$

where J may only take on integer values from 0 to 40 (this is similar to the rotational energy of air molecules at room temperature). Normalize this distribution, plot as a histogram, and determine

(a) The most likely value of J

(b) The expectation value of J

2.3 Consider solution of the time-independent Schrödinger equation for translational motion of a particle with no field forces inside a cubic box of length L. Using the boundary condition that the wave function is zero on the walls of the cube, and the normalization condition, show that the amplitude function is

$$\psi(x,y,z) = \left(\frac{8}{L^3}\right)^{\frac{1}{2}} \sin\left(\frac{n_1\pi x}{L}\right) \sin\left(\frac{n_2\pi y}{L}\right) \sin\left(\frac{n_3\pi z}{L}\right)$$

with eigenvalues

$$\epsilon_i = \frac{h^2}{8mL^2} n_i^2,\ n_i = 1, 2, 3, \ldots$$

Find the *expectation value* of x-position.

2.4 Give the degree of degeneracy and show all possible energy states (by writing down the quantum numbers) for each of the following energy levels:

(a) Translational energy $7h^2/(4mL^2)$

(b) Rotational energy $12\,\hbar^2/(2\mu r_e^2)$ (assuming rigid rotor)

(c) Vibrational energy $7h\nu/2$ (assuming harmonic oscillator)

2.5 (a) For the following un-normalized wave amplitude function, determine the normalized form, the average position, and the average energy for the range $x = [-1, 1]$:

$$\psi(x) = x\sin(\omega t) + ix\cos(\omega t)$$

(b) Write down all information indicated for these electronic states:

- Mg: ^3G
- NO: $F^2\Delta$

2.6 (a) Consider the following one-dimensional wave function for x in [0,1]:

$$\psi = x^2 \exp(i\omega t)$$

Determine the normalized wave function, average position, and average energy.

(b) For N_2 molecules, how many rotational levels exist between the ground and first vibrational levels (assume rigid rotor with $\theta_{\text{rot}} = 2.9$ K and harmonic oscillator with $\theta_{\text{vib}} = 3390$ K)?

(c) Write down all information for the following electronic states:

- N: ^4F
- N_2^-: $X^2\Pi_g$

3 Statistical Mechanics

3.1 Introduction

Similar to *kinetic theory*, the purpose of *statistical mechanics* is to relate molecular-level information to macroscopic gas flow properties. While kinetic theory considers the motions and collisions of particles and can be used to relate molecular behavior to macroscopic fluid dynamics, statistical mechanics considers how particles occupy their allowed quantized energy states and is used to relate molecular behavior to macroscopic thermodynamics.

The first step in making a connection between molecular statistical mechanics and macroscopic thermodynamics is to postulate the following relation:

$$S = k \ln \Omega \tag{3.1}$$

where S is entropy, k is the Boltzmann constant, and Ω is a measure of the degree of molecular randomness of the system. To arrive at this result, we have to consider molecular statistical counting techniques, the distribution of particles across energy states, and internal energy partition functions. We will also extend our ideas to consider a chemically reacting system.

3.2 Molecular Statistical Methods

Consider a gas system consisting of N identical particles, in which each particle i has quantized energy ϵ_i and the total energy of the system is E. In statistical mechanics, we do not analyze particle–particle interactions, i.e., intermolecular collisions, but they are assumed to occur and maintain the distributions of particles across their allowed quantum energies. To formulate an expression for Ω in Eq. 3.1, we need to count the number of different ways that we can arrange the system without changing N and E. Formally, we define the degree of molecular randomness as equating to the number of different ways the N particles can be distributed among their allowable quantum energy states without changing E. Each combination of particles

Table 3.1 Distributions of Particles in Energy States

$\epsilon_0 = 0$	$\epsilon_1 = 2$	$\epsilon_2 = 6$	$\epsilon_3 = 12$	E	Consistent with System?
1	1	1	1	20	No
0	3	1	0	12	Yes
2	0	2	0	12	Yes
1	1	2	0	14	No
3	0	0	1	12	Yes
...	...	...	...	...	...

satisfying this condition is called a *microstate* of the system. So, Ω is the total number of microstates consistent with a gas system defined by N and E. The simplest way to answer our question is by trial and error. Thus, in a systematic way, we try all possible combinations of the N particles in their allowed energy states.

Example 3.1 Consider a system with $N = 4$, and $E = 12$, in which each particle has rigid-rotor-like energy states:

$$\epsilon_J = J(J+1) : J = 0, 1, 2, 3 \tag{3.2}$$

Table 3.1 shows some of the possible distributions, and indicates those that are consistent with the system parameters, and that are therefore microstates.

For Example 3.1, there are only three microstates. Mathematically, we write this procedure as

$$\Omega = \sum 1, \text{ for all possible distributions } N_i \text{ such that}$$
$$\sum_i N_i = N \text{ and } \sum_i N_i \epsilon_i = E \tag{3.3}$$

In this example, we assume that the particles are indistinguishable from each other. Thus, for the combination on line 5 of Table 3.1, we do not obtain additional microstates by reordering the three particles in $J = 0$. We have also assumed that more than one particle can occupy the same quantum state. It may be shown from quantum mechanics that this condition applies only to particles made up of an even number of elementary units: neutrons, protons, and electrons. Such particles are called *bosons* and are analyzed using Bose–Einstein statistics. Examples of bosons include hydrogen atoms, He^4, N_2, and photons. Particles made up of an odd number of elementary units are called fermions and are analyzed using Fermi–Dirac statistics. Examples of fermions include electrons, protons, and He^3. No two fermions may occupy the same quantum state (similar to the Pauli Exclusion Principle). If the particles in Example 3.1 are fermions, there are no microstates at all.

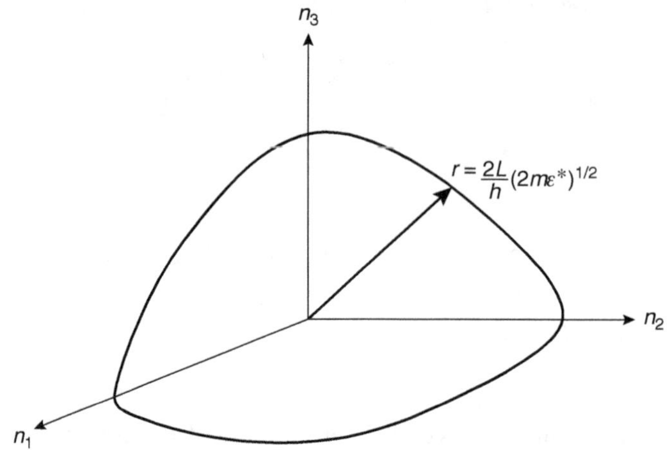

Figure 3.1 Cartesian space for the translational quantum numbers.

For real gas systems, the trial-and-error procedure is too slow for the large numbers of particles and energy states involved. We can use an alternative approach in the special case of translational energy where the spacing between adjacent energy levels is extremely small. Recall our result from Chapter 2 for the quantized translational energy:

$$\epsilon_{tr} = \frac{h^2}{8mL^2}(n_1^2 + n_2^2 + n_3^2) = \frac{h^2 n^2}{8mL^2} \tag{3.4}$$

where we saw that the spacing between adjacent levels $\epsilon_{tr} \approx 10^{-38}$ J for N_2.

Consider the number of translational states with energy less than some value ϵ^*. By plotting all translational energy states on a Cartesian diagram with coordinates (n_1, n_2, n_3), as illustrated in Fig. 3.1, it may be shown that all points lying inside the first octant of a sphere of radius $\frac{2L}{h}(2m\epsilon^*)^{1/2}$ satisfy the condition that $\epsilon < \epsilon^*$. Although the combinations of (n_1, n_2, n_3) are discrete points, since the spacing between adjacent states is very small, the number of states satisfying our condition is approximately given by the total volume of the sphere octant:

$$\Gamma = \frac{1}{8}\frac{4\pi}{3}\left[\frac{2L}{h}(2m\epsilon^*)^{1/2}\right]^3 = \frac{4\pi}{3}\frac{V}{h^3}(2m\epsilon^*)^{3/2} \tag{3.5}$$

Example 3.2 Consider N_2 at a temperature of 293 K in a cube with a side length of 1 cm. Let us find the number of translational energy states that lie below the average translational energy under these conditions. Thus, we use $\epsilon^* = \frac{3}{2}kT$ from kinetic theory to find that $\Gamma = 2 \times 10^{26}$. The total number of particles in a 1 cm^3 volume at standard temperature and pressure is 3×10^{19}. Hence, we find that only about 1 in every 10^7 available translational energy states is actually occupied on average by a particle under these conditions.

Figure 3.2 Energy group structure.

3.2.1 Energy Groups

The special approach described previously for counting translational energy states does not work for the additional internal energy modes, so we need to develop a more general procedure. Because we have a very large number of closely spaced energy states, it is useful to divide the energy spectrum into groups as illustrated in Fig. 3.2.

In this approach, the range of the spectrum covered by each group, $\delta\epsilon_j$, must be smaller than the total system energy E. Each energy group j has the following properties:

ϵ_j: Characteristic energy of the group (a fixed property)
C_j: Number of discrete quantum energy states in the group (a fixed property)
N_j: Number of particles in the group (a variable property).

A *microstate* N_j of the system is a particular distribution of the N particles across the energy groups, as illustrated in Fig. 3.3.

Rearranging the particles occupying the energy states within group j changes the microstate N_i but not the macrostate N_j. Thus, as we saw in our earlier example, a given macrostate may contain many possible microstates.

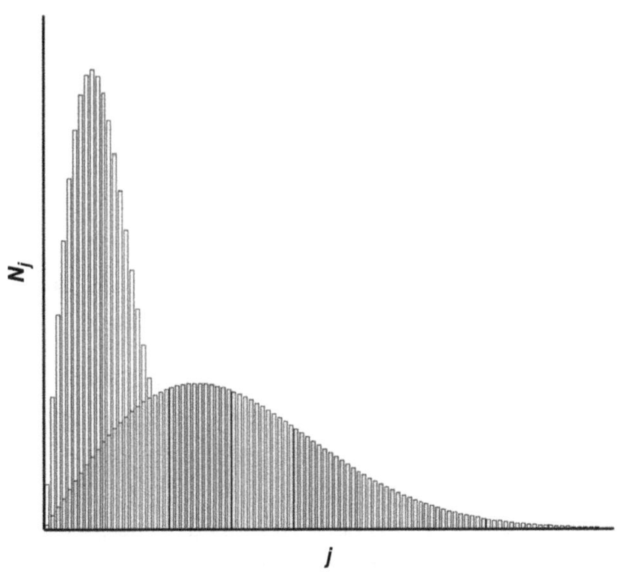

Figure 3.3 Illustration of different macrostates.

This suggests an efficient way to evaluate the total number of microstates, which is used to determine Ω:

(1) Divide the spectrum of energy states into groups.
(2) For each macrostate, determine the number of microstates.
(3) Sum over all macrostates consistent with the system parameters N and E, i.e., $N = \sum_j N_j$ and $E = \sum_j N_j \epsilon_j$.

Example 3.3 Consider a macrostate defined as follows:

$j = 1$	$j = 2$	$j = 3$
ϵ_1	ϵ_2	ϵ_3
$C_1 = 5$	$C_2 = 6$	$C_3 = 3$
$N_1 = 2$	$N_2 = 5$	$N_3 = 1$

The system parameters are evaluated as: $E = 2\epsilon_1 + 5\epsilon_2 + \epsilon_3$ $N = 8$.

We can construct many microstates consistent with the system, for example:

Microstate I

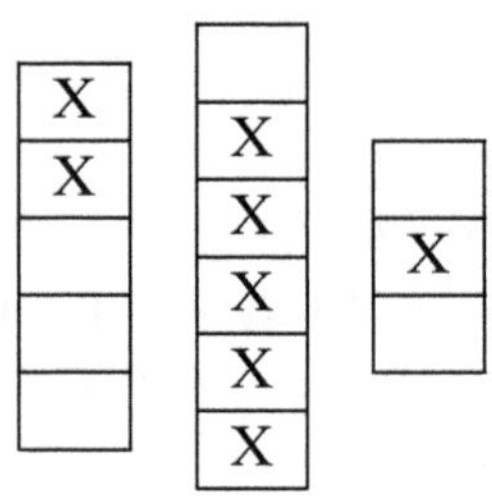

Microstate II

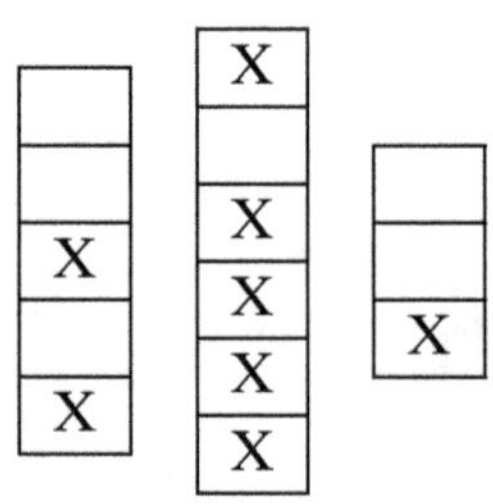

In our proposed counting approach, let us first consider step (2) for a group of C_j states containing N_j indistinguishable particles.

(a) Bose–Einstein Statistics. Recall that for bosons, there is no limit on the number of particles in each energy state. Consider a general group j with parameters ϵ_j, C_j, N_j. We want to find the total number of ways to place the N_j particles into the C_j energy states.

As illustrated in Fig. 3.4, we need to consider the positions of the particles and the locations of the partitions between the energy states such that $\sum_{i=1}^{C_j} N_i = N_j$. We proceed through a process in which at the first step there

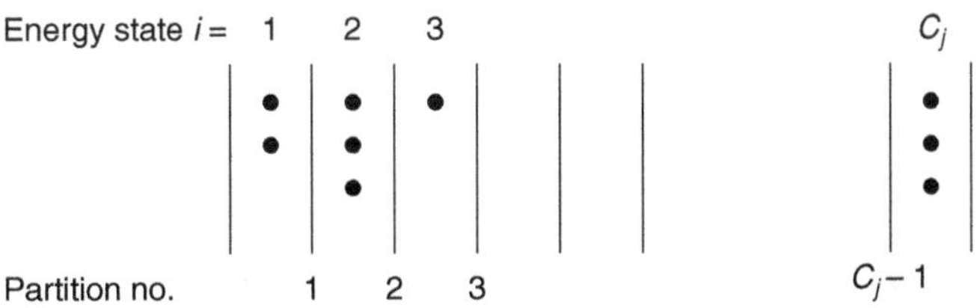

Figure 3.4 Counting of microstates using Bose–Einstein statistics.

is a total of N_j particles and $C_j - 1$ partitions to choose from to arrange. In the next step, having chosen one of these items, there will be one less set of objects to choose from, and so on down to the last step where there is just one item remaining.

Hence, the total number of different arrangements is $(N_j + C_j - 1)!$. However, we do not care about the exact positions of the partitions: Switching the partitions does not affect the counting of microstates. We must therefore divide our result by $(C_j - 1)!$. Similarly, the particles are indistinguishable, so we must also divide by $(N_j)!$. Thus, the total number of ways to arrange particles in energy states without restricting the number of particles in each state is

$$(W_{\mathrm{BE}})_j = \frac{(N_j + C_j - 1)!}{(N_j)!(C_j - 1)!} \tag{3.6}$$

Fermi–Dirac Statistics. For fermions, we must impose a limit of one particle at maximum in each energy state, so that $N_j \leq C_j$

The counting process is illustrated in Fig. 3.5 and consists of first laying down the partitions, then placing a particle one at a time into the available, unoccupied energy states. In the first step there are C_j choices, in the second step there is one less choice, and so on until the final step in which there are $C_j - (N_j - 1)$ choices. Thus, the total number of arrangements may be written

$$\frac{(C_j)!}{(C_j - N_j)!} \tag{3.7}$$

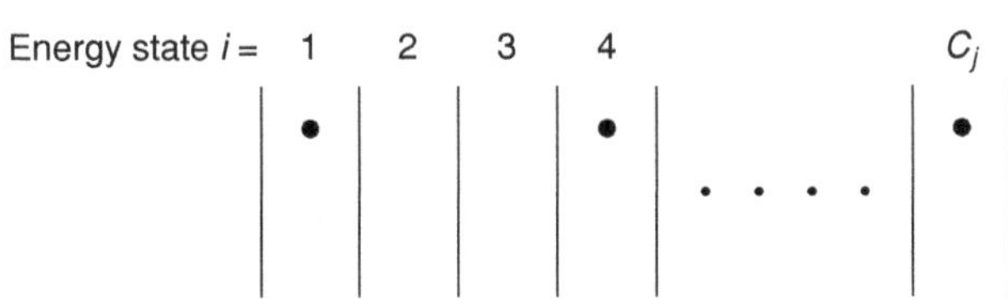

Figure 3.5 Counting of microstates using Fermi–Dirac statistics.

Again, the particles are indistinguishable so we have to divide by $(N_j)!$ to obtain

$$(W_{\mathrm{FD}})_j = \frac{(C_j)!}{(C_j - N_j)!(N_j)!} \tag{3.8}$$

Example 3.4 Consider a group with $C_j = 20$, $N_j = 10$. Using Eqs. 3.6 and 3.8 we find that $W_{\mathrm{BE}} = 2 \times 10^7$ and $W_{\mathrm{FD}} = 2 \times 10^5$. Thus, at least for this specific example, there are significant differences in the results obtained for the degree of randomness for bosons and fermions.

Equations 3.6 and 3.8 provide the number of combinations for a particular group. For a given distribution of particles across energy groups, N_j, the total number of microstates for a system is then the product over all j groups:

$$\text{BE: } W(N_j) = \prod_j \frac{(N_j + C_j - 1)!}{(N_j)!(C_j - 1)!} \tag{3.9}$$

$$\text{FD: } W(N_j) = \prod_j \frac{(C_j)!}{(C_j - N_j)!(N_j)!} \tag{3.10}$$

where $W(N_j)$ indicates the result for a particular macrostate N_j. The final result for the total number of microstates of the system is obtained from

$$\Omega = \sum W(N_j) \tag{3.11}$$

such that $\sum_j N_j = N$ and $\sum_j N_j \epsilon_j = E$.

We next consider part (1) of the overall counting process.

3.3 Distribution of Energy States

Evaluation of Eq. 3.11 is simplified when only the largest term in the summation is significant. We will return to assess this assumption later, and for now write

$$\Omega = W_{\mathrm{max}} \tag{3.12}$$

where W_{max} is the largest term for a particular macrostate. In searching for W_{max}, it is convenient mathematically to consider $\ln(W)$:

$$\text{BE: } \ln(W) = \sum_j \left\{ \ln(N_j + C_j - 1)! - \ln(N_j)! - \ln(C_j - 1)! \right\} \tag{3.13}$$

$$\text{FD: } \ln(W) = \sum_j \left\{ \ln(C_j)! - \ln(C_j - N_j)! - \ln(N_j)! \right\} \tag{3.14}$$

Now, we use Stirling's formula, which is valid for large z:

$$\ln(z)! \approx z \ln z - z \tag{3.15}$$

to obtain

$$\ln(W) = \sum_j \left\{ \pm C_j \ln\left(1 \pm \frac{N_j}{C_j}\right) + N_j \ln\left(\frac{C_j}{N_j} \pm 1\right) \right\} \tag{3.16}$$

where the "+" sign is for Bose–Einsten, for which we further assume that $C_j \gg 1$, and the "−" indicates Fermi–Dirac.

The maximum of this function occurs when

$$\frac{\partial}{\partial N_j} \ln(W) \partial N_j = 0$$

$$\Rightarrow \sum_j \left\{ \frac{\pm C_j \left(\pm \frac{1}{C_j}\right)}{1 \pm \frac{N_j}{C_j}} + \ln\left(\frac{C_j}{N_j} + \pm 1\right) + N_j \frac{\left[-\frac{C_j}{(N_j)^2}\right]}{\frac{C_j}{N_j} \pm 1} \right\} \partial N_j = 0$$

$$\Rightarrow \sum_j \left\{ \ln\left(\frac{C_j}{N_j} \pm 1\right) \right\} \partial N_j = 0 \tag{3.17}$$

Note that to satisfy conservation of particles and energy, the small changes ∂N_j must satisfy

$$\sum_j \partial N_j = 0 \text{ and } \sum_j \epsilon_j \partial N_j = 0 \tag{3.18}$$

An appropriate technique for solution of Eq. 3.17 subject to conditions 3.18 is Lagrange's method of undetermined multipliers, which states that for solution of

$$\sum_j f(x_j) \partial x_j = 0 \tag{3.19}$$

subject to $\sum y_j \partial x_j = 0$ and $\sum z_j \partial x_j = 0$, the most general solution is $f(x_j) = \alpha y_j + \beta z_j$, where α, β are constants to be determined. Applying this method to our case, we find that $x_j = N_j$, $y_j = 1$, $z_j = \epsilon_j$ and so

$$\ln\left(\frac{C_j}{N_j} \pm 1\right) = \alpha + \beta \epsilon_j \tag{3.20}$$

Therefore, the particular macrostate N_j^* that gives $W_{\max}$ is

$$\frac{N_j^*}{C_j} = \frac{1}{\exp(\alpha + \beta \epsilon_j) \mp 1} \tag{3.21}$$

where now the minus sign indicates BE. To evaluate α and β we could use

$$N = \sum_j N_j^* = \sum_j \frac{C_j}{\exp(\alpha + \beta \epsilon_j) \mp 1} \tag{3.22}$$

$$E = \sum_j \epsilon_j N_j^* = \sum_j \frac{C_j \epsilon_j}{\exp(\alpha + \beta \epsilon_j) \mp 1} \tag{3.23}$$

These equations can be solved only for a simplified case that is fortunately of physical relevance.

3.3.1 Boltzmann Limit

We first make the assumption that $C_j \gg N_j$, which we know to be valid from earlier analyses. This condition is called the *Boltzmann Limit*. Then, from Eq. 3.21:

$$\frac{N_j^*}{C_j} = \exp(-\alpha - \beta\epsilon_j) \tag{3.24}$$

and Eq. 3.16 becomes

$$\ln(W) = \sum_j \left[\pm C_j \ln\left(1 \pm \frac{N_j}{C_j}\right) + N_j \ln\left(\frac{C_j}{N_j}\right) \right] \tag{3.25}$$

Using the approximation, $\ln(1+x) \approx \pm x$ when $x \ll 1$, we obtain

$$\ln(W) = \sum_j N_j \left[1 + \ln\left(\frac{C_j}{N_j}\right) \right] \tag{3.26}$$

Equations 3.24 and 3.26 correspond to the *Boltzmann Limit*, $C_j \gg N_j$, i.e., the quantum spacing is very small. Note that these results are now independent of the particle type (bosons or fermions). Physically, this makes sense because in the Boltzmann Limit, the chances of two bosons occupying the same quantum state is very small, and so they essentially behave like fermions.

We may also simplify Eq. 3.22 in the *Boltzmann Limit* to obtain

$$\exp(-\alpha) = \frac{N}{\sum_j C_j \exp(-\beta\epsilon_j)} \tag{3.27}$$

Substitution of Eq. 3.27 into 3.24 gives

$$N_j^* = N \frac{C_j \exp(-\beta\epsilon_j)}{\sum_j C_j \exp(-\beta\epsilon_j)} \tag{3.28}$$

This is the macrostate that yields $W_{\max}$ in the *Boltzmann Limit*. Finally, simplifying Eq. (4.17b) in the *Boltzmann Limit* and using Eq. 3.27, we obtain

$$E = N \frac{\sum_j \epsilon_j C_j \exp(-\beta\epsilon_j)}{\sum_j C_j \exp(-\beta\epsilon_j)} \tag{3.29}$$

From Eq. 3.26:

$$\ln \Omega = \ln(W_{\max}) = \sum_j \left[N_j^* \left(1 + \ln \frac{C_j}{N_j^*} \right) \right] \tag{3.30}$$

Using Eq. 3.28, we obtain

$$\ln \Omega = N \left[1 + \ln \left\{ \frac{\sum_j C_j \exp(-\beta \epsilon_j)}{N} \right\} \right] + \beta E \tag{3.31}$$

This is the total number of microstates for a given system N, E.

We now return to our assumption that $\Omega = W_{\max}$. In the Boltzmann Limit, consider the number of combinations of W in which N_j is close to N_j^*. Introduce a small perturbation $N_j = N_j^* + \Delta N_j$ such that $\frac{\Delta N_j}{N_j^*} \ll 1$ and substitute into Eq. 3.26:

$$\ln(W) = \sum_j (N_j^* + \Delta N_j) \left\{ \ln(C_j) - \ln(N_j^* + \Delta N_j) \right\} + \sum_j N_j \tag{3.32}$$

Since we have $\Delta N_j \ll N_j^*$, we may write

$$\begin{aligned} \ln(N_j^* + \Delta N_j) &= \ln(N_j^*) + \ln \left(1 + \frac{\Delta N_j}{N_j^*} \right) \\ &= \ln(N_j^*) + \frac{\Delta N_j}{N_j^*} - \frac{1}{2} \left(\frac{\Delta N_j}{N_j^*} \right)^2 + \cdots \end{aligned} \tag{3.33}$$

Substitution of 3.33 into 3.32 leads to

$$\begin{aligned} \ln(W) &= \sum_i N_j^* \ln \left[\left(\frac{C_j}{N_j^*} \right) + 1 \right] \\ &\quad - \sum_j \left\{ \Delta N_j + \frac{N_j^*}{2} \left(\frac{\Delta N_j}{N_j^*} \right)^2 + \Delta N_j \ln \left(\frac{C_j}{N_j} \right) \ldots \right\} \\ &= \ln(W_{\max}) - \frac{1}{2} \sum_j \left(\frac{\Delta N_j}{N_j^*} \right)^2 N_j^* \end{aligned} \tag{3.34}$$

where the final expression is obtained using Eqs. 3.17 and 3.18. We see that the right hand side always decreases independent of the sign of ΔN_j so that $W_{\max}$ is indeed a maximum turning point.

Example 3.5 Evaluate $\frac{W}{W_{\max}}$ for N_2 in a 1 cm^3 volume at standard temperature and pressure.

$$N = \sum_j N_j = 2 \times 10^{19} \tag{3.35}$$

Consider an average perturbation of $\frac{\Delta N_j}{N_j^*} = 10^{-3}$

$$\ln\left(\frac{W}{W_{\max}}\right) = -\frac{1}{2} \times 10^{-6} \times 2 \times 10^{19} = -10^{13} \tag{3.36}$$

Thus,

$$W = W_{\max} \exp(-10^{13}) \tag{3.37}$$

that is an incredibly small number. Thus, a small perturbation away from N_j^* contributes only a negligibly small number of additional microstates. This indicates that W is a very peaked function, and so $\Omega = W_{\max}$ is a good assumption. Put another way, the macrostates for which $\Omega = W_{\max}$ must very closely follow $N_j \approx N_j^*$.

3.3.2 Boltzmann Energy Distribution

Recall Eq. 3.28:

$$N_j^* = N \frac{C_j \exp(-\beta \epsilon_j)}{\sum_j C_j \exp(-\beta \epsilon_j)} \tag{3.38}$$

Our prior analysis indicates that the particles in a gas will be distributed across this very special macrostate, N_j^*. For now, we postulate that $\beta = \frac{1}{kT}$ (this will be shown later) to obtain the *Boltzmann energy distribution*:

$$\frac{N_j^*}{N} = \frac{C_j \exp\left(-\frac{\epsilon_j}{kT}\right)}{\sum_j C_j \exp\left(-\frac{\epsilon_j}{kT}\right)} \tag{3.39}$$

The denominator is called the partition function:

$$Q = \sum_j C_j \exp\left(-\frac{\epsilon_j}{kT}\right) = C_1 \exp\left(-\frac{\epsilon_1}{kT}\right) + C_2 \exp\left(-\frac{\epsilon_2}{kT}\right) + \cdots \tag{3.40}$$

From Eq. 3.39,

$$C_j \exp\left(-\frac{\epsilon_j}{kT}\right) = \frac{N_j^*}{N} Q \tag{3.41}$$

Thus, each term in the partition function is proportional to the number of particles in the corresponding energy group, e.g.,

$$C_1 \exp\left(-\frac{\epsilon_1}{kT}\right) = \frac{N_1^*}{N} Q \tag{3.42}$$

If we express the results in terms of the individual quantum energy levels using subscript i in place of j, then the number of states in a group is just the degeneracy, i.e., $C_j = g_i$, and

$$Q = \sum_i g_i \exp\left(-\frac{\epsilon_i}{kT}\right) \tag{3.43}$$

$$\frac{N_i^*}{N} = \frac{g_i \exp\left(-\frac{\epsilon_i}{kT}\right)}{\sum_i g_i \exp\left(-\frac{\epsilon_i}{kT}\right)} = \frac{g_i \exp\left(-\frac{\epsilon_i}{kT}\right)}{Q} \tag{3.44}$$

Equations 3.43 and 3.44 are the most important results from counting microstates: the partition function and the Boltzmann distribution.

3.4 Relation to Thermodynamics

In this section, we consider the plausibility of a relation

$$S = S(\Omega) \tag{3.45}$$

that connects macroscopic thermodynamics and molecular random behavior in terms of how particles can be arranged across their allowed quantum energy states. We review this idea using two simple example configurations.

(a) Isothermal expansion into vacuum. Consider the situation illustrated in Fig. 3.6 in which a gas contained in a chamber A is separated by a partition from a second chamber B of equal volume that is under pure vacuum. A valve in the partition opens at zero time and gas starts to flow from A to B until eventually pressure is equalized across the two chambers at the macroscopic level.

Macroscopically, from classical thermodynamics, the change in entropy is given by

$$dS = \frac{\delta Q}{T} = \frac{dE}{T} + \frac{p}{T}dV - \frac{\tilde{\mu}}{T}dN \tag{3.46}$$

where δQ is the heat transfer, E is the internal energy, p is the pressure, V is the volume, and $\tilde{\mu}$ is the chemical potential. For a perfect gas,

$$p = \rho RT = \frac{m_T}{V}RT \text{ and } E = m_T c_v T \tag{3.47}$$

where m_T is the total mass of gas, c_v is the specific heat at constant volume, and R is the ordinary gas constant. The number of particles is constant so the overall change in entropy between the initial (i) and final (f) states is given by

$$\Delta S = m_T c_v \ln\left(\frac{T_f}{T_i}\right) + m_T R \ln\left(\frac{V_f}{V_i}\right) \tag{3.48}$$

Our process is isothermal, $T_f = T_i$, and $V_f = 2V_i$ so that entropy increases.

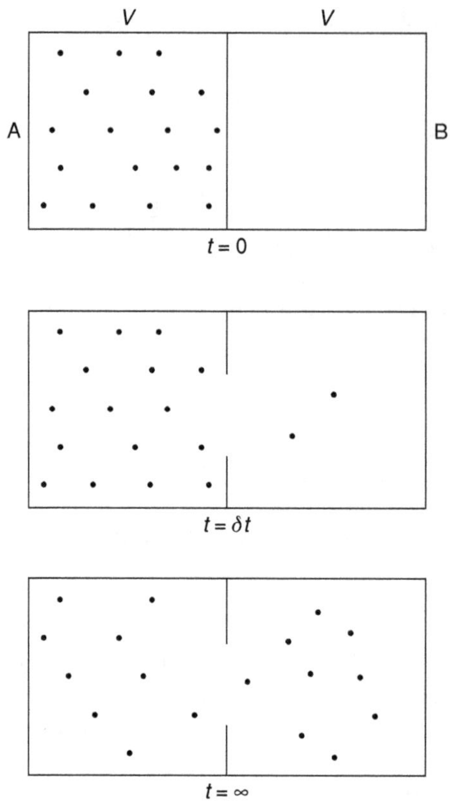

Figure 3.6 Isothermal expansion of a gas into a vacuum.

At the *molecular level*, the degree of randomness also increases since each particle can now place itself within a larger volume of space, thus directly increasing the range of positions it can occupy, but also increasing the number of quantized translational energy levels that it can access. Thus, an increase in Ω occurs simultaneously with an increase in S.

(b) Reversible Heating. Consider reversible heating of a gas of a fixed number of particles at constant volume. *Macroscopically*, from Eq. 3.46, we will see an increase in entropy:

$$\delta S = \frac{\delta Q}{T} \tag{3.49}$$

and the added heat will manifest itself as an increase in temperature. At the *molecular level,* the degree of randomness is increased as each particle can now access a wider range of energy levels due to the increased temperature.

Hence, in each of these two examples, qualitatively we can see that there is a connection between *macroscopic entropy* and the *degree of randomness at the molecular level*, i.e.,

$$S = S(\Omega) \tag{3.50}$$

3.4.1 Boltzmann's Relation

The previous examples illustrated that there is a connection between macroscopic entropy and molecular level random behavior, and let us assume that it is of the form

$$S = \phi(\Omega) \tag{3.51}$$

where Ω is the number of microstates, and ϕ is a universal function. To find an appropriate form for $\phi(\Omega)$ consider two systems with the properties (N_1, S_1, Ω_1) and (N_2, S_2, Ω_2) that are combined such that

$$N_{12} = N_1 + N_2 \tag{3.52}$$

$$S_{12} = S_1 + S_2 \text{(note that entropy is an additive property)} \tag{3.53}$$

$$\Omega_{12} = \Omega_1\Omega_2 \tag{3.54}$$

since microstates Ω_1 coexist with microstates Ω_2. Now, we may also write that $S_{12} = \phi(\Omega_{12})$ and so

$$S_1 + S_2 = \phi(\Omega_1) + \phi(\Omega_2) = \phi(\Omega_1\Omega_2) \tag{3.55}$$

Now, we first differentiate with respect to Ω_1 and then with respect to Ω_2 as follows:

$$\phi(\Omega_1\Omega_2) = \phi(\Omega_1) + \phi(\Omega_2) \tag{3.56}$$

$$\frac{\partial}{\partial\Omega_1} : \Omega_2\frac{\partial\phi}{\partial\Omega_1}(\Omega_1\Omega_2) = \frac{\partial\phi}{\partial\Omega_1}(\Omega_1) \tag{3.57}$$

$$\frac{\partial}{\partial\Omega_2} : \frac{\partial\phi(\Omega_1\Omega_2)}{\partial\Omega_1} + \Omega_1\Omega_2\frac{\partial^2\phi(\Omega_1\Omega_2)}{\partial\Omega_1\partial\Omega_2} = 0 \tag{3.58}$$

Alternatively, returning to Eq. 3.55, we first differentiate with respect to Ω_2 and then with respect to Ω_1 to obtain

$$\phi(\Omega_1\Omega_2) = \phi(\Omega_1) + \phi(\Omega_2)$$

$$\frac{\partial}{\partial\Omega_2} : \Omega_2\frac{\partial\phi}{\partial\Omega_2}(\Omega_1\Omega_2) = \frac{\partial\phi}{\partial\Omega_2}(\Omega_2) = 0$$

$$\frac{\partial}{\partial\Omega_1} : \frac{\partial\phi(\Omega_1\Omega_2)}{\partial\Omega_2} + \Omega_1\Omega_2\frac{\partial^2\phi(\Omega_1\Omega_2)}{\partial\Omega_1\partial\Omega_2} = 0$$

These results may be generalized:

$$\Omega\phi''(\Omega) + \phi'(\Omega) = 0 \tag{3.59}$$

This equation has solutions only of the form:

$$[S =]\, \phi(\Omega) = A(N)\ln(\Omega) + B(N) \tag{3.60}$$

where the constants A, B may be different for each system. Using 3.60 in 3.55:

$$\begin{aligned} &A(N_1+N_2)[\ln(\Omega_1)+\ln(\Omega_2)]+B(N_1+N_2) \\ &= A(N_1)\ln\Omega_1 + B(N_1) + A(N_2)\ln\Omega_2 + B(N_2) \end{aligned} \tag{3.61}$$

This equation only has solutions when $A = k$, an absolute constant, and $B(N) = b \cdot N$ where b is a constant. Thus

$$\begin{aligned} B(N) &= b \cdot N = S_0 \\ S &= k \ln \Omega + S_0 \end{aligned} \tag{3.62}$$

By convention, we set $S = 0$ for a completely ordered system, that is, one for which $\Omega = 1$, so that $S_0 = 0$. Thus, we obtain *Boltzmann's Relation*:

$$S = k \ln \Omega \tag{3.63}$$

3.4.2 Macroscopic Thermodynamic Properties

(a) Entropy. Using Boltzmann's Relation, we now proceed to derive macroscopic thermodynamic properties. Recall Eq. 3.31:

$$S = k\left[N\left\{\ln\left(\frac{\sum C_j \exp(-\beta\epsilon_j)}{N}\right)+1\right\}+\beta E\right] \tag{3.64}$$

We will use this expression to evaluate β. Recall Eq. 3.46:

$$dS = \frac{dE}{T} + \frac{p}{T}dV - \frac{\tilde{\mu}}{T}dN \tag{3.65}$$

and so

$$\left(\frac{\partial S}{\partial E}\right)_{V,N} = \frac{1}{T} \tag{3.66}$$

Recall from Eq. 3.29 that: $\beta = \beta(E)$. Thus

$$\left(\frac{\partial S}{\partial E}\right)_{V,N} = k\left\{\beta + \frac{\partial}{\partial\beta}\left[N\left\{\ln\left(\frac{\sum C_j \exp(-\beta\epsilon_j)}{N}\right)+1\right\}+\beta E\right]\frac{\partial\beta}{\partial E}\right\} \tag{3.67}$$

$$= k\beta + k\left[N\left\{\frac{-\sum \epsilon_j C_j \exp(-\beta\epsilon_j)}{\sum C_j \exp(-\beta\epsilon_j)}\right\}+E\right]\frac{\partial\beta}{\partial E} \tag{3.68}$$

$$= k\beta + k\left[N\left(-\frac{E}{N}\right)+E\right]\frac{\partial\beta}{\partial E} = k\beta = \frac{1}{T} \tag{3.69}$$

$$\therefore \beta = \frac{1}{kT} \text{ as we assumed earlier.} \tag{3.70}$$

Therefore

$$S = Nk\left[\ln\frac{Q}{N}+1\right]+\frac{E}{T} \tag{3.71}$$

where Q is the partition function.

(b) Helmholtz Free Energy. This classical thermodynamic property has the definition

$$F \equiv E - TS \tag{3.72}$$

$$dF = dE - TdS - SdT \tag{3.73}$$

Using Eq. 3.65, Eq. 3.73 becomes

$$dF = -SdT - pdV + \tilde{\mu}dN \tag{3.74}$$

Thus

$$S = -\left(\frac{\partial F}{\partial T}\right)_{V,N}; p = -\left(\frac{\partial F}{\partial V}\right)_{T,N}; \tilde{\mu} = \left(\frac{\partial F}{\partial N}\right)_{T,V} \tag{3.75}$$

Using Eq. 3.72,

$$E = F + TS = F - T\left(\frac{\partial F}{\partial T}\right)_{V,N} = -T^2\left[\frac{\partial(F/T)}{\partial T}\right]_{V,N} \tag{3.76}$$

We now obtain F from Eqs. 3.71 and 3.72:

$$F = E - TS = E - NkT\left(1 + \ln\frac{Q}{N}\right) - E = -NkT\left(1 + \ln\frac{Q}{N}\right) \tag{3.77}$$

Now, assuming that $Q = Q(T, V)$ (to be proved later)

$$S = Nk\left[1 + \ln\left(\frac{Q}{N}\right) + T\frac{\partial}{\partial T}(\ln Q)\right] \tag{3.78}$$

$$E = NkT^2\frac{\partial}{\partial T}(\ln Q) \tag{3.79}$$

$$p = NkT\frac{\partial}{\partial V}(\ln Q) \tag{3.80}$$

$$\tilde{\mu} = kT\ln\frac{Q}{N} \tag{3.81}$$

Thus, the partition function Q determines all of the classical thermodynamic properties.

3.5 Partition Functions

We have seen that all of the macroscopic thermodynamics variables are determined by the molecular partition functions. In this section, we evaluate the partition functions for all of the energy modes.

3.5.1 Translational Energy

Consider a gas of N particles in motion inside a volume V. Let us assume that the particles only have translational energy and interact via collisions.

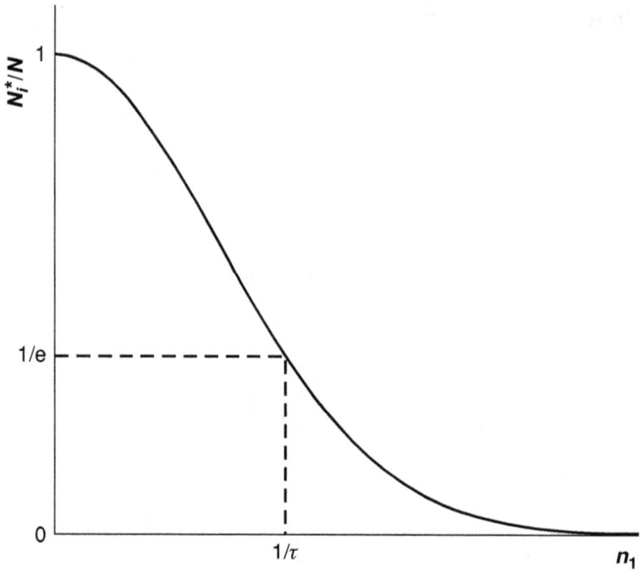

Figure 3.7 Translational energy distribution function.

If they reside inside a cuboid with side lengths a_1, a_2, a_3, then the allowed translational quantum energy states are

$$(\epsilon_{tr})_i = \epsilon_{n_1,n_2,n_3} = \frac{h^2}{8m}\left(\frac{n_1^2}{a_1^2} + \frac{n_2^2}{a_2^2} + \frac{n_3^2}{a_3^2}\right) \tag{3.82}$$

Hence, the partition function for translational motion is

$$\begin{aligned} Q_{tr} &= \sum_i \exp\left[-\frac{(\epsilon_{tr})_i}{kT}\right] = \sum_{n_1}\sum_{n_2}\sum_{n_3} \exp\left[-\frac{h^2}{8mkT}\left(\frac{n_1^2}{a_1^2} + \frac{n_2^2}{a_2^2} + \frac{n_3^2}{a_3^2}\right)\right] \\ &= \sum_{n_1} \exp\left[-\frac{n_1^2}{a_1^2}\frac{h^2}{8mkT}\right] \sum_{n_2} \exp\left[-\frac{n_2^2}{a_2^2}\frac{h^2}{8mkT}\right] \sum_{n_3} \exp\left[-\frac{n_3^2}{a_3^2}\frac{h^2}{8mkT}\right] \end{aligned} \tag{3.83}$$

Consider one of the three summation terms:

$$\sum_{n_1} \exp\left[-\frac{n_1^2}{a_1^2}\frac{h^2}{8mkT}\right] = \sum_{n_1=1}^{\infty} \exp(-\tau^2 n_1^2) \tag{3.84}$$

with $\tau^2 = \frac{h^2}{8ma_1^2kT}$. Now, for N_2 at 293 K and $a = 1$ cm, $\tau^2 = 3 \times 10^{-18}$. Under these conditions, the distribution of energy is shown in Fig. 3.7. Thus, the probability density falls to a value of 1/e at $n_1 = 1/\tau$, indicating again that there are a very large number of translational energy states, this is an energy mode involving very large quantum numbers, and it thus lies in the classical physics regime.

Since the number of translational energy states is so large, we may replace each summation in the partition function by an integral:

$$\sum_{n_1=1}^{\infty} \exp(-\tau^2 n_1^2) = \int_{n_1=1}^{\infty} \exp(-\tau^2 n_1^2) dn_1 = \frac{a_1}{h}\sqrt{2\pi mkT} \tag{3.85}$$

Using the analogous results for n_2 and n_3, we obtain the overall translational partition function

$$Q_{\text{tr}} = V\left(\frac{2\pi mkT}{h^2}\right)^{3/2} \tag{3.86}$$

We now use Eqs. 3.77–3.81 to evaluate the associated thermodynamic properties.

Free Energy

$$\begin{aligned} F_{\text{tr}} &= -NkT\left[\ln\left(\frac{Q_{tr}}{N}\right) + 1\right] \\ &= -NkT\left[\ln\left\{\frac{V}{N}T^{3/2}\left(\frac{2\pi mk}{h^2}\right)^{3/2}\right\} + 1\right] \\ &= -NkT\left[\ln\left(\frac{V}{N}\right) + \frac{3}{2}\ln(T) + \ln\left(\frac{2\pi mk}{h^2}\right)^{3/2} + 1\right] \end{aligned} \tag{3.87}$$

Pressure

$$\begin{aligned} p_{\text{tr}} &= NkT\frac{\partial}{\partial V}\left\{\ln(V) + \ln\left(\frac{2\pi mkT}{h^2}\right)^{3/2}\right\} = NkT/V \\ p_{\text{tr}}V &= NkT \end{aligned} \tag{3.88}$$

Thus, by comparison with the ideal gas law, the constant k in Boltzmann's Relation is the Boltzmann constant from kinetic theory, $k = 1.38 \times 10^{-23}$ J/K.

Entropy

$$\begin{aligned} S_{\text{tr}} &= -\left(\frac{\partial F_{\text{tr}}}{\partial T}\right)_{V,N} \\ &= Nk\left[\ln\left(\frac{V}{N}\right) + \frac{3}{2}\ln(T) + \ln\left(\frac{2\pi mk}{h^2}\right)^{3/2} + 1\right] + \frac{3}{2}Nk \end{aligned} \tag{3.89}$$

Substitute Eq. 3.88 as $\ln\left(\frac{V}{N}\right) = \ln(T) - \ln(p) + \ln(k)$ to obtain

$$S_{\text{tr}} = Nk\left\{\frac{5}{2}\ln(T) - \ln(p) + \ln\left[\left(\frac{2\pi m}{h^2}\right)^{3/2} k^{5/2}\right] + \frac{5}{2}\right\} \tag{3.90}$$

Internal Energy

$$E_{\text{tr}} = F_{\text{tr}} + T \cdot S_{\text{tr}} = \frac{3}{2}NkT \tag{3.91}$$

The associated specific internal energy for translational motion is

$$e_{\text{tr}} = \frac{3}{2}RT \tag{3.92}$$

Chemical Potential

$$\tilde{\mu}_{\text{tr}} = \left(\frac{\partial F_{\text{tr}}}{\partial N}\right)_{V,T} = kT\left[\ln\left(\frac{V}{N}\right) + \frac{3}{2}\ln(T) + \ln\left(\frac{2\pi mk}{h^2}\right)^{3/2}\right] \tag{3.93}$$

Energy Distribution. We now consider the distribution of translational energy. For an energy group j with energy levels in the range $\epsilon \to \epsilon + d\epsilon$, assuming a continuous distribution function, $f(\epsilon)$, the number of particles in a group is

$$N_j^* = Nf(\epsilon)\,d\epsilon \tag{3.94}$$

The Boltzmann energy distribution is

$$\frac{N_j^*}{N} = \frac{C_j \exp\left(-\frac{\epsilon_j}{kT}\right)}{Q} = f(\epsilon)\,d\epsilon \tag{3.95}$$

The number of translational energy states in a group is

$$C_j = \frac{d\Gamma}{d\epsilon}d\epsilon = 2\pi\frac{V}{h^3}(2m)^{3/2}\epsilon^{1/2}d\epsilon \tag{3.96}$$

recalling that the total number of translational energy states was given by Eq. 3.5:

$$\Gamma = \frac{4\pi}{3}\frac{V}{h^3}(2m\epsilon)^{3/2} \tag{3.97}$$

Substituting 3.96 and Q_{tr} into 3.95 we obtain

$$f(\epsilon)\,d\epsilon = \frac{2\pi\epsilon^{1/2}\exp\left(-\frac{\epsilon}{kT}\right)d\epsilon}{(\pi kT)^{3/2}} = \frac{2}{\pi^{1/2}}\exp\left(-\frac{\epsilon}{kT}\right)d\left(\frac{\epsilon}{kT}\right) \tag{3.98}$$

This is the continuous energy distribution at equilibrium for three degrees of freedom. By inspection, the corresponding result for ζ degrees of freedom is

$$f_s(\epsilon)\,d\epsilon = \frac{1}{\Gamma\left(\frac{\zeta}{2}\right)}\left(\frac{\epsilon}{kT}\right)^{\frac{\zeta}{2}-1}\exp\left(-\frac{\epsilon}{kT}\right)d\left(\frac{\epsilon}{kT}\right) \tag{3.99}$$

Returning to 3.98, we can substitute $\epsilon = \frac{1}{2}mC^2$ to obtain:

$$\chi(C)\,dC = 4\pi\left(\frac{m}{2\pi kT}\right)^{3/2}C^2\exp\left(-\frac{mC^2}{2kT}\right)dC \tag{3.100}$$

i.e., the Maxwellian speed distribution. Thus, our results from kinetic theory and statistical mechanics/quantum mechanics are consistent in the classical limit, and so adhere to the *Correspondence Principle.*

3.5.2 Internal Structure

In addition to translational energy, all particles have electronic energy, and molecules also have rotational and vibrational energy. All of these energy modes affect the partition functions and hence the thermodynamic properties.

Let the total particle energy be

$$\epsilon = \epsilon' + \epsilon'' + \epsilon''' + \cdots$$

where ϵ', ϵ'', ϵ''', ... are separate contributions from different modes. The quantized energy states are written

$$\epsilon_{m,n,p,\ldots} = \epsilon'_m + \epsilon''_n + \epsilon'''_p + \cdots$$

where m, n, p...are quantum numbers. The partition function is then

$$\begin{aligned} Q &= \sum_{m,n,p,\ldots} \exp\left(-\frac{\epsilon_{m,n,p,\ldots}}{kT}\right) \\ &= \sum_m \exp\left(-\frac{\epsilon'_m}{kT}\right) \sum_n \exp\left(-\frac{\epsilon''_n}{kT}\right) \sum_p \exp\left(-\frac{\epsilon'_p}{kT}\right) \cdots \\ &= Q' \times Q'' \times Q''' \times \cdots \end{aligned} \tag{3.101}$$

Thus, the total partition function is the product of the individual partition functions of all contributing energy sources. We now apply this idea to a real molecule, for which the total energy is

$$\epsilon = \epsilon_{\text{tr}} + (\epsilon_{\text{rot}} + \epsilon_{\text{vib}} + \epsilon_{\text{el}})_{\text{int}} \tag{3.102}$$

with overall partition function

$$Q = Q_{\text{tr}} Q_{\text{rot}} Q_{\text{vib}} Q_{\text{el}} = Q_{\text{tr}} \times Q_{\text{int}} \tag{3.103}$$

Consider the Helmholtz Free Energy:

$$\begin{aligned} F = F_{\text{tr}} + F_{\text{int}} &= -NkT\left[\ln\left(\frac{Q}{N}\right) + 1\right] \\ &= -NkT\left[\ln\left(\frac{Q_{\text{tr}} Q_{\text{rot}} Q_{\text{vib}} Q_{\text{el}}}{N} + 1\right) + 1\right] \end{aligned} \tag{3.104}$$

However, from Eq. 3.87, we already know that

$$\begin{aligned} F_{\text{tr}} &= -NkT\left[\ln\left(\frac{Q_{tr}}{N}\right) + 1\right] \\ &\Rightarrow F_{\text{int}} = -NkT\ln(Q_{\text{int}}) \end{aligned} \tag{3.105}$$

Thus, by analysis of the Helmholtz Free Energy, it may be shown that for entropy:

$$S = S_{tr} + \sum_{int} S_{int} \tag{3.106}$$

and for specific internal energy

$$e = \frac{E}{mN} = RT^2 \frac{\partial}{\partial T}[\ln(Q)] \tag{3.107}$$

$$e = \underbrace{RT^2 \frac{\partial}{\partial T}[\ln(Q_{tr})]}_{e_{tr}} + \underbrace{\sum_{int} RT^2 \frac{\partial}{\partial T}[\ln(Q_{int})]}_{\sum_{int} e_{int}} \tag{3.108}$$

Using the standard definition of specific heat at constant volume:

$$c_v \equiv \left(\frac{\partial e}{\partial T}\right)_v = c_{v,tr} + \sum_{int} c_{v,int} \tag{3.109}$$

Thus, we find that F, S, e, and c_v are all additive thermodynamic properties. For pressure, we have:

$$p = NkT \frac{\partial}{\partial V} \ln(Q_{tr}) + \sum_{int} NkT \frac{\partial}{\partial V} \ln(Q_{int}) \tag{3.110}$$

We will soon show that Q_{int} is independent of V, so that our earlier result, $pV = NkT$, is unaffected by the internal energy modes.

3.5.3 Monatomic Gas

Let us now apply our results to the particular case of atoms, for which

$$\epsilon = \epsilon_{tr} + \epsilon_{el} \tag{3.111}$$

$$Q = Q_{tr} \times Q_{el} \tag{3.112}$$

where the partition function of the electronic energy mode may be written

$$Q_{el} = \sum_i g_i \exp\left(-\frac{\epsilon_i}{kT}\right) = g_0 \exp\left(-\frac{\epsilon_0}{kT}\right) + g_1 \exp\left(-\frac{\epsilon_1}{kT}\right) + \cdots \tag{3.113}$$

By convention, we set the electronic energy of the ground state, $\epsilon_0 = 0$. Now we introduce characteristic temperatures for electronic excitation,

$$\theta_i \equiv \frac{\epsilon_i}{k} \tag{3.114}$$

$$\Rightarrow Q_{\text{el}} = g_0 + g_1 \exp\left(-\frac{\theta_1}{T}\right) + g_2 \exp\left(-\frac{\theta_2}{T}\right) + \cdots \tag{3.115}$$

For most atoms, the values of θ_i are large compared to the gas temperature, T. So, we need consider only the first few terms in the partition function. We will proceed to illustrate the process for evaluation of the thermodynamic properties for the following simple case:

$$Q_{\text{el}} = g_0 + g_1 \exp\left(-\frac{\theta_1}{T}\right) \tag{3.116}$$

From Eq. 3.106, the specific electronic energy is

$$\begin{aligned} e_{\text{el}} &= RT^2 \frac{\partial}{\partial T}\left\{\ln\left[g_0 + g_1 \exp\left(-\frac{\theta_1}{T}\right)\right]\right\} \\ &= RT^2 \left\{-\frac{g_1\theta_1\left(-\frac{1}{T^2}\exp\left(-\frac{\theta_1}{T}\right)\right)}{g_0 + g_1 \exp\left(-\frac{\theta_1}{T}\right)}\right\} \\ &= \frac{R\theta_1\left(\frac{g_1}{g_0}\right)\exp\left(-\frac{\theta_1}{T}\right)}{1 + \frac{g_1}{g_0}\exp\left(-\frac{\theta_1}{T}\right)} \end{aligned} \tag{3.117}$$

The *specific heat at constant volume* for the electronic energy mode is

$$(c_v)_{\text{el}} = \left(\frac{\partial e_{\text{el}}}{\partial T}\right)_V = R\left(\frac{\theta_1}{T}\right)^2 \frac{\frac{g_1}{g_0}\exp\left(-\frac{\theta_1}{T}\right)}{\left[1 + \frac{g_1}{g_0}\exp\left(-\frac{\theta_1}{T}\right)\right]^2} \tag{3.118}$$

Note that $(c_v)_{\text{el}}$ is a function of T, so that in terms of classical thermodynamics, the gas is no longer calorically perfect, and the *ratio of specific heats*, $\gamma = \gamma(T)$.

Figure 3.8 shows $(c_v)_{\text{el}}$ versus T for two different values of g_1/g_0. These curves are valid only for $\theta_2 \gg T$. At higher temperature, where $T \approx \theta_2$, it is necessary to include additional terms in the evaluation of the partition function.

Example 3.6 (a) Atomic oxygen: from Table 2.3:

$$\begin{array}{llllll} Q_{\text{el}}^{\text{O}} = & 5+ & 3\exp\left(-\frac{228}{T}\right)+ & \exp\left(-\frac{326}{T}\right)+ & \text{O}\left[\exp\left(-\frac{23{,}000}{T}\right)\right] \\ & \uparrow & \uparrow & \uparrow & \\ & {}^3P_2 & {}^3P_1 & {}^3P_0 & \end{array}$$

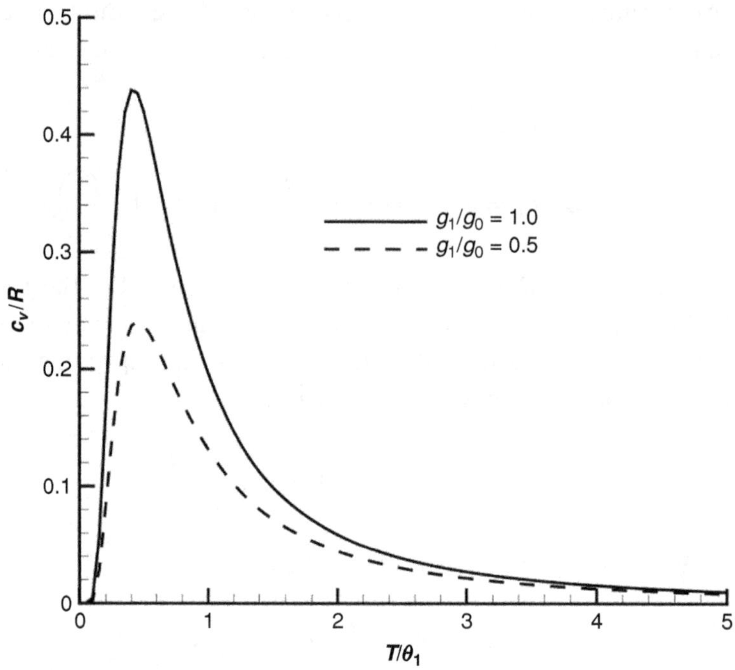

Figure 3.8 Electronic specific heat as a function of temperature.

The first three states are often combined in different ways, depending on the temperature range of interest:

$$Q_{\mathrm{el}}^{\mathrm{O}} \approx 5 + 4\exp\left(-\frac{270}{T}\right) \approx 9 \tag{3.119}$$

$$\therefore\ e_{\mathrm{el}}^{\mathrm{O}} = 0 = (c_v)_{\mathrm{el}}^{\mathrm{O}}$$

Atomic nitrogen: again, from Table 2.3:

$$Q_{\mathrm{el}}^{N} = \underset{\substack{\uparrow \\ {}^4S}}{4} +\ \mathrm{O}\left[\exp\left(-\tfrac{28{,}000}{T}\right)\right]$$

$$\therefore\ e_{\mathrm{el}}^{\mathrm{N}} = 0 = (c_v)_{\mathrm{el}}^{\mathrm{N}}$$

Thus, for temperatures typical of aerospace applications, of a few thousand degrees at the most, for both oxygen and nitrogen atoms, the partition function is simply a number, and there are no contributions to the internal energy and specific heats from the electronic mode. We will find later on, however, that the numerical values of the atomic electronic partition functions are important in determining chemical composition.

3.5.4 Diatomic Gas

Now, we apply our results to the case of a gas consisting of diatomic molecules such as N_2 and O_2.

(a) Rotational Energy: for moderate temperatures, e.g., less than 3000 K, the effects of rotation and vibration are more important than those for electronic energy states. For the rigid rotor:

$$\epsilon_J = k\theta_{\text{rot}} J(J+1)(J = 0, 1, 2, \ldots) \tag{3.120}$$

with a degeneracy, $g_J = 2J + 1$. Thus, the partition function of the rotational mode is

$$Q_{\text{rot}} = \sum_{J=0}^{\infty} (2J+1) \exp\left(-\frac{\theta_r}{T} J(J+1)\right) \tag{3.121}$$

When $\theta_r \gg T$, Eq. 3.121 indicates that $Q_{\text{rot}} = 1 \Rightarrow e_{\text{rot}} = 0 = (c_v)_{\text{rot}}$.

However, this limit is never attained for air molecules since, from Table 2.3:

$$\theta_{\text{rot}}^{N_2} = 2.9K, \theta_{\text{rot}}^{O_2} = 2.1\text{ K}, \theta_{\text{rot}}^{NO} = 2.5K$$

These values indicate that the quantum spacing for rotational energy is small. We may therefore use a continuous approach to evaluate the partition function:

$$Q_{\text{rot}} = \int_0^{\infty} (2J+1) \exp\left[-\frac{\theta_{\text{rot}}}{T} J(J+1)\right] dJ \tag{3.122}$$

Using the transformation $z = J(J+1)$

$$Q_{\text{rot}} = \int_0^{\infty} \exp\left(-\frac{z\theta_{\text{rot}}}{T}\right) dz = \frac{T}{\theta_{\text{rot}}} \tag{3.123}$$

$$e_{\text{rot}} = RT^2 \frac{\partial}{\partial T}\left[\ln\left(\frac{T}{\theta_{\text{rot}}}\right)\right] = RT \tag{3.124}$$

$$(c_v)_{\text{rot}} = \left(\frac{\partial e_{\text{rot}}}{\partial T}\right)_V = R \tag{3.125}$$

We can now evaluate the ratio of specific heats,

$$\gamma = \frac{(c_v)_{\text{tr}} + (c_v)_{\text{rot}} + R}{(c_v)_{\text{tr}} + (c_v)_{\text{rot}}} = \frac{\frac{3}{2}R + R + R}{\frac{3}{2}R + R} = \frac{7}{5} = 1.40 \tag{3.126}$$

as expected for air at room temperature.

Note

Equation 3.123 is valid only for heteronuclear molecules such as NO. Homonuclear molecules (N_2, O_2), have a smaller degeneracy. For example, in a Σ state ($\Lambda = 0$), the degeneracy is the same but every second state is missing. This situation arises from the wave function symmetry properties of homonuclear molecules that are always bosons and have the property that their total eigenfunction preserves sign after coordinate reflection through the origin. Consider the ground state of molecular oxygen, $X^3\Sigma_g^-$. Assuming the total eigenfunction is a superposition of those from the individual energy modes, we have the following situation as a result of coordinate reflection (in which r is replaced by $-r$):

- Translation: No sign change
- Rotation: Sign change for odd values of J, no sign change for even values of J
- Vibration: No sign change
- Electronic: Sign change due to the g and "$-$" properties of the state.

Thus, we have the following situation for the rotational states of oxygen:

$$O_2\text{: } X^3\Sigma_g^-\text{: } J = 1, 3, 5, 7, \ldots$$

The same analysis for nitrogen molecules indicates that:

$$N_2\text{: } X^1\Sigma_g^+\text{: } J = 0, 2, 4, 6, \ldots$$

Thus, for such diatomic molecules, the partition function is reduced on average by a factor of two, and so we write in general:

$$Q_{\text{rot}} = \frac{1}{\sigma}\frac{T}{\theta_{\text{rot}}}$$

$$\sigma = \begin{cases} 1\text{–heteronuclear, e.g., NO} \\ 2\text{–homonuclear, e.g., } N_2 \end{cases}$$

It is straightforward to show that this property has no effect on the results above for internal energy and specific heat.

(b) Vibrational energy: Consider the simplified version of the harmonic oscillator model:

$$\epsilon_{\text{vib}} = vk\theta_{\text{vib}};\ v = 0, 1, 2, \ldots$$

with a degeneracy for all levels of 1. The partition function is

$$Q_{\text{vib}} = \sum_{v=0}^{\infty} \exp\left(-v\frac{\theta_{\text{vib}}}{T}\right)$$

Now, the summation:

$$\sum_{i=0}^{\infty} X^i = 1 + X + X^2 + \cdots = \frac{1}{1-X}(X < 1)$$

$$\Rightarrow Q_{\text{vib}} = \frac{1}{1 - \exp\left(-\frac{\theta_{\text{vib}}}{T}\right)} \tag{3.127}$$

Hence:

$$e_{\text{vib}} = RT^2 \frac{\partial}{\partial T}\left\{-\ln\left[1 - \exp\left(-\frac{\theta_{\text{vib}}}{T}\right)\right]\right\} = \frac{R\theta_{\text{vib}}}{\exp\left(\frac{\theta_{\text{vib}}}{T}\right) - 1} \tag{3.128}$$

$$(c_v)_{\text{vib}} = \frac{\partial e_{\text{vib}}}{\partial T} = R\left[\frac{\theta_{\text{vib}}/2T}{\sinh\left(\frac{\theta_{\text{vib}}}{2T}\right)}\right] \tag{3.129}$$

Equation 3.128 may be written:

$$e_{\text{vib}} = \frac{R\theta_{\text{vib}}}{\frac{\theta_{\text{vib}}}{T} + \frac{1}{2!}\left(\frac{\theta_{\text{vib}}}{T}\right)^2 + \cdots} \tag{3.130}$$

for which we can evaluate two limits:

$$(i)\ \theta_{\text{vib}} \gg T,\ e_{\text{vib}} \to 0$$

$$(ii)\ \theta_{\text{vib}} \ll T,\ e_{\text{vib}} \to RT$$

So, at low temperature, there is no participation in thermodynamics from the vibrational energy mode. Whereas at high temperature, the vibrational mode becomes fully excited with two degrees of freedom. In that same limit:

$$(c_v)_{\text{vib}} \to R,\ \gamma \to \frac{\frac{3}{2}R + R + R + R}{\frac{3}{2}R + R + R} = \frac{9}{7} = 1.28 \tag{3.131}$$

In general, the number of vibrational degrees of freedom may be written

$$\zeta_{\text{vib}} \equiv \frac{e_{\text{vib}}}{RT/2} = \frac{2(\theta_{\text{vib}}/T)}{\exp(\theta_{\text{vib}}/T) - 1} \tag{3.132}$$

Figure 3.9 plots the variation with temperature of the number of vibrational degrees of freedom showing that it approaches two only at high temperature.

Note: From Table 2.3: $\theta_{\text{vib}} = 3390$ K (N_2), 2270 K (O_2), 2740 K (NO)

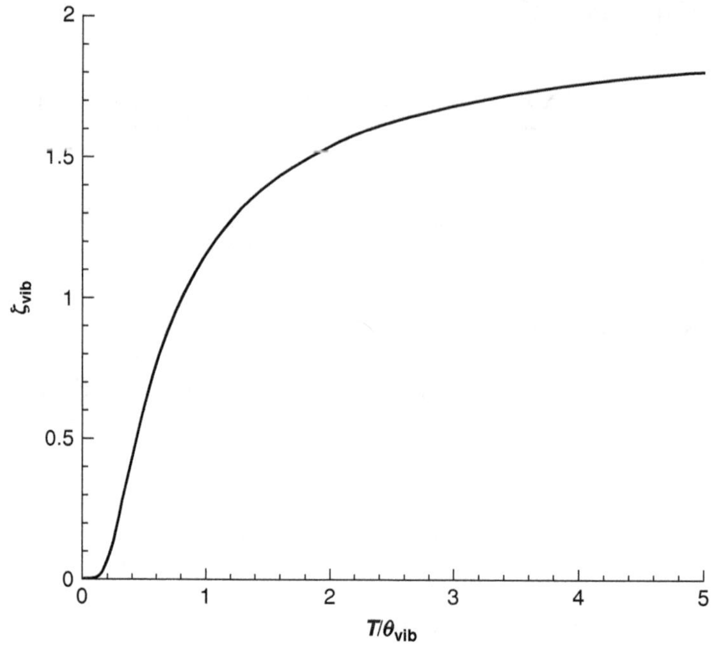

Figure 3.9 Number of vibrational degrees of freedom as a function of temperature.

(c) Electronic Energy: The partition functions for the electronic modes of molecules have the same mathematical form as those for atoms, and spectroscopic measurements give the following data for air species:

$$Q_{el}^{N_2} = 1 + O\left[\exp\left(-\frac{71,000}{T}\right)\right]$$

$$X^1\Sigma_g^+$$

$$Q_{el}^{O_2} = 3 + 2\exp\left(-\frac{11,900}{T}\right) + O\left[\exp\left(-\frac{19,000}{T}\right)\right]$$

$$X^3\Sigma_g^-$$

$$Q_{el}^{NO} = 2 + 2\exp\left(-\frac{174}{T}\right) + O\left[\exp\left(-\frac{65,000}{T}\right)\right]$$

$$X^2\Pi$$

We can model the associated thermodynamic properties using Eqs. 3.92 and 3.93.

Now that we have considered all of the internal energy mode partition functions, we can see by inspection that they are all independent of volume V. Therefore, for pressure

$$p = p_{tr} + NkT\sum_{int}\frac{\partial}{\partial V}(\ln Q_{int}) = p_{tr} = \frac{NkT}{V} = nkT \tag{3.133}$$

that is simply the ideal gas law.

In summary, for the temperature range of interest for most aerospace applications, e.g., 300 K $< T <$ 5000 K, the following is true for air molecules (N_2, O_2, NO):

- The translational mode behaves classically with three degrees of freedom ($\zeta_{tr} = 3$).
- The rotational mode behaves classically with two degrees of freedom ($\zeta_{rot} = 2$).
- The vibrational mode is quantized with $0 \leq \zeta_{vib} \leq 2$.
- Almost all particles lie in their electronic ground state.

3.6 Dissociation–Recombination System

The goal of this section is to use statistical mechanics to analyze the thermodynamics of chemically reacting gases at the molecular level. We will continue to focus here on high-temperature air, for which dissociation–recombination is an important reaction:

$$
\begin{aligned}
a_2 &\overset{\text{d}}{\underset{\text{r}}{\Longleftrightarrow}} a + a \\
a &= \text{O, N}
\end{aligned}
$$

Consider a system containing a fixed number $\tilde{N}_a$ of a-particles that may exist either as atoms or diatoms (a or a_2). The allowed atomic energy states are:

$$\epsilon_1^a, \epsilon_2^a, \epsilon_3^a \ldots \text{Group structure: } \epsilon_j^a, N_j^a, C_j^a$$

and the allowed diatomic energy states are

$$\epsilon_1^{aa}, \epsilon_2^{aa}, \epsilon_3^{aa} \ldots \text{Group structure: } \epsilon_j^{aa}, N_j^{aa}, C_j^{aa}$$

We will therefore have atomic and diatomic macrostates: N_j^a, N_j^{aa}. Thus, the total number of atomic microstates is

$$W^a = W^a(N_j^a) \tag{3.134}$$

and the total number of diatomic microstates is

$$W^{aa} = W^{aa}(N_j^{aa}) \tag{3.135}$$

Hence, the total number of microstates for the system is given by

$$\Omega = \sum W^a(N_j^a) W^{aa}(N_j^{aa}) = \sum W \tag{3.136}$$

where the sets of N_j^a, N_j^{aa} must satisfy two criteria:

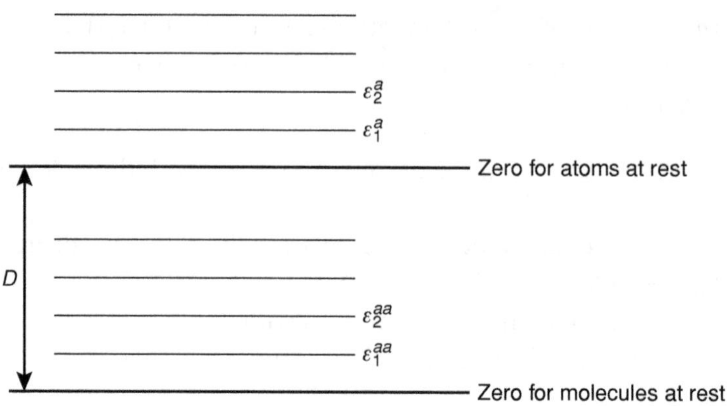

Figure 3.10 Relative energies of atoms and molecules.

(1) Atom conservation:

$$\sum_j N_j^a + 2\sum_j N_j^{aa} = \tilde{N}_a \tag{3.137}$$

(2) Energy conservation
When a molecule dissociates into atoms, an energy D is expended to break the chemical bond. On an energy level diagram, atomic energies lie this energy D above molecular energies. The chemical bond energy D is therefore like a potential energy. When a molecule has energy $\epsilon_i^{aa} > D$, it dissociates into atoms. When atoms recombine, energy D goes into forming the chemical bond. These ideas are illustrated in Fig. 3.10.

By convention, we set the total system energy to zero when we only have atoms, and they are all at rest. Thus:

$$\begin{aligned} \text{total energy of atoms} &= \sum_j N_j^a \epsilon_j^a \\ \text{total energy of molecules} &= \sum_j N_j^{aa}(\epsilon_j^{aa} - D) \end{aligned}$$

Therefore,

$$\sum_j N_j^a \epsilon_j^a + \sum_j N_j^{aa}(\epsilon_j^{aa} - D) = E \tag{3.138}$$

To evaluate the thermodynamics of this system, we follow the same procedure as before:

- Evaluate W^a, W^{aa} using Bose–Einstein and Fermi–Dirac statistics.
- Simplify using Stirling's formula.
- Search for the largest term in Eq. 3.136 at the Boltzmann Limit:

$$\partial[\ln(W)] = \sum_j \ln\left(\frac{C_j^a}{N_j^a}\right)\delta N_j^a + \sum_j \ln\left(\frac{C_j^{aa}}{N_j^{aa}}\right)\delta N_j^{aa} = 0 \tag{3.139}$$

where

$$\sum_j \partial N_j^a + 2\sum_j \partial N_j^{aa} = 0 \tag{3.140}$$

$$\sum_j \epsilon_j^a \partial N_j^a + \sum_j (\epsilon_j^{aa} - D)\partial N_j^{aa} = 0 \tag{3.141}$$

- Solve Eq. 3.139 subject to Eqs. 3.140 and 3.141 using Lagrange's method of undetermined multipliers to obtain the Boltzmann energy distributions for atoms and molecules:

$$\begin{aligned} \frac{N_j^a}{N^{a^*}} &= \frac{C_j^a \exp(-\epsilon_j^a/kT)}{Q^a} \\ \frac{N_j^{aa}}{N^{aa^*}} &= \frac{C_j^{aa} \exp(-\epsilon_j^{aa}/kT)}{Q^{aa}} \end{aligned} \tag{3.142}$$

where N^{a^*} and N^{aa^*} are the numbers of atoms and molecules that exist at equilibrium (the equilibrium composition). The system of equations is closed using

$$\text{atom conservation} : N^{a^*} + 2N^{aa^*} = \tilde{N}_a \tag{3.143}$$

$$\text{and the law of mass action} : \frac{N^{aa^*}}{(N^{a^*})^2} = \frac{Q^{aa}}{(Q^a)^2} \exp\left(\frac{D}{kT}\right) \tag{3.144}$$

that results from the Lagrange analysis. Note that the equilibrium composition depends on $\tilde{N}_a$, T, V. We will use these results in Chapter 4 to analyze equilibrium chemical composition.

Using the Boltzmann Relation for this system, it may be shown that

$$\begin{aligned} S = kN^{a^*}\left[\ln\left(\frac{Q^a}{N^{a^*}}\right) + 1\right] + \frac{1}{T}\sum_j N_j^{a^*}\epsilon_j^a \\ + kN^{aa^*}\left[\ln\left(\frac{Q^{aa}}{N^{aa^*}}\right) + 1\right] + \frac{1}{T}\sum_j N_j^{aa^*}\epsilon_j^{aa} \end{aligned} \tag{3.145}$$

Now, substituting Eq. 3.138:

$$S = kN^{a^*}\left[\ln\left(\frac{Q^a}{N^{a^*}}\right) + 1\right] + kN^{aa^*}\left[\ln\left(\frac{Q^{aa}}{N^{aa^*}}\right) + 1\right] + \frac{1}{T}\left(E + N^{aa^*}D\right) \tag{3.146}$$

The classical thermodynamics result for our reacting system is

$$dS = \frac{\delta Q}{T} = \frac{dE}{T} + \frac{p}{T}dV - \frac{\tilde{\mu}^a}{T}dN^a - \frac{\tilde{\mu}^{aa}}{T}dN^{aa} \tag{3.147}$$

Recalling the Helmholtz Free Energy, $F \equiv E - TS$ we obtain

$$F = -kT\left\{N^{a^*}\left[1 + \ln\left(\frac{Q^a}{N^{a^*}}\right)\right] + N^{aa^*}\left[1 + \ln\left(\frac{Q^{aa}}{N^{aa^*}}\right)\right]\right\} - N^{aa^*}D \tag{3.148}$$

All other thermodynamic properties can then be obtained as derivatives of F.

For example, the pressure for a simple gas is given by

$$p = -\left(\frac{\partial F}{\partial V}\right)_{T,N} \tag{3.149}$$

For our gas mixture, we must be more specific about keeping the number of particles of each species constant, i.e.,

$$p = -\left(\frac{\partial F}{\partial V}\right)_{T,N^a,N^{aa}} \tag{3.150}$$

Also, we can recall that for every species, whether atom or molecule, the translational energy partition function is directly proportional to volume and is independent of particle numbers. Therefore, in the differentiation of the partition functions in Eq. 3.148:

$$\left(\frac{\partial \ln Q}{\partial V}\right)_{T,N^a,N^{aa}} = \frac{d \ln V}{dV} = \frac{1}{V} \tag{3.151}$$

Thus,

$$p = kT\left\{N^{a^*}\frac{1}{V} + N^{aa^*}\frac{1}{V}\right\} = p^a + p^{aa} \tag{3.152}$$

which is simply Dalton's law of partial pressures.

3.7 Summary

In this chapter, a connection was established between the way that particles occupy their allowed, quantized energy states, and the macroscopic thermodynamic properties of a gas. This connection was developed through the study of molecular counting methods and led to the establishment of the Boltzmann energy distribution function. The familiar thermodynamic properties were defined in terms of various derivatives of the partition functions of the particles, and expressions for the partition functions were developed for the four energy modes relevant to a gas. These concepts were extended to account for the complexities of a gas mixture undergoing chemical reactions. Many of the key results will be used in Chapter 4 to study finite rate relaxation processes, and will be seen again in the portions of the molecular simulation material pertaining to internal energy relaxation and chemistry.

3.8 Problems

3.1 Derive an expression for the ratio of the number of particles to the number of energy states below the average energy for the translational energy mode. Plot this ratio for a gas of hydrogen atoms (molecular weight = 1 g/g-mol) in a volume of 1 cm^3 at 1 atm pressure over the temperature range from 1 K to 300 K. Use the plot to assess the validity of the Boltzmann limit for hydrogen under these conditions.

3.2 Consider a system of N identical particles in a variable volume V. For a fixed temperature, the number of microstates can be assessed classically in terms of the position of the particles by dividing the volume into z small cells of identical size such that $z \gg N$. The cell size remains constant even when V is varied.

(a) Write an expression for the number of microstates (assume Fermi–Dirac statistics).

(b) Using Stirling's formula and $\ln(1-x) = -x$ for $x \ll 1$ show that:

$$\ln(\Omega) = N \ln\left(\frac{z}{N}\right) + N$$

(c) If two volumes are in the ratio $r = V_2/V_1$ so that $z_2 = rz_1$, find an expression for $\ln(\Omega_2/\Omega_1)$ in terms of V_2/V_1 and N.

(d) Using macroscopic thermodynamics, write an expression for the entropy difference $S_2 - S_1$ (assume perfect gas, fixed temperature).

(e) By comparing the results of (c) and (d), show for this example that

$$S_2 - S_1 = k\ln(\Omega_2) - k\ln(\Omega_1)$$

and thus $S = k\ln(\Omega)$.

3.3 Consider the Boltzmann energy distributions for N_2 at energies up to the dissociation energy ($D_o = 9.75$ eV, $\theta_{\text{rot}} = 2.9$ K, $\theta_{\text{vib}} = 3390$ K). At temperatures of 300 K and 3000 K, plot the distributions for (1) rigid rotor; and (2) harmonic oscillator. Note: display your results on a semilog plot as

$$\left(\frac{N_i^*}{N}\frac{1}{g_i}\right) \text{ against } \epsilon_i \text{ in eV}$$

where g_i is the degeneracy of level ϵ_i. Comment on your results.

3.4 (a) A diagnostic instrument provides measurements of the number densities of the second and fifth rotational levels of a simple, diatomic gas. Calculate the temperature for nitric oxide when the ratio of the fifth to second number densities is 1.77 (for NO: $\theta_{\text{rot}} = 2.4$ K)

(b) At low temperature, the partition function of a harmonic oscillator may be approximated by retaining only the first two terms:

$$Q_v = 1 + \exp(-\theta_{\text{vib}}/T)$$

Using this equation, determine an expression for the specific internal energy. Evaluate the accuracy of this approach for NO at 300 K by comparing to the result obtained with full treatment of the vibrational partition function (for NO: $\theta_{\text{vib}} = 2740$ K).

3.5 (a) Write the Boltzmann distribution for quantized rotational energy levels given by the rigid rotor model. Evaluate the partition function by assuming a continuous energy distribution. Hence determine the fraction of N_2 molecules in the ground rotational level at a temperature of 300 K. How does this fraction change as the temperature is decreased? ($\theta_{rot} = 2.9$ K)

(b) For molecules with relatively large values of the characteristic temperature for rotation such as H_2 ($\theta_{rot} = 80$ K) an approximate form for the rotational energy partition function is

$$Q = \frac{T}{\theta_{rot}}\left[1 + \frac{1}{3}\frac{\theta_{rot}}{T} + \frac{1}{15}\frac{\theta_{rot}^2}{T^2}\right]$$

Using this equation, determine an expression for the specific energy and evaluate it for H_2 at 100 K. Compare this result with that for full excitation of the rotational mode (normalize your results by RT).

3.6 Consider a perfect monatomic gas constrained to move in the x direction only, so that the permissible energy levels for the system are

$$\epsilon_{n_1} = \frac{h^2 n_1^2}{8ma_1^2}$$

(a) Obtain the partition function, pressure, entropy and internal energy in terms of a_1 and T.

(b) Show that

$$C_j = \frac{a_1}{h}(2m)^{1/2}\epsilon^{-1/2}d\epsilon$$

(c) Obtain the distribution function for molecular speed and find the mean speed.

3.7 (a) Show that the specific rotational entropy when rotation is fully excited is

$$s_{rot} = R\left[\ln\left(\frac{T}{\sigma\theta_{rot}}\right) + 1\right]$$

(b) Show that the specific vibrational entropy is

$$s_{vib} = R\left[-\ln\{1 - \exp(-\theta_{vib}/T)\} + \frac{\theta_{vib}/T}{\exp(\theta_{vib}/T) - 1}\right]$$

What is the limiting form when $T \gg \theta_{vib}$?

(c) Using the two-term approximation given in Eq. 3.116, show that the specific electronic entropy is

$$s_{el} = R\left[\ln(g_0) + \ln\left\{1 + \frac{g_1}{g_0}\exp(-\theta_1/T)\right\} + \frac{(g_1/g_0)(\theta_1/T)\exp(-\theta_1/T)}{1 + (g_1/g_0)\exp(-\theta_1/T)}\right]$$

(d) Plot on a single graph the variation of s/R as a function of temperature from 300 to 10,000 K for each of the translational, rotational, vibrational, and electronic modes, and the total value, for NO at a pressure of 1 atm. Comment on your results.

(For NO: $\theta_{rot} = 2.5$ K; $\theta_{vib} = 2740$ K; $\theta_1 = 174$ K; $g_0 = 2$; $g_1 = 2$).

3.8 (a) For the rotational mode of a diatomic molecule, show that the internal energy and the specific heat at constant volume are independent of whether the molecule is heteronuclear or homonuclear.

(b) For the same conditions as question 3.7, part (d), plot on a single graph the variation of the number of degrees of freedom of NO for each of the translational, rotational, vibrational, and electronic modes, and the total value. Comment on your results.

(c) For the same conditions as part (b), plot on a single graph the variation of c_v/R of NO for each of the translational, rotational, vibrational, and electronic modes, and the total value. Use these results to plot the ratio of specific heats over the same temperature range. Comment on your results.

(For NO: $\theta_{rot} = 2.5$ K; $\theta_{vib} = 2740$ K; $\theta_1 = 174$ K; $g_0 = 2$; $g_1 = 2$).

4 Finite-Rate Processes

4.1 Introduction

The goal in this chapter is to consider different finite-rate processes at the molecular level. Specifically, we consider internal energy mode relaxation and chemical reactions. These studies involve the combination of kinetic theory, quantum mechanics, and statistical mechanics. When particles collide, a variety of processes may result. These intermolecular processes occur at a finite rate, and we may associate a characteristic (or relaxation) time, τ, for each individual process. There are two main types of finite-rate processes:

(1) Energy exchange between the molecular energy modes:
τ_{tr}, τ_{rot}, τ_{vib}, τ_{el} (translational, rotational, vibrational, electronic)
(2) Chemical reactions:
τ_d, τ_i (dissociation, ionization)

It is often informative to express the characteristic times as

$$\tau_x = \tau_{\mathrm{tr}} \times Z_x \tag{4.1}$$

where Z_x is the *collision number* of process x, i.e., the number of collisions required for process x to reach equilibrium, and $\tau_{\mathrm{tr}} = 1/\nu$ is the reciprocal of the collision frequency. From kinetic theory, for a hard-sphere gas:

$$\nu = n\pi d^2 \sqrt{\frac{8kT}{\pi m^*}} \tag{4.2}$$

In general, it is found that

$$Z_{\mathrm{tr}} = 1 < Z_{\mathrm{rot}} < Z_{\mathrm{vib}} < Z_{\mathrm{el}} < Z_d < Z_i \tag{4.3}$$

$$\Rightarrow \tau_{\mathrm{tr}} < \tau_{\mathrm{rot}} < \tau_{\mathrm{vib}} < \tau_{\mathrm{el}} < \tau_d < \tau_i \tag{4.4}$$

In a high-speed gas flow, the gas properties vary significantly with position leading to flow field gradients in macroscopic properties (p, T, ρ, etc.). At each point in a flow field, we may therefore define a characteristic flow time:

$$\tau_f \equiv \frac{\Delta}{u} \tag{4.5}$$

where Δ is a nondimensional gradient length scale, e.g., $\left(\frac{1}{\rho}\frac{\partial \rho}{\partial s}\right)^{-1}$, and s is the distance along a streamline. We can now introduce a *condition for equilibrium* for process x:

$$\tau_x \ll \tau_f \tag{4.6}$$

In this case, a very large number of process x collisions will occur before there are any significant changes in flow properties, thus allowing the process to reach equilibrium. In the following, we first consider the limiting case of equilibrium. We then use results from kinetic theory, quantum mechanics, and statistical mechanics to develop approaches for analysis of nonequilibrium, finite-rate processes.

4.2 Equilibrium Processes

When $\tau_x \ll \tau_f \Rightarrow \tau_x \approx 0 \Rightarrow \nu_x \rightarrow \infty$. This is the equilibrium condition in which the rate of process x is infinitely large, meaning that changes occur instantaneously. Thus, when temperature, pressure, and other properties in a flow field change, under equilibrium conditions this leads to instant changes in other properties such as vibrational energy and chemical composition.

4.2.1 Vibrational Energy

We have discussed before that the energy spacing for the translational and rotational modes is so small that a continuous, classical approach can be used to evaluate the partition functions. This also means that these energy modes respond very rapidly to changes in the state variables, and hence they are usually in equilibrium. Typically, for air molecules, the rotational collision number $Z_{\text{rot}} \approx 5$. By comparison, the vibrational mode of air molecules has larger energy level spacing, responds more slowly to changes in the flow field, and can be in a state of nonequilibrium. For now, recall the important equilibrium properties of the vibrational energy mode for the simple harmonic oscillator:

(a) Specific vibrational energy

$$e_{\text{vib}} = \frac{R\theta_{\text{vib}}}{\exp\left(\frac{\theta_v}{T}\right) - 1} \tag{4.7}$$

(b) Boltzmann distribution of energy

$$\frac{N_v^*}{N} = \frac{\exp\left(-v\frac{\theta_{\text{vib}}}{T}\right)}{Q_{\text{vib}}} = \exp\left(-v\frac{\theta_{\text{vib}}}{T}\right)\left[1 - \exp\left(-\frac{\theta_{\text{vib}}}{T}\right)\right] \tag{4.8}$$

where v is the vibrational quantum number and Q_{vib} is the partition function. Figure 4.1 compares the translational, rotational, and vibrational

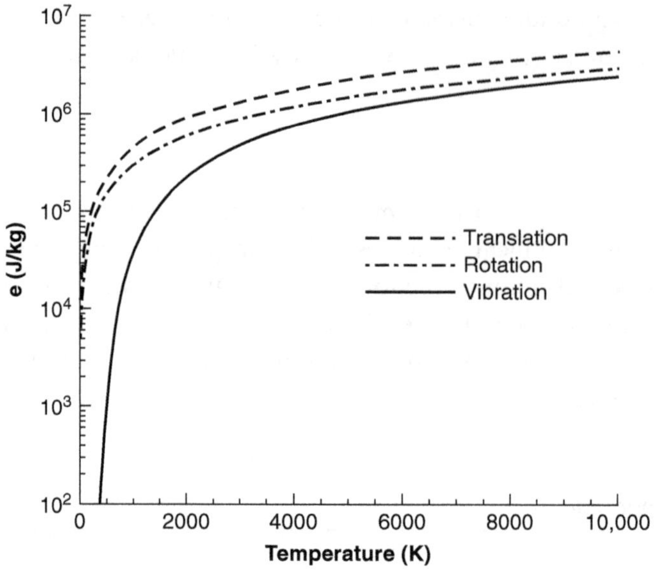

Figure 4.1 Specific internal energies as a function of temperature for N_2.

specific energies for N_2 as a function of temperature. At room temperature, there is a small amount of vibrational energy and this mode does not participate significantly in thermodynamics. By comparison, at high temperature, the vibrational energy approaches that contained in the rotational mode and is therefore an important contributor to the thermodynamic state of the gas. Figure 4.2 shows the vibrational energy distribution functions at two different temperatures again for N_2. The plot illustrates how a significant

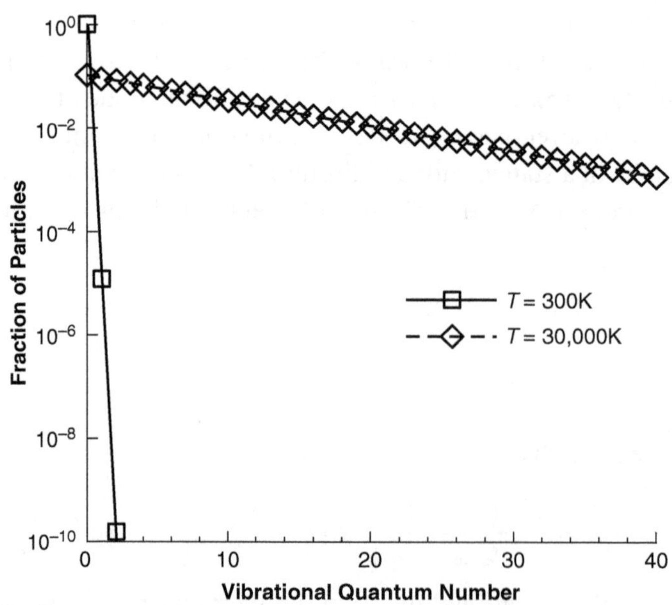

Figure 4.2 Vibrational energy distributions for N_2.

fraction of molecules occupies high vibrational quantum number states at high temperatures.

4.2.2 Equilibrium Chemistry

We continue our focus on high-temperature air in which the dissociation–recombination system for a fixed number of a particles, $\tilde{N}_a$:

$$a_2 \Leftrightarrow a + a$$

is an important chemical reaction. We now introduce the degree of dissociation (that is also the atomic mass fraction):

$$\alpha \equiv \frac{N^a}{N^a + 2N^{aa}} = \frac{N^a}{\tilde{N}_a} = \frac{m_a N^a}{m_a N^a + m_{aa} N^{aa}} \tag{4.9}$$

$$\begin{cases} \alpha = 0 \Rightarrow N^a = 0 \\ \alpha = 1 \Rightarrow N^{aa} = 0 \end{cases} \tag{4.10}$$

We can write Eq. 4.9 as

$$N^a = \alpha \tilde{N}_a \tag{4.11}$$

Then, since $N^a + 2N^{aa} = \tilde{N}_a = \alpha \tilde{N}_a + 2N^{aa}$

$$\Rightarrow N^{aa} = \frac{1-\alpha}{2} \tilde{N}_a \tag{4.12}$$

These are general results independent of whether we are in a state of equilibrium or nonequilibrium conditions. Now, we introduce the equilibrium condition using Eq. 3.123, the law of mass action:

$$\frac{N^{aa^*}}{(N^{a^*})^2} = \frac{Q^{aa}}{(Q^a)^2} \exp\left(\frac{D}{kT}\right) \tag{4.13}$$

Using Eqs. (4.11) and (4.12) and defining the characteristic temperature for dissociation, $\theta_d \equiv \frac{D}{k}$, we obtain

$$\frac{1-\alpha^*}{(\alpha^*)^2} \frac{1}{2\tilde{N}_a} = \frac{Q^{aa}}{(Q^a)^2} \exp\left(\frac{\theta_d}{T}\right) \tag{4.14}$$

where α^* is the equilibrium degree of dissociation.

The total number of a atoms may be written

$$\tilde{N}_a = N^a + 2N^{aa} = \frac{1}{m_a}(m_a N^a + m_{aa} N^{aa}) = \frac{\rho V}{m_a} \tag{4.15}$$

where m_a is the mass of an atom. Thus, we can write

$$\frac{(\alpha^*)^2}{1-\alpha^*} = \frac{m_a}{2\rho V} \frac{(Q^a)^2}{Q^{aa}} \exp\left(-\frac{\theta_d}{T}\right) \tag{4.16}$$

From statistical mechanics:

$$Q^a = Q^a_{tr} Q^a_{el} \tag{4.17}$$

$$Q^{aa} = Q^{aa}_{tr} Q^{aa}_{rot} Q^{aa}_{vib} Q^{aa}_{el} \tag{4.18}$$

Remember that each translational partition function is proportional to volume, $Q_{tr} \propto V$, so that volume is canceled in Eq. 4.16, and thus $\alpha^* = \alpha^*(\rho, T)$, which means that α^* (and more generally the chemical composition) is a thermodynamic state variable as it is described as a function of two other state variables.

Example 4.1

We evaluate α^* for the $N_2 \Leftrightarrow N + N$ dissociation–recombination system.

$$m_a = 14/6.023 \times 10^{26}\ \mathrm{kg} = 2.32 \times 10^{-26}\ \mathrm{kg} \tag{4.19}$$

$$Q^i_{tr} = V\left(\frac{2\pi m_i kT}{h^2}\right)^{3/2} \tag{4.20}$$

$$Q_{rot} = \frac{1}{2}\frac{T}{\theta_{rot}}\ (\text{homonuclear, } \theta_{rot} = 2.9\ \mathrm{K}) \tag{4.21}$$

$$Q_{vib} = \left(1 - \exp\left(-\frac{\theta_{vib}}{T}\right)\right)^{-1}\ (\theta_{vib} = 3390\ \mathrm{K}) \tag{4.22}$$

Recall the law of mass action:

$$\frac{(\alpha^*)^2}{1-\alpha^*} = \frac{\exp(-\theta_d/T)}{\rho}\left\{m_a\left(\frac{\pi m_a k}{h^2}\right)^{3/2}\theta_{rot}\sqrt{T}\left[1 - \exp\left(-\frac{\theta_{vib}}{T}\right)\right]\frac{(Q^a_{el})^2}{Q^{aa}_{el}}\right\} \tag{4.23}$$

Now, from our discussions in statistical mechanics:

$$Q^{N}_{el} \approx 4 \tag{4.24}$$

$$Q^{N_2}_{el} \approx 1 \tag{4.25}$$

$$\Rightarrow \frac{(\alpha^*)^2}{1-\alpha^*} = \frac{\exp\left(-\frac{\theta_d}{T}\right)}{\rho}\left\{3700\sqrt{T}\left[1 - \exp\left(-\frac{\theta_{vib}}{T}\right)\right]\right\} \tag{4.26}$$

The term in { } is sometimes called the *characteristic density for dissociation* ρ_d, and is plotted in Fig. 4.3 for nitrogen.

We now evaluate the degree of dissociation as a function of pressure, by making use of the ideal gas law, and temperature: note that $\theta_d = 113{,}000$ K for N_2 and $\theta_d = 59{,}500$ K for O_2. The results are shown in Fig. 4.4. At the conditions considered, we see in general that

- For fixed pressure, higher temperature gives higher α^* that occurs because there is more energy available for dissociation.

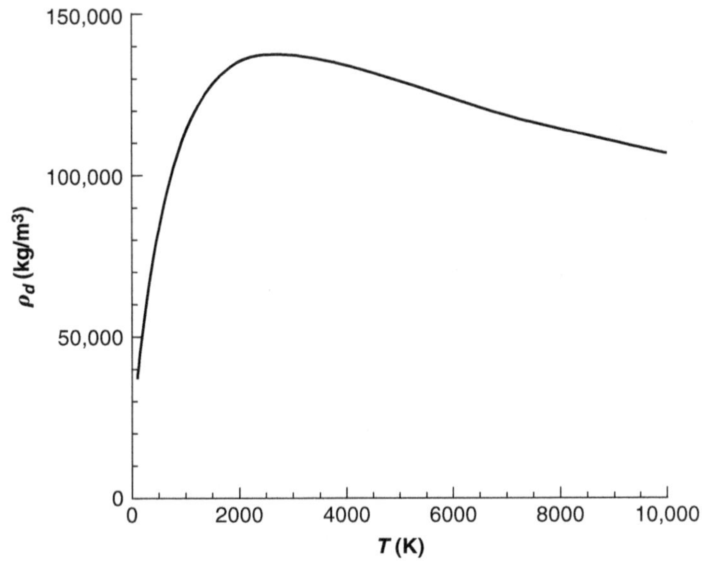

Figure 4.3 Characteristic density for dissociation of N_2.

- For fixed temperature, higher pressure gives lower α^* that is explained by the fact that recombination is a three-body process that occurs more frequently at higher pressure. The effect can be seen directly from the presence of density in the denominator on the right hand side of Eq. 4.26. At a given temperature, a higher value of density will give a higher pressure and a lower value of α^*.

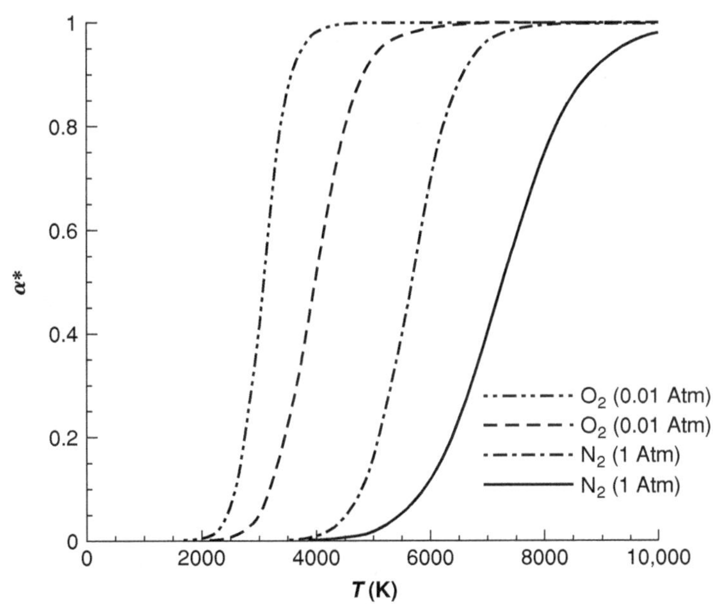

Figure 4.4 Equilibrium degree of dissociation.

Also shown in Fig. 4.4 are data for oxygen under the same conditions. The lower characteristic temperature for dissociation of oxygen, because of its weaker chemical bond, means that oxygen dissociates at lower temperatures than nitrogen.

4.2.3 Equilibrium Constant

The law of mass action may be written

$$\frac{(\alpha^*)^2}{1-\alpha^*} = \frac{G(T)}{\rho} \tag{4.27}$$

Using Eq. 4.9, the left-hand side becomes

$$\frac{(N^a)^2}{N^{aa}}\frac{1}{2\tilde{N}_a} = \frac{(N^a)^2}{N^{aa}}\frac{m_a}{2\rho V} \tag{4.28}$$

Now, introduce the partial pressure for species i:

$$p_i V = N^i kT \tag{4.29}$$

So, Eq. 4.28 becomes

$$\frac{(p_a)^2}{p_{aa}}\frac{V}{kT}\frac{m_a}{2\rho V} = \frac{G(T)}{\rho} \tag{4.30}$$

Thus,

$$\frac{(p_a)^2}{p_{aa}} = \frac{2kT}{m_a}G(T) = K_p(T) \tag{4.31}$$

where $K_p(T)$ is the equilibrium constant, a function only of temperature.

Equilibrium constants can be evaluated for all chemical reactions occurring in a system. In general for a reaction: $A + B + C + \cdots \Leftrightarrow \alpha + \beta + \chi + \cdots$

$$K_p(T) = \frac{p_\alpha p_\beta p_\chi \cdots}{p_A p_B p_C \cdots} = \frac{Q_\alpha Q_\beta Q_\chi \cdots}{Q_A Q_B Q_C \cdots}\exp\left(-\frac{\epsilon_{\text{act}}}{kT}\right) \tag{4.32}$$

where p_x is the partial pressure of species x, Q_x is the total partition function of species x, and ϵ_{act} is the activation energy for the reaction.

4.2.4 Equilibrium Composition

For a simple homonuclear dissociation–recombination system, $\alpha^*(T,\rho)$ represents the equilibrium composition. For high-temperature air, the composition may involve five species (N_2, O_2, NO, N, and O), and is conveniently expressed in terms of partial pressures. We therefore require five equations for the five unknowns.

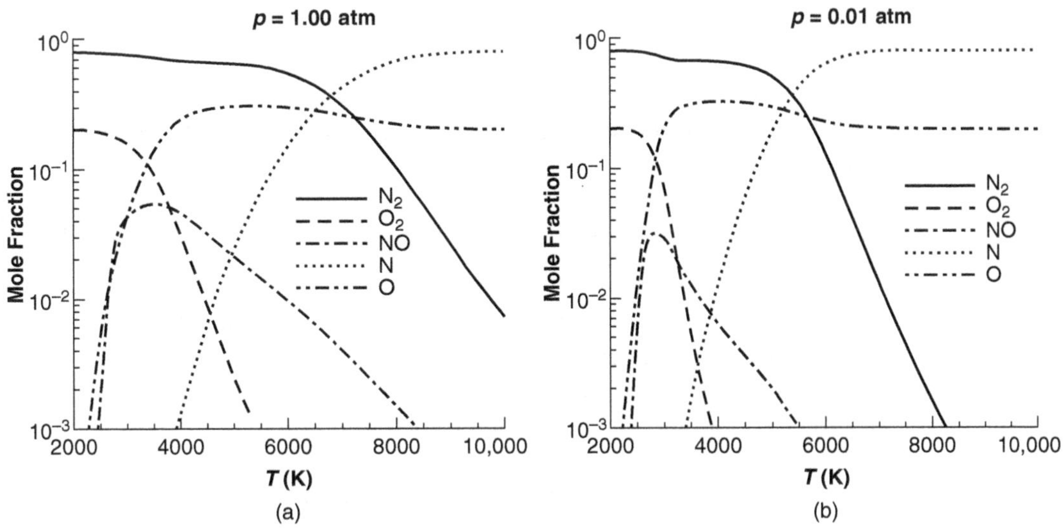

Figure 4.5 Equilibrium composition of air (a) at 1 atm; (b) at 0.01 atm.

Equilibrium constants are defined for the dissociation–recombination of each molecule:

$$\frac{(p_{\rm N})^2}{(p_{\rm N_2})} = K_{\rm N_2}(T) \quad \text{(i)}$$

$$\frac{(p_{\rm O})^2}{(p_{\rm O_2})} = K_{\rm O_2}(T) \quad \text{(ii)}$$

$$\frac{p_{\rm N}p_{\rm O}}{p_{\rm NO}} = K_{\rm NO}(T) \quad \text{(iii)}$$

Dalton's law of partial pressures:

$$p_{\rm N_2} + p_{\rm O_2} + p_{\rm N} + p_{\rm O} + p_{\rm NO} = p \quad \text{(iv)}$$

Atom conservation:

$$\frac{2p_{\rm N_2} + p_{\rm N} + p_{\rm NO}}{2p_{\rm O_2} + p_{\rm O} + p_{\rm NO}} = \frac{\tilde{N}_{\rm N}}{\tilde{N}_{\rm O}} \quad \text{(v)}$$

Equations i–v represent a nonlinear system that depends on T, p, and $\frac{\tilde{N}_{\rm N}}{\tilde{N}_{\rm O}}$. Solutions are shown in Fig. 4.5 as a function of temperature for pressures of 1 and 0.01 atm, and for $\frac{\tilde{N}_{\rm N}}{\tilde{N}_{\rm O}} = 4$ that is the approximate elemental ratio of air.

Equilibrium flow, in which $\tau_x \ll \tau_f$, represents a limiting case where changes in flow properties lead to instant changes in internal energy and chemical composition. The other extreme case, $\tau_x \gg \tau_f$, is called *frozen flow*. In this case, $\tau_x \to \infty$, $\upsilon_x \to \infty$. Thus, there is no change at all in process x as other flow properties change.

For the remainder of this chapter, we consider the situation in which $\tau_x \approx \tau_f$, for which the finite rate of change, also called relaxation, leads to a nonequilibrium state for internal energy and chemical composition.

4.3 Vibrational Relaxation

The process by which the vibrational energy changes in a nonequilibrium situation is called *vibrational relaxation*. In such cases, the vibrational relaxation time

$$\tau_{\text{vib}} \gg \tau_{\text{tr}}, \tau_{\text{rot}} \text{ and } \tau_{\text{vib}} \approx \tau_f \tag{4.33}$$

At the molecular level, the vibrational energy of particles changes via transfer of energy between modes during an intermolecular collision, such as, for example, between the translational and vibrational modes. At the macroscopic level, the process is modeled using the Landau–Teller vibrational relaxation equation:

$$\frac{dE_{\text{vib}}}{dt} = \frac{E^*_{\text{vib}}(T) - E_{\text{vib}}(t)}{\tau_{\text{vib}}} \tag{4.34}$$

where $E^*_{\text{vib}}(T)$ is the total equilibrium vibrational energy at temperature T, evaluated using Eq. 4.7, and $E_{\text{vib}}(t)$ is the total system vibrational energy at time t. As an aside, the analogous equation for relaxation of the rotational mode is called the Jeans equation:

$$\frac{dE_{\text{rot}}}{dt} = \frac{E^*_{\text{rot}}(T) - E_{\text{rot}}(t)}{\tau_{\text{rot}}} \tag{4.35}$$

that can usually be written in terms of temperatures because the rotational mode is generally fully excited:

$$\frac{dT_{\text{rot}}}{dt} = \frac{T^*_{\text{rot}}(T) - T_{\text{rot}}(t)}{\tau_{\text{rot}}} \tag{4.36}$$

In general, the vibrational relaxation time, $\tau_{\text{vib}} = \tau_{\text{vib}}(T, p)$ and this complicates integration of Eq. 4.34. However, a heat bath is a special situation in which the chemical species of interest, say N_2, is introduced in a very small quantity, such as 1% by mole, into a bath of an inert gas such as argon. In this situation, the temperature and pressure are essentially constants so that E^*_{vib}, τ_{vib} are also constants. In this case, the vibrational relaxation equation can be integrated analytically to obtain

$$\int_{E^0_{\text{vib}}}^{E_{\text{vib}}} \frac{dE_{\text{vib}}}{E^*_{\text{vib}} - E_{\text{vib}}} = \int_0^t \frac{dt}{\tau_{\text{vib}}} \tag{4.37}$$

where E^0_{vib} is the initial vibrational energy of the molecules. Evaluation of the integrals leads to

$$\begin{aligned} \left[-\ln(E^*_{\text{vib}} - E_{\text{vib}})\right]^{E_{\text{vib}}}_{E^0_{\text{vib}}} &= \frac{t}{\tau_{\text{vib}}} \\ \frac{E^*_{\text{vib}} - E_{\text{vib}}(t)}{E^*_{\text{vib}} - E^0_{\text{vib}}} &= \exp\left(-\frac{t}{\tau_{\text{vib}}}\right) \end{aligned} \tag{4.38}$$

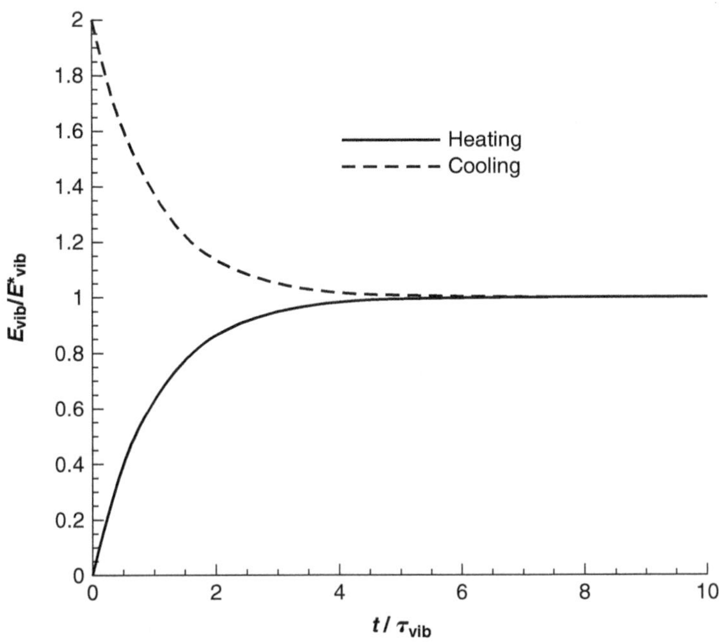

Figure 4.6 Vibrational relaxation in a heat bath.

As illustrated in Fig. 4.6, this solution indicates that vibrational relaxation involves exponential decay in which $|E^*_{\text{vib}} - E_{\text{vib}}|$ is changed by a factor of 1/e over each time interval τ_{vib}.

Note that when $(E^0_{\text{vib}}) > E^*_{\text{vib}}$ the relaxation process corresponds to a cooling process that resembles an expanding gas in a nozzle or jet, and when $(E^0_{\text{vib}}) < E^*_{\text{vib}}$ this corresponds to a heating case as experienced by a compressed gas such as in a shock wave.

4.3.1 Vibrational Relaxation Time

At the molecular level, the vibrational energy of particles changes through collisions, e.g., through vibrational activation and deactivation:

$$\mathrm{N_2}(v) + \mathrm{M} \Leftrightarrow \mathrm{N_2}(v+1) + \mathrm{M} \tag{4.39}$$

Activation requires a finite amount of collision energy to be transferred into the vibrational mode for the higher vibrational energy level to be reached, e.g.,

$$\epsilon_{\text{coll}} = \frac{1}{2} m^* g^2 > k\theta_{\text{vib}} \tag{4.40}$$

Landau and Teller (1936) proposed that the vibrational activation probability for a collision is given by

$$P_{\text{vib}} \propto \exp\left(-\frac{g^*}{g}\right) \tag{4.41}$$

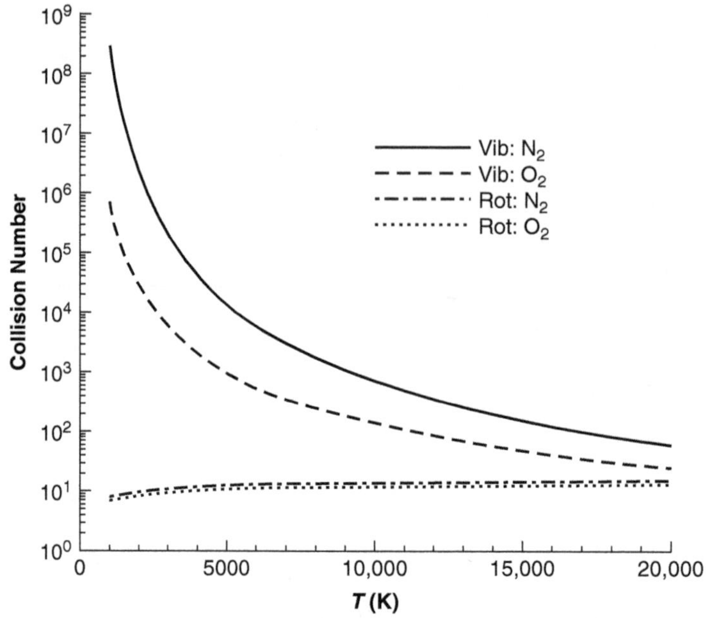

Figure 4.7 Vibrational and rotational collision numbers for air molecules.

where g^* is a *characteristic velocity for vibration.* The vibrational relaxation time is then obtained from the integration

$$\langle P_{\text{vib}} \rangle = \frac{1}{Z_{\text{vib}}} = \frac{\tau_{\text{tr}}}{\tau_{\text{vib}}} = \int_0^\infty P_{\text{vib}}(g) f(g)\, dg \tag{4.42}$$

where $f(g)$ is the Maxwellian distribution function for relative velocity. The final result is

$$\tau_{\text{vib}} = \frac{K_1 T^{5/6} \exp(K_2/T)^{1/3}}{p\left[\left(1 - \exp\left(-\frac{\theta_{\text{vib}}}{T}\right)\right)\right]} \tag{4.43}$$

where K_1, K_2 depend on molecular constants.

Millikan and White (1963) used Eq. 4.43 as a model to fit many sets of experimental measurements and proposed the following modified expression:

$$p\tau_{\text{vib}} = \exp[A(T^{-1/3} - 0.015\mu^{1/4}) - 18.42][\text{atm sec}] \tag{4.44}$$

where μ is the reduced mass in atomic units (e.g., μ for N_2 is 14) and $A = 1.16 \times 10^{-3}\mu^{1/2}\theta_{\text{vib}}^{4/3}$. Despite its empiricism, this model is quite accurate and widely used.

Recall the vibrational collision number:

$$Z_{\text{vib}} = \frac{\tau_{\text{vib}}}{\tau_{\text{tr}}} \tag{4.45}$$

Figure 4.7 shows the vibrational collision numbers for nitrogen and oxygen as a function of temperature, and illustrates the significant temperature variation. Also note that the vibrational collision number of oxygen is always smaller than that of nitrogen owing to its lower characteristic temperature for vibration. Included for comparison are the rotational collision numbers for nitrogen and oxygen evaluated using Parker's model (Park 1990). Note that the rotational collision numbers increase with temperature. This opposite trend to the vibrational mode is explained by the fact that the rotational quantum energy spacing increases with quantum number, so that it becomes increasingly difficult to excite the more energetic rotational states.

Example 4.2
An arcjet is a small rocket used for in-orbit spacecraft propulsion. Consider a hydrogen arcjet with a nozzle length of 1 cm and the following spatially averaged values: $p = 15{,}000$ N/m^2, $T = 5000$ K, $u = 5$ km/s.

For molecular hydrogen (H_2): $\mu = 1, \theta_{\text{vib}} = 6345$ K.

We can evaluate a characteristic flow time based on the velocity change:

$$\tau_f = \left[\frac{1}{u}\frac{du}{ds}\right]^{-1}\frac{1}{u} = 10^{-6}\text{sec} \tag{4.46}$$

For evaluation of the vibrational relaxation time:

$$A = 1.16 \times 10^{-3} \times 1 \times (6345)^{4/3} \tag{4.47}$$

$$\tau_{\text{vib}} = 5 \times 10^{-6} \text{ sec } > \tau_f \Rightarrow \text{nonequilibrium process} \tag{4.48}$$

Indeed, it is found that vibrational nonequilibrium is an important performance degradation mechanism in arcjets that is known as a *frozen flow loss*. The energy that remains trapped in the vibrational mode is not available for translation into directed kinetic motion, and thus the thrust generated is lower than that predicted theoretically by equilibrium analysis.

4.4 Finite-Rate Chemistry

We now consider the situation where $\tau_f \approx \tau_{\text{chemistry}}$ so that finite-rate chemical processes are important. Consider a reacting system of s different species. The changes in concentrations of reactants and products are expressed for each reaction:

$$\nu_1' X_1 + \nu_2' X_2 + \cdots \rightarrow \nu_1'' X_1 + \nu_2'' X_2 + \cdots \tag{4.49}$$

where ν_s' and ν_s'' are the stoichiometric coefficients for reactants and products, respectively.

Example 4.3
Consider air (N_2, N, O_2, O, NO) and the particular nitrogen dissociation reaction occurring through interaction with another nitrogen molecule:

$$N_2 + N_2 \rightarrow 2N + N_2 \tag{4.50}$$

$$\nu_{N_2}' = 2,\ \nu_{O_2}' = \nu_{NO}' = \nu_N' = \nu_O' = 0 \tag{4.51}$$

$$\nu_N'' = 2,\ \nu_{N_2}'' = 1,\ \nu_{O_2}'' = \nu_{NO}'' = \nu_O'' = 0 \tag{4.52}$$

Note that, unlike in equilibrium chemistry analysis, we must now consider the catalyst particle M in dissociation–recombination reactions:

$$N_2 + M \Leftrightarrow N + N + M \tag{4.53}$$

because, at the molecular level, two particles must collide for dissociation to occur.

To describe the chemical composition, we will use *species concentrations*:

$$[X_s] \text{ mol/m}^3 \tag{4.54}$$

The rate of formation of a product is given by

$$\frac{d[X_s]}{dt} = \nu_s'' k_f(T) \prod_s [X_s]^{z_s'} \tag{4.55}$$

where $k_f(T)$ is the forward rate coefficient and is usually expressed in modified Arrhenius form:

$$k_f(T) = C_f T^{\eta_f} \exp\left(-\frac{\theta_d}{T}\right) \tag{4.56}$$

The rate of depletion of a reactant is similarly given by

$$\frac{d[X_s]}{dt} = -\nu_s' k_f(T) \prod_s [X_s]^{z_s'} \tag{4.57}$$

Note that the order of reaction, $z_1' + z_2' + \cdots = z'$ that means that k_f has units $\left(\frac{\text{mol}}{\text{m}^3}\right)^{1-z'} \frac{1}{\text{sec}}$.

As a specific example, consider dissociation–recombination of nitrogen.

(a) Dissociation:

$$N_2 + M \quad \overset{k_f}{\longrightarrow} \quad N + N + M$$

Here, $z' = 2$, so we have a second-order reaction. The time rates of change of the atom and molecule concentrations can be written

$$\frac{d[N]}{dt} = 2k_f(T)[N_2][M] \tag{4.58}$$

$$\frac{d[N_2]}{dt} = -k_f(T)[N_2][M] \tag{4.59}$$

(b) Recombination:

$$2N + M \quad \overset{k_b}{\longrightarrow} \quad N_2 + M$$

Considering both dissociation and recombination, the overall rate of change in concentration of atomic nitrogen is

$$\frac{d[N]}{dt} = \left(\frac{d[N]}{dt}\right)_f + \left(\frac{d[N]}{dt}\right)_b = 2k_f(T)[N_2][M] + \left(\frac{d[N]}{dt}\right)_b \tag{4.60}$$

Under conditions of chemical equilibrium, the net rate of change of all variables including species concentrations is zero:

$$\Rightarrow \left(\frac{d[\mathrm{N}]}{dt}\right)_b^* = -2k_f(T)[\mathrm{N_2}]^*[\mathrm{M}]^* \tag{4.61}$$

where we use * to indicate equilibrium. Recall the equilibrium constant, which we here express in terms of concentrations instead of pressures,

$$K_c(T) \equiv \frac{([\mathrm{N}]^*)^2}{[\mathrm{N_2}]^*} \tag{4.62}$$

Thus,

$$\left(\frac{d[\mathrm{N}]}{dt}\right)_b^* = -2\frac{k_f(T)}{K_c(T)}[\mathrm{N}]^{*2}[\mathrm{M}]^* \tag{4.63}$$

We now make the assumption that chemical processes in any one direction proceed at rates determined by the temperature alone and are independent of whether or not equilibrium exists. This same approach is employed in experiments when changes in concentrations are measured and used to deduce rate coefficients. Thus

$$\left(\frac{d[\mathrm{N}]}{dt}\right)_b^* = -2\frac{k_f(T)}{K_c(T)}[\mathrm{N}]^2[\mathrm{M}] = -2k_b(T)[\mathrm{N}]^2[\mathrm{M}] \tag{4.64}$$

where the backward rate coefficient is

$$k_b(T) = \frac{k_f(T)}{K_c(T)} \tag{4.65}$$

In general, we can write

$$K_c(T) = C_c T^{\eta_c} \exp\left(-\frac{\theta_d}{T}\right) \tag{4.66}$$

so that

$$k_b(T) = \frac{C_f}{C_c} T^{\eta_f - \eta_c} = C_b T^{\eta_b} \tag{4.67}$$

Note that the activation energy for recombination is zero, and that recombination has order $z' = 3$, so the units of k_b are $\frac{\mathrm{m}^6}{\mathrm{mol}^2 \cdot \mathrm{sec}}$.

(c) Combined system

Let us now consider the effects of all reactions leading to concentration changes in a relatively simple system consisting of molecular (N_2) and atomic (N) nitrogen.

$$\mathrm{N_2} + \mathrm{M} \underset{k_b}{\overset{k_f}{\longleftrightarrow}} 2\mathrm{N} + \mathrm{M}$$

$$\frac{d[\mathrm{N}]}{dt} = 2k_f(T)[\mathrm{M}]\left\{[\mathrm{N_2}] - \frac{1}{K_c(T)}[\mathrm{N}]^2\right\} \tag{4.68}$$

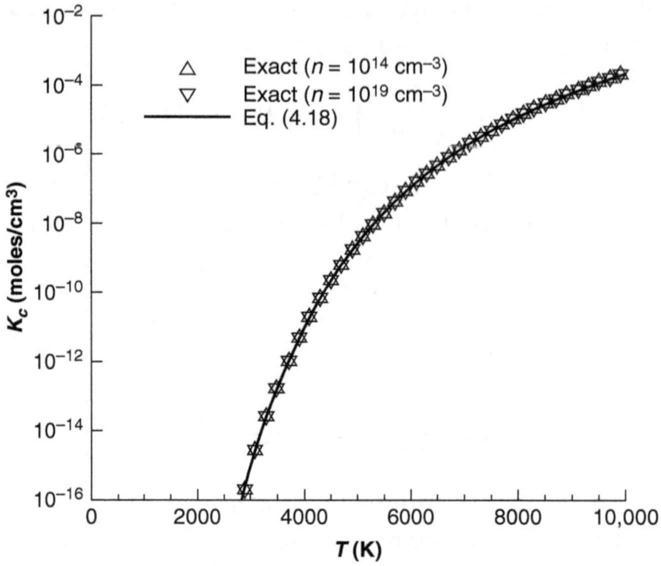

Figure 4.8 Equilibrium constant for nitrogen dissociation–recombination.

By inspection:

$$\text{If } [\mathrm{N}] > [\mathrm{N}]^* \quad \Rightarrow \tfrac{[\mathrm{N}]^2}{K_c} > [\mathrm{N}]_2 \quad \Rightarrow \tfrac{d[\mathrm{N}]}{dt} < 0 \text{ because } K_c(T) \equiv \tfrac{([\mathrm{N}]^*)^2}{[\mathrm{N}_2]^*}$$

$$\text{If } [\mathrm{N}] < [\mathrm{N}]^* \quad \Rightarrow \tfrac{d[\mathrm{N}]}{dt} > 0$$

Therefore, the system always moves toward equilibrium.

In the overall analysis, we must account for all possible catalysts: $M = N_2, N$; therefore

$$\frac{d[\mathrm{N}]}{dt} = 2\{k_{f1}(T)[\mathrm{N}_2] + k_{f2}(T)[\mathrm{N}]\}\left\{[\mathrm{N}_2] - \frac{[\mathrm{N}]^2}{K_c(T)}\right\} \tag{4.69}$$

where we use

$$\mathrm{N}_2 + \mathrm{N}_2 \underset{k_{b1}}{\overset{k_{f1}}{\longleftrightarrow}} \mathrm{N}_2 + 2\mathrm{N}, \quad \text{and} \quad \mathrm{N}_2 + \mathrm{N} \underset{k_{b2}}{\overset{k_{f2}}{\longleftrightarrow}} 3\mathrm{N}$$

The equilibrium constant is evaluated from

$$K_c(T) = \frac{K_p(T)}{kT\hat{N}} \approx C_c T^{\eta_c} \exp\left(-\frac{\theta_d}{T}\right) \tag{4.70}$$

As shown in Fig. 4.8, the following approximate expression for nitrogen agrees well with more exact calculations that are based on detailed analysis of the partition functions (Park 1990).

$$K_c(T) = 18 \exp\left(-\frac{113{,}000}{T}\right) \text{ mol/cm}^3 \tag{4.71}$$

4.4.1 Rate Coefficient

Our next goal is to develop expressions for the forward rate coefficients based on molecular ideas. We must determine (1) the collision conditions that lead to reaction and (2) the frequency that these conditions occur.

Consider a general dissociation reaction:

$$AB + M \xrightarrow{k_f} A + B + M$$

Let us begin with the incorrect assumption that every AB–M dissociation–recombination collision leads to dissociation. The reaction frequency is then simply the bimolecular collision rate from kinetic theory:

$$Z_{AB,M} = \frac{n_{AB}n_M}{1+\delta_{AB,M}}\sigma_{AB,M}\left(\frac{8kT}{\pi m^*_{AB,M}}\right)^{1/2} \tag{4.72}$$

Writing in terms of concentrations, we obtain a rate coefficient:

$$k_f(T) = \frac{\hat{N}}{1+\delta_{AB,M}}\sigma_{AB,M}\left(\frac{8kT}{\pi m^*_{AB,M}}\right)^{1/2} \tag{4.73}$$

since $n_a = \hat{N}[X_a]$ and $\hat{N}$ is Avogadro's number. Our simple model represented by Eq. 4.73 is evaluated using a hard-sphere diameter of 4×10^{-10} m and compared with experimental data in Fig. 4.9, from which we conclude that only a small fraction of all collisions actually lead to dissociation. In addition, the simple model gives the wrong dependence of the rate coefficient on temperature.

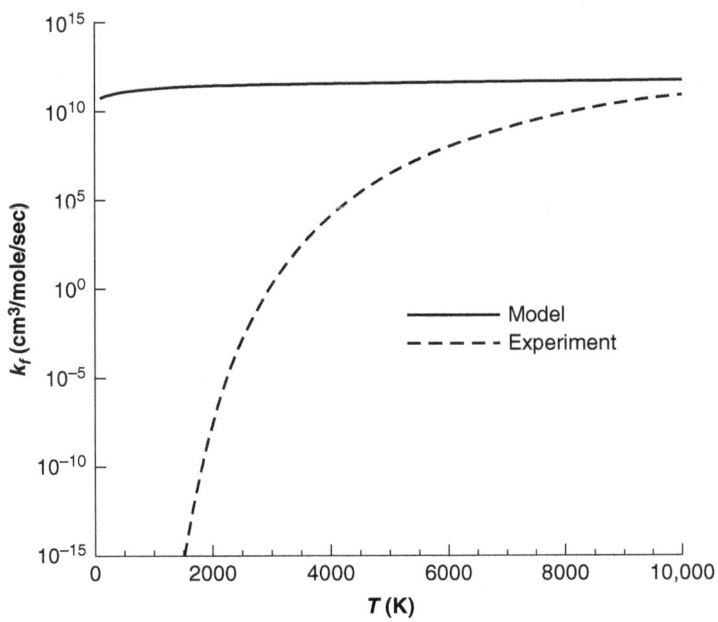

Figure 4.9 Nitrogen dissociation rate as a function of temperature.

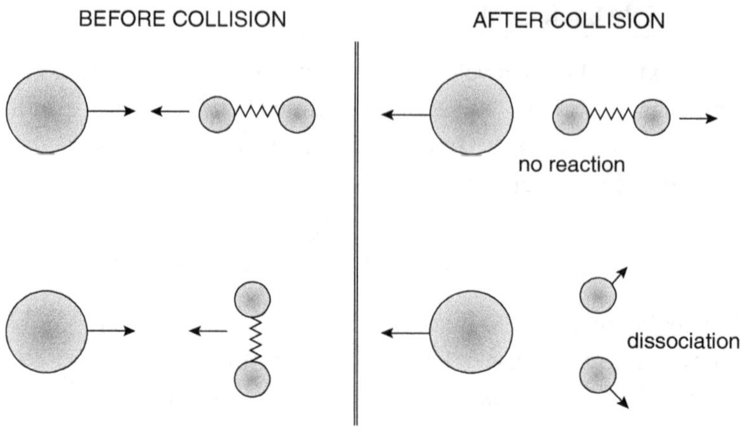

Figure 4.10 Effect of collision orientation on reaction likelihood.

Conditions for Reaction

Let us consider in more detail the properties of a collision that may lead to a chemical reaction. These are

(1) Collision energy: It must be large enough to break the chemical bond.
(2) Orientation: As illustrated in Fig. 4.10, for the same collision energy, some particle orientations are more likely to lead to reaction than others.

Based on these ideas, we write

$$\text{reaction rate} = \{\text{collision rate} \times F\} \times P \tag{4.74}$$

where F is the fraction of collisions with sufficient energy to react, and P is the fraction of sufficiently energetic collisions that actually do react, and this quantity is sometimes called the steric factor. We will take the approach of modeling F and then inferring P from comparison to measured data.

To determine F, we need to think about the energy available in a collision to break the chemical bond, and we begin by consideration of translational energy only. Further, we make the strong assumption of considering only the component of translational energy along the line of centers (LOC) of the colliding particles, as illustrated in Fig. 4.11. The other components of translational energy are considered later.

We aim to find the fraction of collisions with translational energy along the LOC exceeding some threshold energy, ϵ_0. From kinetic theory, the rate of hard-sphere collisions with relative velocity in the range $[g, g + dg]$ and angle between $\bar{g}$ and the LOC in the range $[\psi, \psi + d\psi]$ may be shown to be

$$dZ = 8\pi n_{AB} n_M \left(\frac{m^*_{AB,M}}{2\pi kT} \right)^{3/2} \sigma_{AB,M} g^3 \exp\left(-\frac{m^*_{AB,M} g^2}{2kT} \right) \sin\psi \cos\psi \, d\psi \, dg \tag{4.75}$$

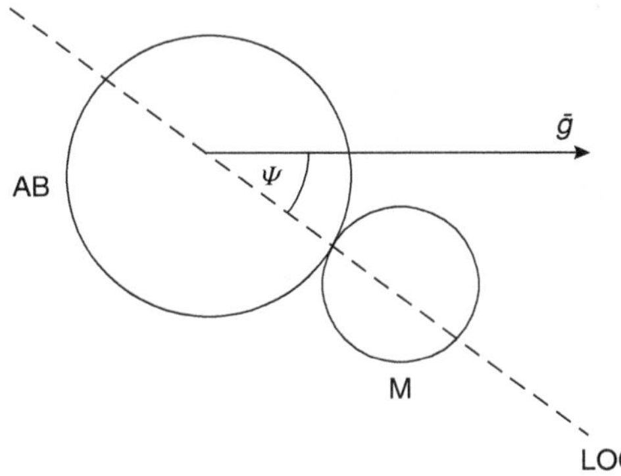

Figure 4.11 Illustration of the LOC of a collision.

Now, dividing by total collision rate, we obtain the fraction of collisions in $[g, g + dg]$ and $[\psi, \psi + d\psi]$:

$$dF = 4\pi^2 \left(\frac{m^*_{\mathrm{AB,M}}}{2\pi kT}\right)^2 g^3 \exp\left(-\frac{m^*_{\mathrm{AB,M}} g^2}{2kT}\right) \sin\psi \cos\psi \, d\psi \, dg \tag{4.76}$$

The fraction of these collisions with velocity component along the LOC greater than a threshold v_0 is obtained by integrating for $g\cos\psi \geq v_0$ as follows:

First, over ψ from 0 to $\cos^{-1}(v_0/g)$

$$= \int_{\cos\psi=1}^{\cos\psi=v_0/g} \sin\psi \cos\psi \, d\psi = \left[-\frac{1}{2}\cos^2\psi\right]_{\cos\psi=1}^{\cos\psi=v_0/g} = \frac{1}{2}\left(1 - \frac{v_0^2}{g^2}\right) \tag{4.77}$$

then, over g from v_0 to ∞:

$$F = 2\pi^2 \left(\frac{m^*_{\mathrm{AB,M}}}{2\pi kT}\right)^2 \int_{v_0}^{\infty} g^3 \left(1 - \frac{v_0^2}{g^2}\right) \exp\left(-\frac{m^*_{\mathrm{AB,M}} g^2}{2kT}\right) dg \tag{4.78}$$

Using the substitution $u^2 = g^2 - v_0^2$ and standard integrals, it is found that

$$F(g\cos\psi \geq v_0) = \exp\left(-\frac{m^*_{\mathrm{AB,M}} v_0^2}{2kT}\right) \tag{4.79}$$

Note that $F = 1$ when $v_0 = 0$, as required. Now, using $\epsilon_0 = \frac{1}{2} m^*_{\mathrm{AB,M}} v_0^2$, the collision rate of interactions with translational energy along the LOC greater than the threshold, $\epsilon \geq \epsilon_0$, is

$$Z(\epsilon_0) = Z_{\mathrm{AB,M}} \exp\left(-\frac{\epsilon_0}{kT}\right) \tag{4.80}$$

where $Z_{\mathrm{AB,M}}$ is the bimolecular collision rate.

4.4.2 Effects of Internal Energy

We now extend our analysis to include effects of the internal energy modes. Recall from statistical mechanics, the Boltzmann distribution of energy for ζ degrees of freedom is given by Eq. 3.99:

$$\frac{N_\zeta}{N} = f_\zeta(\epsilon)\,d\epsilon = \frac{1}{\Gamma(\zeta/2)}\left(\frac{\epsilon}{kT}\right)^{\frac{\zeta}{2}-1}\exp\left(-\frac{\epsilon}{kT}\right)d\left(\frac{\epsilon}{kT}\right) \tag{4.81}$$

where the gamma function $\Gamma(z+1) = z!$ for positive integer z.

Hence

$$\frac{\mathrm{N}_2}{\mathrm{N}} = \exp\left(-\frac{\epsilon}{kT}\right)d\left(\frac{\epsilon}{kT}\right) \text{ for } \zeta = 2 \tag{4.82}$$

Various forms of particle energy can be represented by a "square term" that also corresponds to a degree of freedom. For example,

Translational energy associated with one coordinate direction, $(\epsilon_{\mathrm{tr}})_x = \frac{1}{2}mC_x^2$

Rotational energy associated with rotation about one axis, $(\epsilon_{\mathrm{rot}})_x = \frac{1}{2}I\omega_x^2$

Vibrational energy associated with the spring potential, $(\epsilon_{\mathrm{vib}})_{\mathrm{rot}} = \frac{1}{2}kx^2$

So, the fraction of particles for which two degrees of freedom of energy is greater than ϵ_0 is, using Eq. 4.80:

$$\int_{\epsilon_0/kT}^{\infty}\exp\left(-\frac{\epsilon}{kT}\right)d\left(\frac{\epsilon}{kT}\right) = \exp\left(-\frac{\epsilon_0}{kT}\right) = \frac{Z(\epsilon_0)}{Z_{\mathrm{AB,M}}} \tag{4.83}$$

Thus, we may infer that the component of translational energy along the LOC corresponds to two degrees of freedom. Extending these ideas, the energy distribution associated with 2ζ degrees of freedom is

$$\frac{N_{2\zeta}}{N} = \frac{1}{\Gamma(\zeta)}\left(\frac{\epsilon}{kT}\right)^{\zeta-1}\exp\left(-\frac{\epsilon}{kT}\right)d\left(\frac{\epsilon}{kT}\right)$$

$$\Rightarrow \frac{Z_{2\zeta}(\epsilon_0)}{Z_{\mathrm{AB,M}}} = \frac{1}{\Gamma(\zeta)}\int_{z_0/kT}^{\infty}\left(\frac{\epsilon}{kT}\right)^{\zeta-1}\exp\left(-\frac{\epsilon}{kT}\right)d\left(\frac{\epsilon}{kT}\right)$$

This integral is evaluated by repeated integration by parts:

$$\frac{Z_{2\zeta}(\epsilon_0)}{Z_{\mathrm{AB,M}}} = \exp\left(-\frac{\epsilon_0}{kT}\right)\left[\frac{1}{\Gamma(\zeta)}\left(\frac{\epsilon_0}{kT}\right)^{\zeta-1} + \frac{1}{\Gamma(\zeta-1)}\left(\frac{\epsilon_0}{kT}\right)^{\zeta-1} + \cdots + 1\right] \tag{4.84}$$

In the limit of $\epsilon_0 \gg (\zeta-1)kT$, which is usually true for air molecules, we obtain

$$F = \frac{Z_{2\zeta}(\epsilon_0)}{Z_{\mathrm{AB,M}}} \approx \frac{1}{\Gamma(\zeta)}\left(\frac{\epsilon_0}{kT}\right)^{\zeta-1}\exp\left(-\frac{\epsilon_0}{kT}\right) \tag{4.85}$$

Note, for an odd number of degrees of freedom, e.g., $2\zeta + 1$, the mathematical analysis is more complicated, so for simplicity we will proceed with Eq. 4.85.

We now apply Eq. 4.85 and obtain the steric factor for the dissociation-recombination system of nitrogen:

$$\mathrm{N_2 + M} \overset{k_f}{\longrightarrow} \mathrm{N + N + M}$$

The rate of change of molecular and atomic number densities are given by

$$\left(\frac{dn_{\mathrm{N_2}}}{dt}\right)_f = -P\frac{Z(\epsilon_0)}{Z_{\mathrm{N_2,M}}}Z_{\mathrm{N_2,M}} \tag{4.86}$$

$$\left(\frac{dn_{\mathrm{N}}}{dt}\right)_f = 2P\frac{Z(\epsilon_0)}{Z_{\mathrm{N_2,M}}}Z_{\mathrm{N_2,M}} \tag{4.87}$$

where

$$Z_{\mathrm{N_2,M}} = \frac{n_{\mathrm{N_2}}n_{\mathrm{M}}}{1+\delta_{\mathrm{N_2,M}}}\sigma_{\mathrm{N_2,M}}\left(\frac{8kT}{\pi m^*_{\mathrm{N_2,M}}}\right)^{1/2} \tag{4.88}$$

In terms of concentrations, we may write the rate of change of molecular nitrogen as

$$\frac{d[\mathrm{N_2}]}{dt} = -k_f[\mathrm{N_2}][\mathrm{M}] \tag{4.89}$$

where

$$k_f(T) = \frac{\hat{N}}{1+\delta_{\mathrm{N_2,M}}}\sigma_{\mathrm{N_2,M}}\left(\frac{8kT}{\pi m^*_{\mathrm{N_2,M}}}\right)^{1/2} F \times P \tag{4.90}$$

Now, using Eq. 4.85 in 4.90:

$$k_f(T) = P\frac{\hat{N}\sigma_{\mathrm{N_2,M}}}{1+\delta_{\mathrm{N_2,M}}}\left(\frac{8k}{\pi m^*_{\mathrm{N_2,M}}}\right)^{1/2}\frac{1}{\Gamma(\zeta)}\theta_d^{\zeta-1}T^{3/2-\zeta}\exp\left(-\frac{\theta_d}{T}\right) \tag{4.91}$$

where $\theta_d \equiv \frac{\epsilon_0}{k}$ is the characteristic temperature for dissociation. Recall the modified Arrhenius form:

$$k_f(T) = C_f T^{\eta_f}\exp\left(-\frac{\theta_d}{T}\right)$$

$$\Rightarrow C_f = P\frac{\hat{N}\sigma_{\mathrm{N_2,M}}}{1+\delta_{\mathrm{N_2,M}}}\left(\frac{8k}{\pi m^*_{\mathrm{N_2,M}}}\right)^{1/2}\frac{1}{\Gamma(\zeta)}\theta_d^{\zeta-1} \tag{4.92}$$

$$\eta_f = \frac{3}{2} - \zeta \tag{4.93}$$

Note that C_f depends on molecular properties, the catalyst M, and ζ, whereas η_f depends on ζ only.

Table 4.1 Sources of Energy for Nitrogen Dissociation Reactions

Energy Mode	N_2–N_2	N_2–N
$(\epsilon_{tr})_{LOC}$	2	2
$(\epsilon_{tr})_{\perp}$	0 or 2	0 or 2
(ϵ_{rot})	2	0
(ϵ_{vib})	2 or 4	2
Total $(= 2\zeta)$	6 or 8 or 10	4 or 6

For the reverse recombination step:

$$2\mathrm{N} + \mathrm{M} \overset{k_b}{\longrightarrow} \mathrm{N}_2 + \mathrm{M}$$

the backward rate is, as discussed earlier, evaluated using our model for the forward rate coefficient together with the equilibrium constant.

4.4.3 Calculation of Dissociation Rates

Our goal here is to use our model to calculate nitrogen dissociation rates. The remaining unknowns in our model are the total number of degrees of freedom participating in the reaction, ζ, and the steric factor, P. These will be obtained for specific reactions by comparing to measured reaction rates coefficients. First, we consider the sources of degrees of freedom in our reactions of interest, as shown in Table 4.1.

In terms of the fidelity of our model to describe the complex physical processes involved in chemical reactions, there is considerable uncertainty in determination of the number of degrees of freedom. With respect to Table 4.1, all we can say is that

- For $(\epsilon_{tr})_{\perp}$, this is uncertain, but must either be 0 or 2.
- For ϵ_{rot}, by conservation of angular momentum, only rotational energy from the catalyst particle M can contribute.
- For ϵ_{vib}, 2 degrees of freedom from the dissociating molecule must contribute, and there is uncertainty about any contribution from the catalyst particle, M.

We now use experimentally measured rates to determine ζ and then P. There are many sets of measurements reported in the literature for nitrogen dissociation, although most of them were taken more than 40 years ago. We choose the following rates due to Hanson and Baganoff (1972):

$$\mathrm{N}_2\text{–}\mathrm{N}_2 : (k_f)_1 = 7 \times 10^{21} T^{-1.5} \exp\left(-\frac{113{,}000}{T}\right) \left[\frac{\mathrm{cm}^3}{\mathrm{mol} \cdot \mathrm{sec}}\right] \quad (4.94)$$

$$\mathrm{N}_2\text{–}\mathrm{N} : (k_f)_2 = 3 \times 10^{22} T^{-1.5} \exp\left(-\frac{113{,}000}{T}\right) \left[\frac{\mathrm{cm}^3}{\mathrm{mol} \cdot \mathrm{sec}}\right] \quad (4.95)$$

In both cases, $\eta_f = -1.5 = \frac{3}{2} - \zeta \Rightarrow \zeta_1 = \zeta_2 = 3 \Rightarrow 2\zeta_1 = 2\zeta_2 = 6$.

So, for each reaction, there are 6 degrees of freedom of energy involved in breaking the chemical bond. In terms of a physical interpretation of this result, we might say that 2 come from $(\epsilon_{tr})_{LOC}$ and 2 from (ϵ_{vib}) of the dissociating molecule, and 2 from $(\epsilon_{tr})_{\perp}$.

The interpretation is not unique, and it is best not to employ our relatively simple model to make strong conclusions about the complex chemical processes.

With these results, we can now evaluate the steric factor using

$$P_{N_2,M} = \frac{1+\delta_{N_2,M}}{\hat{N}\sigma_{N_2,M}}\left(\frac{\pi m^*_{N_2,M}}{8k}\right)^{1/2}\frac{\Gamma(\zeta)}{\theta_d^{\zeta-1}}C_f \tag{4.96}$$

N_2–N_2

$$\delta = 1,\, m^* = \frac{1}{2}m_{N_2},\, \sigma = \pi(4\times10^{-10})^2\ \mathrm{m}^2 \tag{4.97}$$

$$\begin{aligned} P_1 &= \frac{2}{6.023\times10^{23}\pi(4\times10^{-10})^2}\left(\frac{\pi\frac{1}{2}m_{N_2}}{8k}\right)^{1/2}\frac{\Gamma(3)\times7\times10^{21}}{(113{,}000)^2}10^{-6} \\ &= 0.19 \end{aligned} \tag{4.98}$$

N_2–N

$$\delta = 0,\, m^* = \frac{1}{3}m_{N_2},\, \sigma = \pi(4\times10^{-10})^2\ \mathrm{m}^2 \tag{4.99}$$

$$P_2 = 0.45 \tag{4.100}$$

Thus, molecule–atom dissociation has a higher geometric reaction efficiency.

Prediction of Rate Coefficients

Our dissociation model can also be used to predict rate coefficients for reactions where measurements are not available, by assuming values of ζ and P.

Consider NO–NO dissociation as an example, for which

$$\delta = 1,\, m^* = \frac{1}{2}m_{NO},\, \theta_d = 75{,}500\ \mathrm{K} \tag{4.101}$$

Note that the dissociation temperature may be obtained from the value for D_0 in Table 2.4. Let us assume $P = 1/3$ as a reasonable value. Then, we compare different rate coefficients that are generated by assuming values of $2\zeta = 6, 8$, and 10.

To illustrate the performance of the model, we compare with the measured rate of Koshi et al. (1978):

$$k_f(T) = 1.1\times10^{17}\exp\left(-\frac{75{,}500}{T}\right)\left[\frac{\mathrm{cm}^3}{\mathrm{mol}\cdot\mathrm{sec}}\right] \tag{4.102}$$

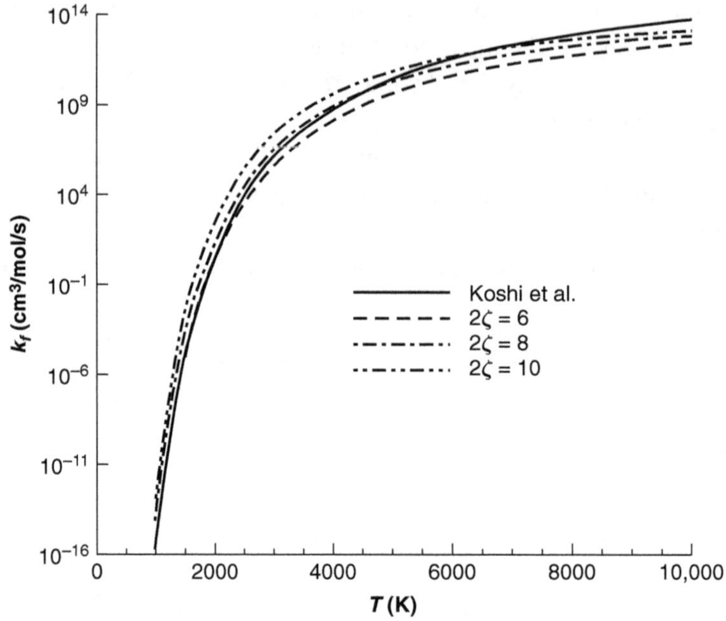

Figure 4.12 NO–NO dissociation rates as a function of temperature.

Figure 4.12 shows the comparisons of the model predictions and the measured rate from which we can see that the rate varies by many orders of magnitude over the temperature range considered, and that no single value of ζ offers best agreement with the data. This observation perhaps suggests that ζ and P themselves may vary with temperature, and of course there is always some uncertainty in the measured data.

4.4.4 Finite-Rate Relaxation

Now that we have models for the rate coefficients, we are able to fully analyze finite rate chemical relaxation processes. We consider two cases.

The N_2–N System

First, we express the molecular and atomic concentrations in terms of the degree of dissociation α:

$$[\mathrm{N}] = \frac{\rho\alpha}{M_\mathrm{N}}, \quad [\mathrm{N}_2] = \frac{\rho(1-\alpha)}{2M_\mathrm{N}} \tag{4.103}$$

where $M_\mathrm{N} = 14$. We can now write Eq. 4.69 as

$$\frac{d\alpha}{dt} = \frac{\rho}{M_\mathrm{N}}\left(k_{f1}\frac{1-\alpha}{2} + k_{f2}\alpha\right)\left\{(1-\alpha) - \frac{2\rho\alpha^2}{M_\mathrm{N}K_c}\right\} \tag{4.104}$$

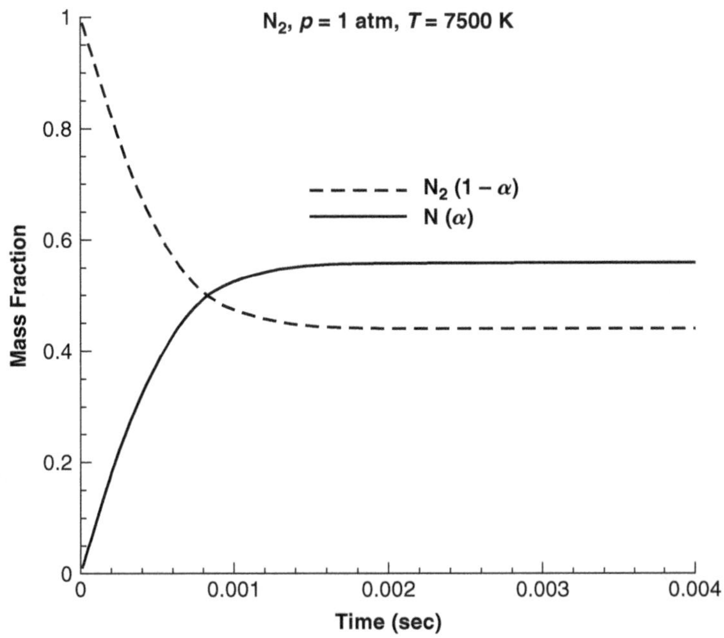

Figure 4.13 Species mass fractions as a function of time for the (N_2, N) system for a fixed temperature.

We solve this ordinary differential equation subject to initial conditions, for example:

$$p = 1 \text{ atm}, T = 7500 \text{ K}, \alpha = 0 \tag{4.105}$$

Figure 4.13 shows that indeed it takes a finite time to reach steady state, and the steady state is exactly that predicted by our earlier equilibrium theory (see Fig. 4.4). Under these conditions, if the characteristic flow time is less than 10^{-3} sec, then we will experience chemical nonequilibrium.

An important consideration that we have neglected so far is the effect of chemistry on the energy and temperature of the gas. The analysis used to generate Fig. 4.13 assumes the temperature is constant, whereas we know that energy is required to break chemical bonds and so the temperature should be reduced. We can write the total energy of our system as follows:

$$E = m_{N_2} N_{N_2} e_{N_2} + m_N N_N e_N + 0.5 m_N N_N k \theta_d \tag{4.106}$$

where the final term accounts for the difference in the energy levels of the molecules and atoms. Using the standard results for the internal energies of molecules and atoms that are functions of temperature, we may solve this equation for temperature as the chemical composition of the gas changes. The results of this analysis are provided in Fig. 4.14 that shows that the temperature does indeed decrease and that this leads to a lower level

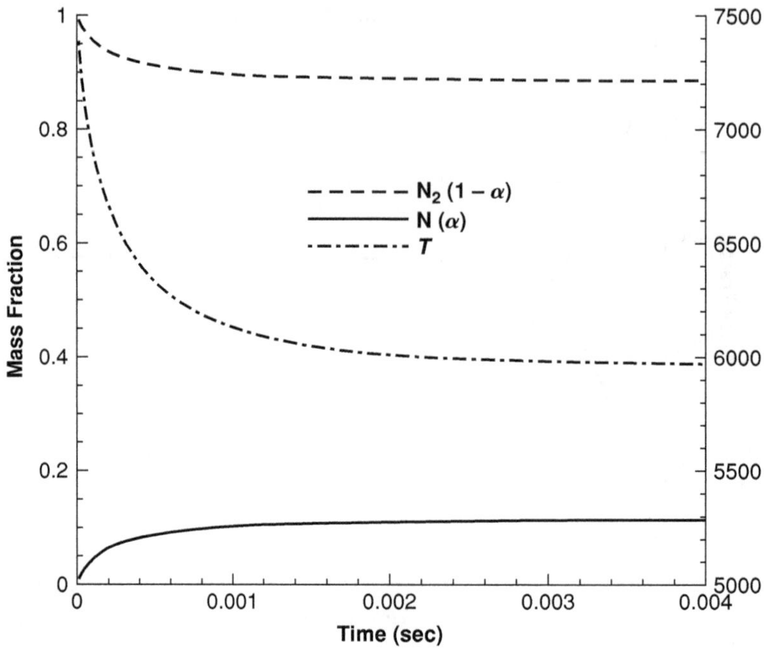

Figure 4.14 Species mass fractions and temperature as a function of time for the (N_2, N) system.

of dissociation in comparison to the fixed temperature analysis shown in Fig. 4.13.

(2) Air System (N_2, O_2, NO, N, O)

Finally, we consider a more complex, five-species reacting air system.

$$\left.\begin{array}{ccc} N_2 + M & \overset{k_{f1}}{\underset{k_{b1}}{\longleftrightarrow}} & 2N + M \\ O_2 + M & \overset{k_{f2}}{\underset{k_{b2}}{\longleftrightarrow}} & 2O + M \\ NO + M & \overset{k_{f3}}{\underset{k_{b3}}{\longleftrightarrow}} & N + O + M \end{array}\right\} \begin{array}{l} \text{dissociation–recombination reactions} \\ M = \{N_2, O_2, NO, N, O\} \end{array}$$

$$\left.\begin{array}{ccc} NO + O & \overset{k_{f4}}{\underset{k_{b4}}{\longleftrightarrow}} & N + O_2 \\ N_2 + O & \overset{k_{f5}}{\underset{k_{b5}}{\longleftrightarrow}} & N + NO \end{array}\right\} \text{exchange reactions}$$

In describing the production and depletion of each species, all possible reactions must be accounted for. For example, in the case of molecular

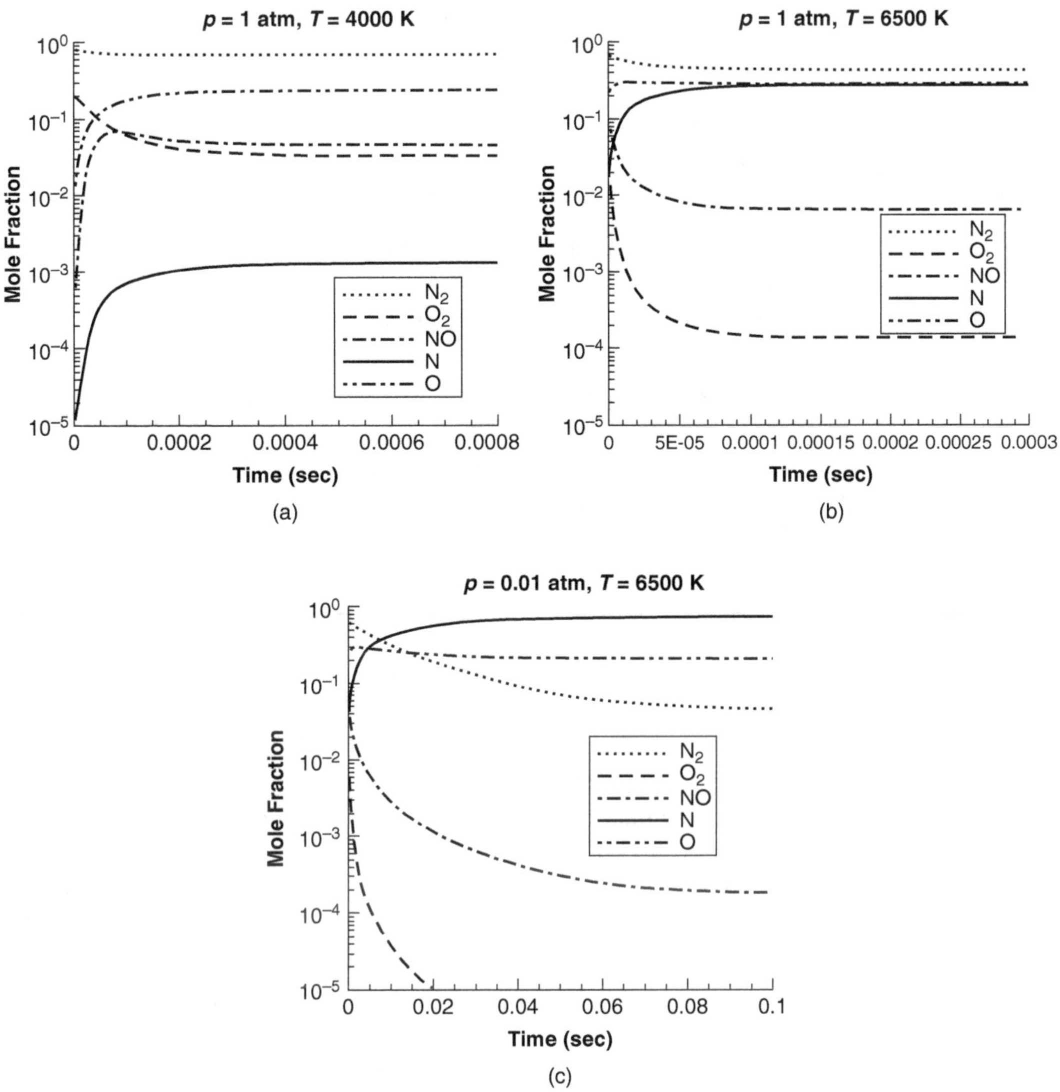

Figure 4.15 Species mole fractions as a function of time for air at fixed temperature.

nitrogen:

$$\frac{d[\mathrm{N}_2]}{dt} = -\{k_{f1}^{\mathrm{N}_2}[\mathrm{N}_2] + k_{f1}^{\mathrm{O}_2}[\mathrm{O}_2] + k_{f1}^{\mathrm{NO}}[\mathrm{NO}] + k_{f1}^{\mathrm{N}}[\mathrm{N}] + k_{f1}^{\mathrm{O}}[\mathrm{O}]\}\left\{[\mathrm{N}_2] - \frac{[\mathrm{N}]^2}{K_{c1}}\right\}$$
$$- k_{f_5}[\mathrm{N}_2][\mathrm{O}] + \frac{k_{f_5}}{K_{c_5}}[\mathrm{NO}][\mathrm{N}] \qquad (4.107)$$

We therefore form a set of five ordinary differential equations that must be solved simultaneously. Figure 4.15 provides solutions for three different conditions showing steady-state results that agree with our earlier equilibrium analysis. In addition, we see significant sensitivity of the profiles to temperature. There is also significant variation in profiles and time scales for the two different pressures. The lower pressure condition is more likely to lead to chemical nonequilibrium. Similar to our first example, accounting for the effects of chemistry on the gas temperature leads to reductions in the level of dissociation.

4.5 Summary

This chapter began with an overview of the equilibrium properties of a gas in terms of its vibrational energy and chemical composition. The results presented were derived from our earlier work in Chapter 3 on statistical mechanics. Using ideas from kinetic theory and quantum mechanics, we subsequently analyzed finite rate relaxation of the vibrational energy of a gas. Similarly, finite rate chemical processes were studied. The main results developed in this chapter will be used to illustrate the molecular simulation models employed to describe these important processes.

4.6 Problems

4.1 An experiment is conducted in which a small population of diatomic molecules, initially at a temperature of 300 K, is introduced into a heat bath of inert argon atoms ($W = 40$ g/gmol) at a pressure of 1 atm. At time $t = 0$, the bath is instantaneously heated to a temperature of 5000 K. Up to a maximum time of 4×10^{-5} sec, and ignoring any chemical reactions, plot on a single graph the variation of $E_{\text{vib}}/E^*_{\text{vib}}$ for molecular nitrogen ($\theta_{\text{vib}} = 3390$ K, $W = 28$ g/g-mol) and molecular hydrogen ($\theta_{\text{vib}} = 6345$ K, $W = 2$ g/g-mol). Comment on and explain any differences in the profiles.

4.2 (a) For the dissociation-recombination system of iodine, suggest a value for the characteristic density for dissociation, ρ_d, by plotting it over the temperature range $T = 100$–1500 K. Assume the atomic and molecular electronic partition functions are represented by the following simple two-term expressions respectively:

$$Q^e_{\text{A}} = g^0_{\text{A}} + g^1_{\text{A}} \exp(-10{,}900/T)$$

$$Q^e_{\text{M}} = g^0_{\text{M}} + g^1_{\text{M}} \exp(-22{,}600/T)$$

Degeneracies are obtained from the first two electronic excited states of

(1) Atomic iodine: $^2P_{3/2}$ (ground state) and $^2P_{1/2}$ (at $\theta_{\text{A}} = 10{,}900$ K); and

(2) Molecular iodine: X $^1\Sigma^+_g$ (ground state) and B $^3\Pi_u$ (at $\theta_{\text{M}} = 22{,}600$ K).

(b) Using part (a), plot the equilibrium degree of dissociation of iodine over the temperature range of $T = 100$ to 1500 K and $\rho = 1.225$ kg/m^3.

(c) Plot the variation of the equilibrium constant for the iodine system over the temperature range of $T = 100$ to 1500 K.

(Additional information: $\theta_{\mathrm{rot}} = 0.053$ K; $\theta_{\mathrm{vib}} = 310$ K; $\theta_d = 17{,}900$ K. Weight of iodine atom =127 g mol^{-1})

4.3 For analysis of a nitrogen plasma, evaluate the backward rate coefficient at a temperature of 6000 K for the following reaction:

$$\mathrm{N} + e^- \Leftrightarrow \mathrm{N}^+ + e^- + e^-$$

Use the following information:

- For electrons, the mass is 9.11×10^{-31} kg, and the internal partition function is 2 (due to spin).
- Assume a collision cross section of 50×10^{-20} m^2.
- Assume that all of the translational collision energy participates in the reaction and that the steric factor is 0.5.
- The partition functions of the atom and ion electronic states may be described by

$$Q^e = g^0 + O[\exp(-\theta_1/T)]$$

and the degeneracies are obtained from
(1) Nitrogen atom: ^4S (ground state) with $\theta_1 = 28{,}000$ K
(2) Nitrogen ion: ^3P (ground state) with $\theta_1 = 22{,}000$ K

- The equilibrium constant (mol/m^3) for the reaction is given by

$$K_c = \frac{1}{\hat{N}V}\frac{Q^{N^+}Q^{e^-}}{Q^N}\exp\left(-\frac{\theta_i}{T}\right)$$

where $\hat{N}$ is Avogadro's number, V is volume, and $\theta_i = 161{,}000$ K.

4.4 (a) The rotational relaxation equation is

$$\frac{dE_{\mathrm{rot}}}{dt} = \frac{E^*_{\mathrm{rot}} - E_{\mathrm{rot}}}{\tau}$$

where * indicates the equilibrium value. The rotational relaxation time τ of H_2 at 300 K is 10^{-8} sec. At this equilibrium temperature, find the time taken for the rotational energy of H_2 to relax to within 1% of the equilibrium value from an initial value of twice the equilibrium value.

(b) Determine the equilibrium degree of dissociation of Cl_2 at $T = 3000$ K and $\rho = 1.225$ kg/m^3. Calculate the value when the density is doubled and explain the answer. Assume the following forms for the atomic and molecular electronic partition functions:

$$Q^e_{\mathrm{A}} = g^0_{\mathrm{A}} + g^1_{\mathrm{A}}\exp(-\theta_{\mathrm{A}}/T)$$
$$Q^e_{\mathrm{M}} = g^0_{\mathrm{M}} + g^1_{\mathrm{M}}\exp(-\theta_{\mathrm{M}}/T)$$

The degeneracies are obtained from the electronic excited states of

(1) Atomic chlorine: $^2P_{3/2}$ (ground state) and $^4P_{1/2}$ ($\theta_A = 103{,}000$ K) and

(2) Molecular chlorine: X $^1\Sigma_g^+$ (ground state) and A $^3\Pi_u$ ($\theta_M = 26{,}325$ K).

Additional information: $\theta_{rot} = 0.35$ K; $\theta_{vib} = 808$ K; $\theta_d = 28{,}700$ K.

4.5 Consider dissociation processes of the N_2 molecule.

(a) Assuming a steric factor of one, calculate a dissociation rate coefficient for N_2–N_2 collisions at 5000 K assuming energy contributions from the translational energy along the LOC, and from the vibrational energy of the reacting molecule. ($\theta_d = 113{,}000$ K, $d = 4 \times 10^{-10}$ m).

(b) An experiment measured the following rate for N_2 dissociation:

$$N_2{-}N_2 : 2.3 \times 10^{29} T^{-3.5} \exp(-113{,}000/T) \text{ cm}^3/\text{mol/sec}$$

Using the model based on available energy, determine the number of degrees of freedom contributing to the reactions and discuss the physical interpretation of this value. Determine the steric factor.

Part II Numerical Simulation

5 Relations Between Molecular and Continuum Gas Dynamics

In Part I of this textbook, the theoretical foundations of nonequilibrium gas dynamics were presented. In the second part of this textbook, computational approaches based on the material from Part I are presented that enable solutions to practical nonequilibrium flow problems. Since the continuum Navier–Stokes equations are widely used to obtain accurate gas flow solutions in the limit of near equilibrium, Part II of this textbook begins by analyzing the relations between molecular and continuum gas dynamics.

5.1 Introduction

The purpose of this chapter is to establish a rigorous mathematical link between molecular and continuum descriptions of a nonequilibrium gas. Specifically, the relation between the Boltzmann equation and Navier–Stokes equations will be presented. Carrying out this analysis is important for a number of reasons. In establishing this link, a more fundamental understanding of the Navier–Stokes model equations is gained and the mathematical theory is able to provide quantitative limits for the validity of the Navier–Stokes model.

This chapter contains the equations that connect interatomic forces to collision cross sections to the transport properties of gases. In fact, as we will see, the interatomic potential energy surface (PES) is the model input for molecular dynamics (MD) calculations, the collision cross section is the model input for direct simulation Monte Carlo (DSMC), and transport property coefficients are the model input for computational fluid dynamics (CFD) calculations. Thus, as depicted in Fig. 5.1, the theory presented in this chapter rigorously establishes consistency between these numerical methods.

The procedure begins with the molecular description of a nonequilibirum gas presented earlier in Chapter 1. By taking moments of the Boltzmann equation, a set of averaged equations called the conservation equations are obtained. In the limit of near-equilibrium flow, the conservation equations reduce to the same form as the well-known Navier–Stokes equations. By comparing the two sets of equations, rigorous expressions for macroscopic state

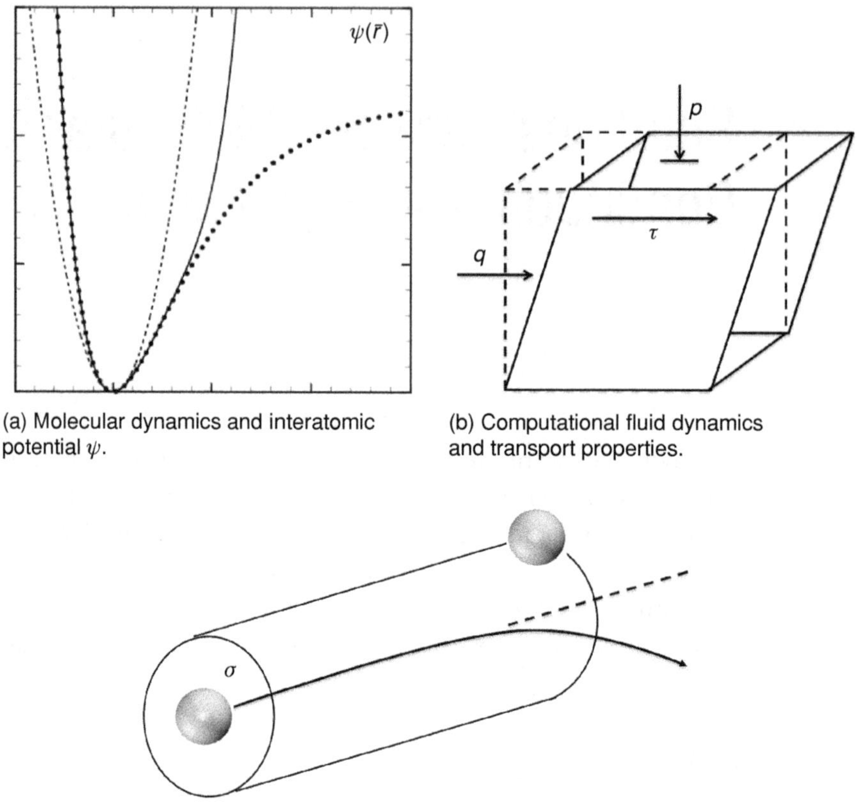

(a) Molecular dynamics and interatomic potential ψ.

(b) Computational fluid dynamics and transport properties.

(c) Direct Simulation Monte Carlo and cross section σ.

Figure 5.1 Various numerical methods and associated model parameters.

properties and transport properties are obtained in terms of the properties of gas molecules.

In particular, the collision cross section becomes a convenient, physically meaningful quantity and general expressions for determining relevant collision cross sections are presented. The collision cross section is the most appropriate model for molecular simulations of dilute gases in nonequilibrium, and the equations in this chapter are referenced in Chapters 6 and 7, which describe the DSMC method.

Finally, a main goal of this chapter is to mathematically link the Boltzmann equation (and therefore the DSMC method) to the Navier–Stokes equations including commonly used models for viscosity, thermal conductivity, and diffusivity for monatomic and polyatomic gas mixtures that are used widely in the field of CFD.

5.2 The Conservation Equations

As described in Chapter 1, one can take moments of the Boltzmann equation (Eq. 1.77) to obtain Maxwell's equation of change (Eq. 1.179), which is rewritten here in index notation for a monatomic simple gas with no external forces,

$$\frac{\partial}{\partial t}(n\langle Q\rangle) + \frac{\partial}{\partial x_j}(n\langle C_j Q\rangle) = \Delta[Q] \tag{5.1}$$

Here $Q = Q(C_i)$ is some function of the particle velocity vector (C_i) and is therefore defined independent of x_j and t. Furthermore, we can separate the particle velocity vector into a mean value $\langle C_i\rangle$, also referred to as the bulk gas velocity, and the remaining thermal value, C_i', defined as

$$C_i' \equiv C_i - \langle C_i\rangle \tag{5.2}$$

It is important to note that $\langle C_i\rangle$ represents an average over the local velocity distribution function (VDF) (Eq. 1.50). Finally, by setting $Q = m, mC_i, 1/2mC_iC_i$, we obtain the mass, momentum, and energy conservation equations. Noting from Eq. 5.2 that $\langle C_i'\rangle = 0$ and also that $\Delta[Q] = 0$ since mass, momentum, and energy are conserved during collisions, the conservation equations become

$$\frac{\partial \rho}{\partial t} + \frac{\partial}{\partial x_j}\left[\rho\langle C_j\rangle\right] = 0 \tag{5.3}$$

$$\frac{\partial}{\partial t}\left[\rho\langle C_i\rangle\right] + \frac{\partial}{\partial x_j}\left[\rho\langle C_i\rangle\langle C_j\rangle\right] = -\frac{\partial}{\partial x_j}\left[\rho\langle C_i'C_j'\rangle\right] \tag{5.4}$$

$$\begin{aligned} &\frac{\partial}{\partial t}\left[\frac{1}{2}\rho\langle C\rangle^2 + \frac{1}{2}\rho\langle C'^2\rangle\right] + \frac{\partial}{\partial x_j}\left\{\langle C_j\rangle\left[\frac{1}{2}\rho\langle C\rangle^2 + \frac{1}{2}\rho\langle C'^2\rangle\right]\right\} \\ &\qquad = -\frac{\partial}{\partial x_j}\left[\frac{1}{2}\rho\langle C_j'C'^2\rangle + \rho\langle C_j'C_k'\rangle\langle C_k\rangle\right] \end{aligned} \tag{5.5}$$

The preceding equations use standard index notation, where single indices correspond to multiple equations (such as the momentum equations for $i = x$, y, and z coordinate directions), and repeated indices correspond to multiple terms. In the case of the final term in Eq. 5.5, which contains two sets of repeated indices, this corresponds to nine terms. Finally, similar to the notation used in Vincenti and Kruger (1967), the velocity vector averages appearing in squared terms without subscript follow the shortened notation:

$$\langle C\rangle^2 \equiv \langle C_x\rangle^2 + \langle C_y\rangle^2 + \langle C_z\rangle^2 \tag{5.6}$$

$$\langle C'^2\rangle \equiv \langle C_x'^2\rangle + \langle C_y'^2\rangle + \langle C_z'^2\rangle \tag{5.7}$$

Note that since the averages in the conservation equations (Eqs. 5.3–5.5) are obtained by proper integration over an arbitrary velocity distribution function (Eq. 1.50), these equations are valid for any degree of nonequilibrium (any Knudsen number regime). In the reminder of this section, we compare these conservation equations (Eqs. 5.3–5.5), which are simply moments of

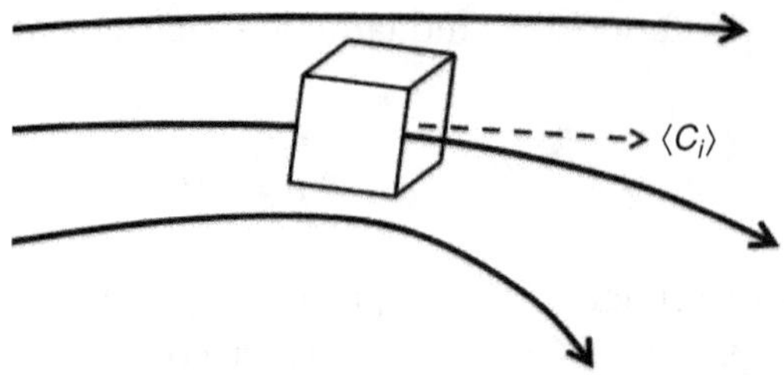

(a) Fluid volume convecting with the bulk flow.

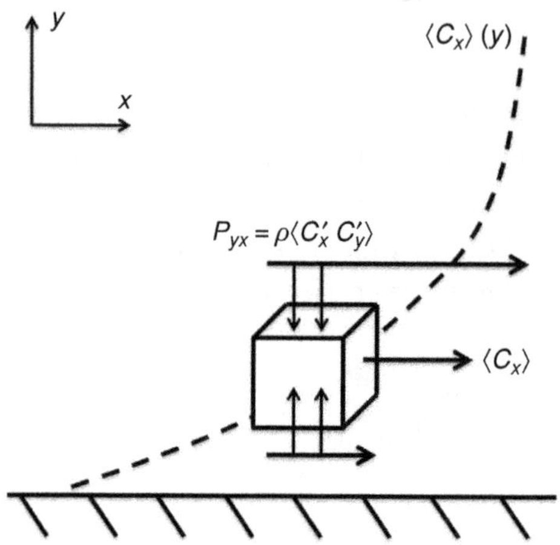

(b) Boundary layer flow.

Figure 5.2 Momentum transfer due to thermal molecular motion relative to the bulk flow velocity.

the Boltzmann equation, with the continuum Navier–Stokes equations. In doing so, we can determine the relationship between molecular and continuum quantities and determine under what conditions the Boltzmann equation reduces to the Navier–Stokes equations.

To begin, it is useful to recast the conservation equations in the frame of reference of a small gas volume convecting with the bulk flow. This is accomplished using the substantial derivative, which for the momentum equations leads to the following:

$$\rho \frac{D}{Dt} \langle C_i \rangle = -\frac{\partial}{\partial x_j} P_{ij}^T \tag{5.8}$$

where the substantial derivative is defined by

$$\frac{D}{Dt} \equiv \frac{\partial}{\partial t} + \langle C_j \rangle \frac{\partial}{\partial x_j} \tag{5.9}$$

and the *pressure tensor* is defined as

$$P_{ij}^T \equiv \rho \langle C_i' C_j' \rangle \tag{5.10}$$

As evident from Eq. 5.8 and portrayed in Fig. 5.2(a), the pressure tensor represents the flux of momentum relative to the bulk gas motion. Indeed, it is the thermal motion of molecules that transports mass, momentum, and energy relative to the bulk flow leading to diffusivity, viscosity and thermal conductivity. For example, Fig. 5.2(b) depicts a boundary layer flow where the bulk flow is predominantly parallel to the wall, in the x-coordinate direction. In this case, an important element of the pressure tensor would be $P_{yx} = P_{xy} = \rho \langle C'_x C'_y \rangle$, corresponding to a flux of x-momentum carried in the y-direction, due to thermal molecular motion. On average, molecules transporting toward the wall $(-y)$ carry more x-momentum than molecules transporting away from the wall $(+y)$, setting up a net transfer of momentum toward the wall. This transport of momentum results purely from molecular motion relative to the bulk flow.

The full pressure tensor, which determines the entire right-hand side of the momentum equations (Eq. 5.4), is given by

$$P_{ij}^T = \begin{pmatrix} P_{xx} = \rho \langle C_x'^2 \rangle & P_{yx} = \rho \langle C'_x C'_y \rangle & P_{zx} = \rho \langle C'_x C'_z \rangle \\ P_{xy} = P_{yx} & P_{yy} = \rho \langle C_y'^2 \rangle & P_{zy} = \rho \langle C'_y C'_z \rangle \\ P_{xz} = P_{zx} & P_{yz} = P_{zy} & P_{zz} = \rho \langle C_z'^2 \rangle \end{pmatrix} \tag{5.11}$$

This tensor provides a fundamental description of all momentum transport due to the thermal motion of molecules relative to the bulk flow motion and, since averages are taken over arbitrary velocity distribution functions, is general to a gas in any degree of nonequilibrium.

Likewise we can define the *heat flux vector* as

$$q_j \equiv \frac{1}{2} \rho \langle C'_j C'^2 \rangle = \frac{1}{2} \rho \langle C'_j (C_x'^2 + C_y'^2 + C_x'^2) \rangle \tag{5.12}$$

Notice that the pressure tensor appears in the momentum equations (Eq. 5.4) and both the pressure tensor and heat flux vector appear in the energy equation (Eq. 5.5).

We can now compare this molecular description (i.e., the conservation equations obtained by taking moments of the Boltzmann equation) with the classical macroscopic Navier–Stokes equations. We start by separating the isotropic portion of the pressure tensor and defining a scalar pressure,

$$p \equiv \frac{1}{3} \rho \langle C'^2 \rangle = \frac{1}{3} \rho \left(\langle C_x'^2 \rangle + \langle C_y'^2 \rangle + \langle C_z'^2 \rangle \right) \tag{5.13}$$

which is simply the average of the three diagonal (isotropic) terms of the pressure tensor. With this definition for scalar pressure, we call the remaining (non-isotropic) portion of the pressure tensor the *viscous stress tensor*:

$$\tau_{ij} \equiv -\left(\rho \langle C'_i C'_j \rangle - \delta_{ij} p \right) \tag{5.14}$$

where δ_{ij} is the Kronecker delta. Later in this chapter, in the limit of near-equilibrium velocity distribution functions, we will relate q_j and τ_{ij} to gradients of temperature and bulk velocity; however, the definitions in Eqs. 5.12 and 5.14 make no such assumption.

Since the average translational energy, per unit mass, associated with the thermal motion of molecules is $e_{tr} = \frac{1}{2}\langle C'^2 \rangle$ (see Eqs. 1.5, 1.6, and 1.12), then the scalar pressure defined in Eq. 5.13 is directly related to e_{tr} by

$$p = \frac{2}{3}\rho e_{tr} \tag{5.15}$$

Finally, we can compare this molecular definition of scalar pressure with the well-established empirical ideal gas law ($p = \rho RT$) to determine how the classical quantity T relates to molecular quantities. The result

$$\frac{3}{2}RT = e_{tr} = \frac{1}{2}\langle C'^2 \rangle \tag{5.16}$$

shows that temperature in the ideal gas law is simply the average kinetic energy per unit mass associated with the thermal motion of molecules. We refer to it as the *translational* temperature,

$$T_{tr} = \frac{\langle C'^2 \rangle}{3R} = \frac{m}{3k}\left(\langle C_x^2 \rangle + \langle C_y^2 \rangle + \langle C_z^2 \rangle - \langle C_x \rangle^2 - \langle C_y \rangle^2 - \langle C_z \rangle^2\right) \tag{5.17}$$

which is seen to be proportional to the variance of the velocity distribution function.

For the ideal gas law, we can write more specifically that

$$p = \rho RT_{tr} \tag{5.18}$$

This expression makes it clear that the defined quantity of scalar pressure is related to the *translational* temperature (i.e., center-of-mass motion of molecules), and has no dependence on internal energy stored within molecules. This is important to note for gases in thermal nonequilibrium, characterized by different temperatures for each internal energy mode (translation, rotational, and vibrational modes). Unlike the formulation of the ideal gas law for an equilibrium gas contained in a box (Chapter 1), the above formulation, resulting in Eq. 5.18, is general to any point in a dilute gas flow in any degree of nonequilibrium (any local VDF). For highly nonequilibrium flows, such as free molecular flow or the interior of a shock wave, the quantities p and T_{tr} may lose their intuitive meaning; however, the general relation between them (Eq. 5.18) holds.

In the preceding comparison with the ideal gas law, the specific gas constant R is seen to act as a conversion factor between molecular and classical variable definitions (Eq. 5.16). In fact, the two fundamental conversion factors are the Boltzmann constant, $k = 1.38065 \times 10^{-23}$ J/K, and Avogadro's number of molecules in 1 kmol of gas, $\hat{N} = 6.02214 \times 10^{26} \frac{\text{molecules}}{\text{kmol}}$. These constants combine to give $R = \hat{R}/M_w$, where $\hat{R} = k\hat{N}$ is the universal gas constant and $M_w = m\hat{N}$ is the molecular weight corresponding to molecules of mass m. In this manner, we can view R, $\hat{R}$, and k as the same conversion factor per unit mass of gas, per kmol of gas, and per particle, respectively.

Finally, for a monatomic simple gas, using the definition of enthalpy ($h \equiv e_{\text{tr}} + p/\rho$) the conservation equations (Eqs. 5.3–5.5) can be written in the following form:

$$\frac{\partial \rho}{\partial t} + \frac{\partial}{\partial x_j}\left[\rho\langle C_j\rangle\right] = 0 \tag{5.19}$$

$$\frac{\partial}{\partial t}\left[\rho\langle C_i\rangle\right] + \frac{\partial}{\partial x_j}\left[\rho\langle C_i\rangle\langle C_j\rangle\right] = -\frac{\partial p}{\partial x_i} + \frac{\partial \tau_{ij}}{\partial x_j} \tag{5.20}$$

$$\frac{\partial}{\partial t}\left[\rho e_{\text{tr}} + \frac{1}{2}\rho\langle C\rangle^2\right] + \frac{\partial}{\partial x_j}\left\{\rho\langle C_j\rangle\left[h + \frac{1}{2}\langle C\rangle^2\right]\right\} = \frac{\partial}{\partial x_j}\left[\tau_{jk}\langle C_k\rangle - q_j\right] \tag{5.21}$$

Equivalently, using the substantial derivative to write the momentum and energy conservation equations, we have

$$\rho\frac{D\langle C_i\rangle}{Dt} = -\frac{\partial p}{\partial x_i} + \frac{\partial \tau_{ij}}{\partial x_j} \tag{5.22}$$

$$\rho\frac{D}{Dt}\left[h + \frac{1}{2}\langle C\rangle^2\right] = \frac{\partial p}{\partial t} + \frac{\partial}{\partial x_j}\left[\tau_{jk}\langle C_k\rangle - q_j\right] \tag{5.23}$$

With the defined relations between molecular quantities and classical macroscopic gas properties, the conservation equations appear fully consistent with the Navier–Stokes equations, except for the transport terms. In the next section we switch focus to analysis of the transport terms involving q_j and τ_{ij}.

5.3 Chapman–Enskog Analysis and Transport Properties

A significant contribution to the theory of nonequilibrium gas dynamics was made independently by Chapman and Enskog, commonly referred to as Chapman–Enskog analysis and described in Chapman and Cowling (1952). This mathematical analysis determined the precise velocity distribution function that reduces thc Boltzmann equation to the Navier–Stokes equations. A different approach leading to the same result was contributed by Grad (1963). This mathematical framework connecting the Boltzmann and Navier–Stokes equations provides theory linking interatomic forces to coefficients of viscosity, thermal conductivity, and diffusivity. The Chapman–Enskog analysis rigorously verifies the Newtonian and Fourier transport models that relate the shear stress tensor and heat flux vector to linear functions of velocity and temperature gradients, respectively. Most notably, Chapman–Enskog analysis establishes quantitative limits on the accuracy of these transport models and therefore on the accuracy of the Navier–Stokes equations for nonequilibrium flows. In this section, we describe the Chapman–Enskog procedure and present the main results relevant to the compressible Navier–Stokes equations for gas mixtures.

5.3.1 Analysis for the BGK Equation

To begin, we present Chapman–Enskog analysis applied to a simplification of the Boltzmann equation developed by Bhatnagar, Gross, and Krook (1954). The analysis of this equation is less involved than for the full Boltzmann equation, yet the process is identical and the results are surprisingly similar.

The BGK model replaces the full collision integral (Eq. 1.76) with a uniform relaxation of the distribution function,

$$\left[\frac{\partial}{\partial t}(nf)\right]_{\text{coll}} = n\nu\left(f_0 - f\right) \tag{5.24}$$

which reduces the Boltzmann equation (Eq. 1.77) to

$$\frac{\partial}{\partial t}(nf) + C_j \frac{\partial}{\partial x_j}(nf) = n\nu\left(f_0 - f\right) \tag{5.25}$$

where external forces on particles, caused by an electric or gravitational field for example, are neglected. Here we follow the same notation and derivation steps as presented in Vincenti and Kruger (1967), which also includes the external force term in the analysis. For the remainder of this chapter, we refer to Eq. 5.25 as the "BGK equation." In this equation, ν is a specified collision rate, n is the number density, f is the velocity distribution function (VDF), and the variable f_0 is an equilibrium VDF that requires further discussion to define properly. Essentially, the BGK collision operator relaxes f toward f_0 according to the prescribed collision rate ν. Typically, ν is modeled as a function of local average properties such as n and possibly T; therefore $\nu = \nu(x, y, z)$ only. Since ν is not a function of C_j, the BGK collision operator relaxes the entire VDF at the same rate.

In the BGK model, f_0 is defined by a Maxwell–Boltzmann velocity distribution (Eq. 1.112), written in terms of local average quantities $\langle C_j \rangle$ and $\langle C'^2 \rangle$,

$$f_0 \equiv \left(\frac{3}{2\pi\langle C'^2\rangle}\right)^{3/2} \exp\left\{-\frac{3}{2\langle C'^2\rangle}\left[\left(C_x - \langle C_x\rangle\right)^2 + \left(C_y - \langle C_y\rangle\right)^2 + \left(C_z - \langle C_z\rangle\right)^2\right]\right\} \tag{5.26}$$

where,

$$\langle C'^2 \rangle = \int_{-\infty}^{+\infty} C'^2 f\, dC \tag{5.26a}$$

$$\langle C_j \rangle = \int_{-\infty}^{+\infty} C_j f\, dC \tag{5.26b}$$

This equilibrium distribution (f_0) is defined locally at any given point in a flow field where the local VDF is represented by f. By construction, the BGK equation (Eq. 5.25) has $f = f_0$ in the limit of equilibrium. Also, by taking moments of Eq. 5.25, we find that $\Delta[Q] = 0$ for $Q = m, mC_j, 1/2mC^2$, by the definition of f_0. Specifically,

$$\Delta[Q = m] = mn\nu \left[\int_{-\infty}^{+\infty} f_0 \, dC - \int_{-\infty}^{+\infty} f \, dC \right] = mn\nu \, [1 - 1] = 0 \quad (5.27)$$

$$\begin{aligned} \Delta[Q = mC_j] &= mn\nu \left[\int_{-\infty}^{+\infty} C_j f_0 \, dC - \int_{-\infty}^{+\infty} C_j f \, dC \right] \\ &= mn\nu \left[\langle C_j \rangle - \langle C_j \rangle \right] = 0 \end{aligned} \quad (5.28)$$

$$\begin{aligned} \Delta[Q = \frac{1}{2} mC^2] &= \frac{1}{2} mn\nu \left[\int_{-\infty}^{+\infty} \left(C_j' + \langle C_j \rangle \right)^2 f_0 \, dC \right. \\ &\left. - \int_{-\infty}^{+\infty} \left(C_j' + \langle C_j \rangle \right)^2 f \, dC \right] = \frac{1}{2} mn\nu \left[\langle C'^2 \rangle - \langle C'^2 \rangle \right] = 0 \end{aligned} \quad (5.29)$$

Therefore, by taking moments of the BGK equation, we obtain the same conservation equations as Eqs. 5.3–5.5. If f were equal to f_0 everywhere in the flow, then the conservation equations reduce to the Euler equations (refer to Chapter 1, Section 1.3.7). If f is not equal to f_0, then the averages appearing in the heat flux vector (Eq. 5.12) and shear stress tensor (Eq. 5.14) are not zero and require proper integration over the local distribution function, f. Therefore, the conservation equations will contain transport terms and we now use Chapman–Enskog analysis to determine precisely what form of f reduces the conservation equations to the Navier–Stokes equations with Newtonian and Fourier transport models. For this analysis, we follow the derivation and notation of Vincenti and Kruger (1967).

To start, we nondimensionalize the variables in Eq. 5.25 with reference values (subscript r) and a reference length scale (L) as

$$\widehat{C}_j = \frac{C_j}{C_r}, \quad \widehat{x}_j = \frac{x_j}{L}, \quad \widehat{t} = \frac{t}{L/C_r}, \quad \widehat{\nu} = \frac{\nu}{\nu_r}, \quad \widehat{n} = \frac{n}{n_r}, \quad \widehat{f} = \frac{f}{1/C_r^3} \quad (5.30)$$

The values n_r, C_r, and ν_r could be chosen as the number density, mean thermal speed, and modeled collision rate in the free-stream flow region, for example. Their precise values are not relevant and are chosen simply so that nondimensional varaibles are of order unity. The nondimensional form of the BGK equation becomes

$$\xi \left[\frac{\partial}{\partial \widehat{t}} (\widehat{n}\widehat{f}) + \widehat{C}_j \frac{\partial}{\partial \widehat{x}_j} (\widehat{n}\widehat{f}) \right] = \widehat{n}\widehat{\nu} \left(\widehat{f}_0 - \widehat{f} \right) \quad (5.31)$$

where

$$\xi \equiv \frac{C_r}{L\nu_r} \quad (5.32)$$

In Eq. 5.31, all variables and all terms are of order unity, except for ξ, which can also be written in a more physically meaningful manner,

$$\xi = \left(\frac{C_r}{\langle C\rangle}\right)\left(\frac{1/\nu_r}{L/\langle C\rangle}\right) \approx \left(\frac{\text{mean-collision-time}}{\text{characteristic flow time}}\right) \tag{5.33}$$

or equivalently, using the relation between collision frequency and mean free path in Eq. 1.22, we can write

$$\xi = \left(\frac{C_r}{\sqrt{8kT/\pi m}}\right)\left(\frac{\lambda_r}{L}\right) \approx \left(\frac{\text{mean free path}}{\text{characteristic length}}\right) = \text{Kn} \tag{5.34}$$

Typically, for near-equilibrium conditions $\xi \approx \text{Kn} \ll 1$ and in the limit of equilibrium (where $f \to f_0$) it is evident from Eq. 5.31 that $\xi \to 0$. Chapman–Enskog analysis exploits this result to determine a formulation for f that reduces Eqs. 5.3–5.5 to the Navier–Stokes equations in the near-equilibrium limit. Specifically, f is written as an expansion about an equilibrium VDF (f_0) for small values of ξ:

$$\widehat{f} = \widehat{f_0}\left(1 + \xi\phi_1 + \xi^2\phi_2 + \cdots\right) \tag{5.35}$$

Substituting this expansion into Eq. 5.31, and neglecting all higher-order terms (O(ξ^2) and higher), gives

$$\xi\left[\frac{\partial}{\partial \widehat{t}}(\widehat{n}\widehat{f_0}) + \widehat{C}_j\frac{\partial}{\partial \widehat{x}_j}(\widehat{n}\widehat{f_0})\right] = -\widehat{n}\widehat{\nu}\widehat{f_0}\xi\phi_1 \tag{5.36}$$

Using the quantities defined in Eq. 5.30, it follows that f_0, T_{tr}, and the ideal gas law are nondimensionalized as

$$\widehat{f_0} = \frac{1}{\left(2\pi\widehat{T_{\text{tr}}}C_r^2\right)^{3/2}} e^{-\left(\widehat{C}_i - \langle\widehat{C}_i\rangle\right)^2/2\widehat{T_{\text{tr}}}} \tag{5.37}$$

with

$$\widehat{T_{\text{tr}}} = \frac{T_{\text{tr}}}{mC_r^2/k}, \quad \text{and} \quad \widehat{p} = \frac{p}{n_r m C_r^2} = \widehat{n}\widehat{T_{\text{tr}}} \tag{5.38}$$

where p and T_{tr} are defined by averages over the local velocity distribution function as given earlier in Eqs. 5.13 and 5.17, respectively. With the preceding definition for $\widehat{f_0}$, we seek an expression for ϕ_1 that satisfies Eq. 5.36. Note, for the remainder of the derivation, the nondimensional $(\widehat{\ })$ notation is dropped for convenience.

To evaluate the partial derivative terms in Eq. 5.36, one can rewrite the functional dependence of f_0 in the following way:

$$f_0 = f_0(x_j, t) = f_0\left(\langle C_i\rangle(x_j, t), T_{\text{tr}}(x_j, t)\right) = f_0\left(\langle C_i\rangle, T_{\text{tr}}\right) \tag{5.39}$$

The partial derivatives become

$$\frac{\partial}{\partial t}(nf_0) = \left[\frac{\partial (nf_0)}{\partial n}\right]\frac{\partial n}{\partial t} + \left[\frac{\partial (nf_0)}{\partial \langle C_i\rangle}\right]\frac{\partial \langle C_i\rangle}{\partial t} + \left[\frac{\partial (nf_0)}{\partial T_{\text{tr}}}\right]\frac{\partial T_{\text{tr}}}{\partial t} \tag{5.40}$$

$$\frac{\partial}{\partial x_j}(nf_0) = \left[\frac{\partial (nf_0)}{\partial n}\right]\frac{\partial n}{\partial x_j} + \left[\frac{\partial (nf_0)}{\partial \langle C_i\rangle}\right]\frac{\partial \langle C_i\rangle}{\partial x_j} + \left[\frac{\partial (nf_0)}{\partial T_{\text{tr}}}\right]\frac{\partial T_{\text{tr}}}{\partial x_j} \tag{5.41}$$

Using Eq. 5.37 we can directly evaluate the derivatives as follows:

$$\left[\frac{\partial (nf_0)}{\partial n}\right] = f_0 \tag{5.42}$$

$$\left[\frac{\partial (nf_0)}{\partial \langle C_i\rangle}\right] = nf_0\left[\frac{1}{T_{\text{tr}}}(C_i - \langle C_i\rangle)\right] = nf_0\frac{C_i'}{T_{\text{tr}}} \tag{5.43}$$

$$\left[\frac{\partial (nf_0)}{\partial T_{\text{tr}}}\right] = nf_0\left[\frac{C_i'^2}{2T_{\text{tr}}^2} - \frac{3}{2T_{\text{tr}}}\right] \tag{5.44}$$

Chapman–Enskog analysis then solves for the perturbation, ϕ_1, assuming higher-order terms can be neglected ($O(\xi^2)$ and higher) for near-equilibrium conditions. Specifically, substituting Eqs. 5.40 and 5.41 into Eq. 5.36 gives

$$\begin{aligned} -nf_0\nu\xi\phi_1 = {} & \left[\frac{\partial (nf_0)}{\partial n}\right]\xi\left\{\frac{\partial n}{\partial t} + C_j\frac{\partial n}{\partial x_j}\right\} \\ & + \left[\frac{\partial (nf_0)}{\partial \langle C_i\rangle}\right]\xi\left\{\frac{\partial \langle C_i\rangle}{\partial t} + C_j\frac{\partial \langle C_i\rangle}{\partial x_j}\right\} \\ & + \left[\frac{\partial (nf_0)}{\partial T_{\text{tr}}}\right]\xi\left\{\frac{\partial T_{\text{tr}}}{\partial t} + C_j\frac{\partial T_{\text{tr}}}{\partial x_j}\right\} \end{aligned} \tag{5.45}$$

Next, the definition of mean and thermal velocities (Eq. 5.2) can be combined with the conservation equations (Eqs. 5.3–5.5 or equivalently Eqs. 5.19–5.21), which are valid for any degree of nonequilibrium, to simplify the expressions contained in curled brackets above. For now, it is noted that the heat flux vector and shear stress tensor (q_i and τ_{ij}) are already proportional to ξ, therefore becoming proportional to ξ^2 in Eq. 5.45, and can be neglected due to the first-order expansion used in this analysis. Using the conservation equations eliminates the time derivatives in Eq. 5.45, and by further using Eqs. 5.42–5.44, we arrive at the following expression:

$$\begin{aligned} -nf_0\nu\phi_1 = {} & f_0\left\{-n\frac{\partial \langle C_j\rangle}{\partial x_j} + C_j'\frac{\partial n}{\partial x_j}\right\} \\ & + nf_0\frac{C_i'}{T_{\text{tr}}}\left\{-\frac{\partial T_{\text{tr}}}{\partial x_i} - \frac{T_{\text{tr}}}{n}\frac{\partial n}{\partial x_i} + C_j'\frac{\partial \langle C_i\rangle}{\partial x_j}\right\} \\ & + nf_0\frac{C_i'C_i'}{2T_{\text{tr}}^2}\left\{-\frac{2T_{\text{tr}}}{3}\frac{\partial \langle C_j\rangle}{\partial x_j} + C_j'\frac{\partial T_{\text{tr}}}{\partial x_j}\right\} \\ & - nf_0\frac{3}{2T_{\text{tr}}}\left\{-\frac{2T_{\text{tr}}}{3}\frac{\partial \langle C_j\rangle}{\partial x_j} + C_j'\frac{\partial T_{\text{tr}}}{\partial x_j}\right\} \end{aligned} \tag{5.46}$$

A number of terms cancel leading to an explicit expression for ϕ_1,

$$-\phi_1 = \frac{1}{\nu T_{\rm tr}}\left\{C_i'C_j'\frac{\partial\langle C_i\rangle}{\partial x_j} - \frac{C_i'C_i'}{3}\frac{\partial\langle C_j\rangle}{\partial x_j}\right\} - \frac{1}{\nu T_{\rm tr}}\left\{C_i'\frac{\partial T_{\rm tr}}{\partial x_i} - \frac{C_j'C_i'C_i'}{2T_{\rm tr}}\frac{\partial T_{\rm tr}}{\partial x_j} + C_j'\frac{3}{2}\frac{\partial T_{\rm tr}}{\partial x_j}\right\} \tag{5.47}$$

Finally, recall that the above equation is in terms of nondimensional quantities, where the $(\widehat{\ })$ notation was dropped for convenience. Switching back to dimensional quantities we can write the first-order perturbed VDF with the following compact expression:

$$f = f_0\left(1 + \xi\phi_1\right) \tag{5.48}$$

where

$$\phi_1 = -\frac{1}{\xi\nu}\left[C_j'\left(\frac{mC'^2}{2kT_{\rm tr}} - \frac{5}{2}\right)\frac{\partial}{\partial x_j}\left(\ln T_{\rm tr}\right) + \frac{m}{kT_{\rm tr}}\left(C_i'C_j' - \frac{C'^2}{3}\delta_{ij}\right)\frac{\partial\langle C_i\rangle}{\partial x_j}\right] \tag{5.49}$$

Equation 5.49 represents a first-order perturbation to an equilibrium velocity distribution function (f_0), where the departure from equilibrium is proportional to local gradients of temperature and bulk velocity.

Proceeding further, one can actually write f in terms of a local *gradient-length* Knudsen number. Combining Eqs. 1.118, 1.119, and 1.143 from Chapter 1, we can write the mean free path in terms of the most probable thermal speed, $C_{\rm mp}$, and the collision rate as

$$\lambda = \frac{1}{\nu}\sqrt{\frac{8kT}{\pi m}} = \frac{2}{\sqrt{\pi}}\frac{C_{\rm mp}}{\nu} \tag{5.50}$$

Also, we can define a local velocity ratio,

$$s_i \equiv \frac{\langle C_i\rangle}{C_{\rm mp}} = \sqrt{\frac{\gamma}{2}}\frac{\langle C_i\rangle}{a} = \sqrt{\frac{\gamma}{2}}M_i \tag{5.51}$$

where a is the local speed of sound in a gas with ratio of specific heats γ, and therefore M_i is the local Mach number corresponding to the component of bulk velocity in the x_i direction. Combining the preceding expressions with the solution for f in Eq. 5.49, we have

$$f = f_0\left\{1 - \frac{\sqrt{\pi}}{2}\frac{C_j'}{C_{\rm mp}}\left(\frac{C'^2}{C_{\rm mp}^2} - \frac{5}{2}\right){\rm Kn}_{GL-T_{\rm tr}} - \sqrt{\pi}\left(\frac{C_i'C_j'}{C_{\rm mp}^2} - \frac{1}{3}\frac{C'^2}{C_{\rm mp}^2}\delta_{ij}\right)\left(s\,{\rm Kn}_{GL-\langle C\rangle}\right)_i\right\} \tag{5.52}$$

where we have defined the local *gradient-length* Knudsen number as

$$\mathrm{Kn}_{GL-Q} \equiv \lambda \left[\frac{1}{Q} \frac{\partial Q}{\partial x_j} \right] \tag{5.53}$$

with Q being a macroscopic flow variable.

This result is quite insightful as it shows that the perturbation terms are proportional to $\mathrm{Kn}_{GL-T_{\mathrm{tr}}}$ and $M_i \mathrm{Kn}_{GL-\langle C_i \rangle}$. For flow regions where λ is small and gradients are relatively weak, these terms will become small and the assumption to neglect higher-order terms is accurate. Therefore, Eq. 5.52 gives a quantitative means of determining the accuracy limit of the Navier–Stokes equations. The gradient-length Knudsen number can be thought of as the percentage change in a macroscopic flow variable that occurs over a length scale of one mean free path. For example, for flow regions where $|\mathrm{Kn}_{GL-Q}| < 0.05$ the Navier–Stokes equations have been shown to provide an accurate model, whereas, in flow regions where $|\mathrm{Kn}_{GL-Q}| > 0.05$ there are noticeable differences between solutions of the Navier–Stokes equations and the Boltzmann equation using DSMC (Boyd, Chen and Candler 1995; Wang and Boyd 2003; Schwartzentruber and Boyd 2006; Schwartzentruber, Scalabrin and Boyd 2007, 2008a, 2008b, 2008c). Next we analytically relate this departure from equilibrium to the transport terms in the Navier–Stokes equations.

To summarize, f (Eqs. 5.48 and 5.49) and f_0 (Eq. 5.26) can be substituted into Eq. 5.25 and moments of the resulting equation can be taken. As determined in Eqs. 5.27–5.29, by construction of the BGK collision operator (i.e., the definition of f_0), the moments of the collision operator vanish. Furthermore, the remaining moment terms reduce to the conservation equations as shown earlier by Eqs. 5.3–5.5. All terms in the resulting conservation equations can be written in terms of bulk velocity (Eq. 5.2), scalar pressure (Eq. 5.13), and translational temperature (Eq. 5.17), except for the transport terms involving q_j and τ_{ij} defined in Eqs. 5.12 and 5.14. We can now use the result for f to analytically determine the transport terms and complete the analysis.

Specifically, an evaluation of the average quantities in Eqs. 5.12 and 5.14 by proper integration over the velocity distribution function f gives

$$\begin{aligned} \tau_{ij} &= -\left[\int_{-\infty}^{+\infty} mnC_i'C_j'(1+\xi\phi_1) f_0 \, dC - p\delta_{ij} \right] \\ &= -\xi mn \int_{-\infty}^{+\infty} C_i'C_j'\phi_1 f_0 \, dC \end{aligned} \tag{5.54}$$

$$q_i = \xi mn \int_{-\infty}^{+\infty} \frac{1}{2} C_i' C'^2 \phi_1 f_0 \, dC \tag{5.55}$$

As mentioned earlier, both quantities are proportional to ξ and, by substituting the perturbation term ϕ_1 (Eq. 5.49), after a number of steps we recover the following analytical expressions for the shear stress tensor and heat flux vector:

$$\tau_{ij} = \frac{nkT_{\text{tr}}}{\nu}\left[\frac{\partial\langle C_i\rangle}{\partial x_j} + \frac{\partial\langle C_j\rangle}{\partial x_i} - \frac{2}{3}\frac{\partial\langle C_k\rangle}{\partial x_k}\delta_{ij}\right] \tag{5.56}$$

$$q_i = -\frac{5}{2}\left(\frac{k}{m}\right)\frac{nkT_{\text{tr}}}{\nu}\frac{\partial T_{\text{tr}}}{\partial x_i} \tag{5.57}$$

These expressions are identical to the shear stress tensor and heat flux vector appearing in the Navier–Stokes equations with Newtonian and Fourier transport models; however, we have now derived the coefficients of viscosity (μ) and thermal conductivity (κ). By comparison with Newtonian and Fourier transport models, the transport coefficients specific to the BGK equation are

$$\mu^{\text{BGK}} = \frac{nkT_{\text{tr}}}{\nu} \tag{5.58}$$

$$\kappa^{\text{BGK}} = \frac{5}{2}\left(\frac{k}{m}\right)\frac{nkT_{\text{tr}}}{\nu} \tag{5.59}$$

These results are analogous to those derived earlier in Chapter 1 (Eqs. 1.34 and 1.36), however, by performing Chapman–Enskog analysis there are no longer unknown coefficients. Rather, the transport terms and transport coefficients are fully determined according to the collision operator in the Boltzmann equation (in this case the BGK operator). It is important to note that the Prandtl number ($\text{Pr} \equiv c_p\mu/\kappa$) predicted by the BGK model is unity. This is a well-known deficiency of the BGK collision operator since $\text{Pr} \approx 2/3$ for most monatomic gases. It should be noted that there are "extended" BGK equations that enable more generality, leading to more realistic values for Pr (originally discussed by Gross and Jackson (1959) with many variations in the present literature). Readers are referred to the literature for such models.

5.3.2 Analysis for the Boltzmann equation

The procedure to apply Chapman–Enskog analysis including the full Boltzmann collision integral (Eq. 1.77) is almost identical to the procedure applied above for the BGK equation. In fact, the results only differ in the value of the coefficients appearing in the expression for f and in the coefficients of the transport properties. Whereas for the BGK collision operator the coefficients could be analytically determined, when the full Boltzmann collision integral is included the coefficients become integral expressions with no closed-form solution. The mathematical procedure used to derive these coefficient expressions can be found in Chapman and Cowling (1952) and is also presented in Chapter 10 of Vincenti and Kruger (1967). In this section, we summarize the procedure and present the main results using the notation and derivation from Vincenti and Kruger (1967).

The Boltzmann equation (Eq. 1.77) can be nondimensionalized using the quantities in Eq. 5.30, where now it is the quantity $ng\sigma$ that is normalized by the reference collision rate ν_r. Substitution of a first-order perturbed Maxwell–Boltzmann VDF (Eq. 5.35) leads to the following analogous expression as Eq. 5.36,

$$\xi\left[\frac{\partial}{\partial \hat{t}}(\widehat{n f_0})+\widehat{C}_j\frac{\partial}{\partial \widehat{x}_j}(\widehat{n f_0})\right]$$
$$=\xi\int_{-\infty}^{+\infty}\int_0^{4\pi}\widehat{n}^2\left[\phi_1(\widehat{C}_i^*)-\phi_1(\widehat{C}_i)+\phi_1(\widehat{Z}_i^*)-\phi_1(\widehat{Z}_i)\right]f_0(C_i)f_0(Z_i)\widehat{g\sigma}\,d\Omega\,d\widehat{Z} \tag{5.60}$$

Using the same procedure detailed by Eqs. 5.35–5.49, the conservation equations can be used to simplify this expression and the result for ϕ_1 is analogous to the BGK result in Eq. 5.49. Specifically, the result corresponding to the Boltzmann equation becomes

$$nf_0\left[C_j'\left(\frac{mC'^2}{2kT_{\text{tr}}}-\frac{5}{2}\right)\frac{\partial}{\partial x_j}(\ln T_{\text{tr}})+\frac{m}{kT_{\text{tr}}}\left(C_i'C_j'-\frac{C'^2}{3}\delta_{ij}\right)\frac{\partial\langle C_i\rangle}{\partial x_j}\right]$$
$$=\xi\int_{-\infty}^{+\infty}\int_0^{4\pi}n^2\left[\phi_1(C_i^*)-\phi_1(C_i)+\phi_1(Z_i^*)-\phi_1(Z_i)\right]f_0(C_i)f_0(Z_i)g\sigma\,d\Omega\,dZ \tag{5.61}$$

Whereas the BGK collision operator led to an explicit expression for ϕ_1 (Eq. 5.49), in Eq. 5.61 ϕ_1 appears inside the collision integral term. For this reason, obtaining the solution for the function ϕ_1 is more difficult mathematically. However, one can reason (and show by substitution) that the functional form of ϕ_1 remains unchanged. Specifically,

$$\phi_1=-\frac{1}{\xi n}\left[\sqrt{\frac{2kT_{\text{tr}}}{m}}A_j\frac{\partial}{\partial x_j}(\ln T_{\text{tr}})+B_{jk}\frac{\partial\langle C_j\rangle}{\partial x_k}+\Gamma\right] \tag{5.62}$$

where the coefficients A_j, B_{jk}, and Γ are yet unknown functions of C_i and T_{tr}. These coefficients have complicated integral constraints, which must satisfy Eq. 5.61. Chapman and Enskog developed an approximate solution method using an expansion of the function ϕ_1 in Sonine polynomials. This particular expansion converges quickly and accurate solutions are typically obtained even when only the first term is maintained. Derivation of the expressions for coefficients A_j, B_{jk}, and Γ can be found in Chapman and Cowling (1952) and also in Chapter 10 of Vincenti and Kruger (1967).

After the coefficients are determined, the final solution for ϕ_1 is obtained. When the resulting expression for ϕ_1 is used to evaluate the shear-stress tensor (Eq. 5.54) and heat flux vector (Eq. 5.55), the Newtonian and Fourier models are recovered:

$$\tau_{ij} = \mu \left[\frac{\partial \langle C_i \rangle}{\partial x_j} + \frac{\partial \langle C_j \rangle}{\partial x_i} - \frac{2}{3} \frac{\partial \langle C_k \rangle}{\partial x_k} \delta_{ij} \right] \tag{5.63}$$

$$q_i = -\kappa \frac{\partial T_{\rm tr}}{\partial x_i} \tag{5.64}$$

The resulting transport expressions are identical to those derived for the BGK equation (Eqs. 5.56 and 5.57) except the coefficients are different. Specifically, maintaining only the first term in the Sonine polynomial solution, the final expressions for the "first approximation" to the coefficients of viscosity and thermal conductivity, corresponding to the full Boltzmann equation, become

$$\mu = \frac{5}{8} \sqrt{\pi m k T_{\rm tr}} \left\{ \left(\frac{m}{4kT_{\rm tr}} \right)^4 \int_0^\infty \int_0^{4\pi} g^7 e^{-(mg^2/4kT_{\rm tr})} \sin^2 \chi \sigma \, d\Omega \, dg \right\}^{-1} \tag{5.65}$$

$$\kappa = \frac{15}{4} \left(\frac{k}{m} \right) \mu. \tag{5.66}$$

The Prandtl number is now, $\mathrm{Pr} = 2/3$, which is physically accurate.

Finally, the above equations for μ and κ include the integral expressions that appear in the coefficients A_j and B_{jk}. This enables Eq. 5.62 and the final Chapman–Enskog VDF to be written directly in terms of the shear-stress tensor and heat flux vector. Following the notation of Garcia and Alder (1998), the Chapman–Enskog VDF can be generally written as

$$f(\mathcal{C}) = f_0(\mathcal{C}) \{1 + \Phi(\mathcal{C})\} \tag{5.67}$$

where

$$\Phi(\mathcal{C}) = \xi \phi_1(\mathcal{C}) = \frac{1}{p} \left[\sqrt{\frac{2m}{kT_{\rm tr}}} \left(q_x \mathcal{C}_x + q_y \mathcal{C}_y + q_z \mathcal{C}_z \right) \left(\frac{2}{5} \mathcal{C}^2 - 1 \right) \right]$$
$$- \frac{2}{p} \left[\tau_{xy} \mathcal{C}_x \mathcal{C}_y + \tau_{xz} \mathcal{C}_x \mathcal{C}_z + \tau_{yz} \mathcal{C}_y \mathcal{C}_z + \frac{1}{2} \tau_{xx} \left(\mathcal{C}_x^2 - \mathcal{C}_z^2 \right) + \frac{1}{2} \tau_{yy} \left(\mathcal{C}_y^2 - \mathcal{C}_z^2 \right) \right] \tag{5.68}$$

Here, the thermal velocity has been normalized with the most probable velocity as

$$\mathcal{C} \equiv \frac{C'}{\sqrt{2kT_{\rm tr}/m}} = \frac{C'}{C_{\rm mp}} \tag{5.69}$$

and therefore

$$f_0(\mathcal{C}) = \frac{1}{\pi^{3/2}} e^{-\mathcal{C}^2} \tag{5.70}$$

This is analogous to the result for the BGK equation (Eqs. 5.48 and 5.49), where only the coefficients in front of the gradient terms are different. Therefore, the perturbation terms in the Chapman–Enskog VDF, corresponding to the full Boltzmann equation, are also proportional to $\mathrm{Kn}_{GL-T_{\mathrm{tr}}}$ and $(s\,\mathrm{Kn}_{GL-\langle C\rangle})_i$. As discussed in the previous section, this provides a quantitative limit on the accuracy of the Navier–Stokes equations.

In summary, on substitution of the near-equilibrium velocity distribution function in Eqs. 5.67–5.70 into the Boltzmann equation, moments of the Boltzmann equation reduce exactly to the Navier–Stokes equations. The momentum and energy transport terms in the resulting equations have the Netwonian and Fourier forms and the expressions for the coefficients of viscosity and thermal conductivity are determined. The resulting transport coefficient expressions (Eqs. 5.65 and 5.66) are derived in terms of molecular parameters and therefore Chapman–Enskog analysis provides a rigorous mathematical connection between the Boltzmann equation (molecular description) and the Navier–Stokes equations (continuum description). We note that the equations in this section were presented for a simple, monatomic gas. We next generalize the relations for gas mixtures.

5.3.3 Analysis for Gas Mixtures

The purpose of this section is to present the results of Chapman–Enskog analysis applied to a polyatomic gas mixture. In this manner, the link between the Boltzmann equation and the most commonly used forms of the compressible Navier–Stokes equations for gas mixtures will be established. The resulting equations quantify how interatomic potential functions can be used to compute collision cross sections, which in turn can be used to compute the viscosity, thermal conductivity, and diffusion coefficients used in the continuum description of gas mixtures. As we will see, a number of assumptions are inherent in commonly used forms of the Navier–Stokes equations for such mixtures. The theory and equations presented in this section are useful when determining consistency in the limit of near-equilibrium flow between numerical solutions of the Boltzmann equation (using DSMC) and numerical solutions of the Navier–Stokes equations (using CFD).

The Chapman–Enskog analysis for a gas mixture proceeds in a similar manner as the approach applied to the BGK and Boltzmann equations for a monatomic simple gas presented in prior sections. Here we present only the main results for gas mixtures. More detailed derivations are presented in Vincenti and Kruger (1967), Chapman and Cowling (1952), and Hirschfelder, Curtiss, Bird and Mayer (1954).

For a dilute gas mixture, a separate velocity distribution function (f_s) must be written for each species s and the evolution of each f_s is described by a separate Boltzmann equation. Analogous to the above derivations, moments of each Boltzmann equation can be determined and it becomes appropriate to sum the resulting conservation equations for each species together to derive momentum and energy conservation equations for the mixture.

A number of mixture properties appear in the final set of equations. The *species velocity* vector, for species s, is defined as

$$\langle C_i \rangle_s \equiv \int_{-\infty}^{+\infty} C_i f_s \, dC \tag{5.71}$$

It is also useful to define the *mixture mass velocity* (C_{0_i}) by

$$\rho C_{0_i} \equiv \sum_s \rho_s \langle C_i \rangle_s \tag{5.72}$$

since this quantity appears frequently in the mixture conservation equations. For this reason, it is most convenient to define the "peculiar" velocity with respect to the mixture mass velocity,

$$C_i' \equiv C_i - C_{0_i} \tag{5.73}$$

With this definition it is important to note that, unlike for a single-species gas, the average peculiar velocity is no longer zero, and we call this quantity the *diffusion velocity* of species s,

$$\langle C_i' \rangle_s = \langle C_i \rangle_s - C_{0_i} \tag{5.74}$$

which appears directly in the conservation equations for a gas mixture. Finally, note that although individual diffusion velocities are nonzero, by Eq. 5.72 the sum of all diffusion velocities is zero as required by mass conservation,

$$\sum_s \rho_s \langle C_i' \rangle_s = 0 \tag{5.75}$$

After taking moments of each Boltzmann equation corresponding to f_s, summing the equations together, and using the mixture quantities defined previously, the following set of conservation equations are obtained,

$$\frac{\partial \rho_s}{\partial t} + \frac{\partial}{\partial x_j}\left[\rho_s C_{0_j} + \rho_s \langle C_j' \rangle_s\right] = 0 \tag{5.76}$$

$$\rho \frac{DC_{0_i}}{Dt} = -\frac{\partial p}{\partial x_i} + \frac{\partial \tau_{ij}}{\partial x_j} \tag{5.77}$$

$$\rho \frac{D}{Dt}\left(h + \frac{C_0^2}{2}\right) = \frac{\partial p}{\partial t} + \frac{\partial}{\partial x_j}\left(\tau_{kj} C_{0_k} - q_j\right) \tag{5.78}$$

Therefore, the set of equations consists of one mass-conservation equation for each species, three momentum equations (one in each coordinate direction) for the mixture, and one energy equation for the mixture. Further definitions for mixture quantities include

$$n \equiv \sum n_s, \quad \rho \equiv \sum \rho_s, \quad h \equiv \sum h_s \tag{5.79}$$

$$\frac{3}{2}kT \equiv \frac{1}{n}\sum_s \frac{1}{2} n_s m_s \langle C'^2 \rangle_s \tag{5.80}$$

$$p \equiv \sum_s p_s \equiv \frac{1}{3}\sum_s n_s m_s \langle C'^2 \rangle_s = nkT \tag{5.81}$$

Applying Chapman–Enskog analysis by assuming f_s to be a first-order perturbation to a Maxwell–Boltzmann velocity distribution function, solving for the perturbation ϕ_{1s}, and evaluating the transport terms, results in the following expressions for the shear-stress tensor and heat flux vector (appearing in Eqs. 5.77 and 5.78),

$$\tau_{ij} = \mu_{\text{mix}} \left(\frac{\partial C_{0_i}}{\partial x_j} + \frac{\partial C_{0_j}}{\partial x_i} \right) - \frac{2}{3}\mu_{\text{mix}} \frac{\partial C_{0_k}}{\partial x_k} \delta_{ij} \tag{5.82}$$

$$q_i = -\kappa_{\text{mix}} \frac{\partial T}{\partial x_i} + \sum_s n_s h_s \langle C_i' \rangle_s + \underbrace{\frac{kT}{n} \sum_s \sum_{t \neq s} \frac{n_t D_s^T}{m_s \mathscr{D}_{st}} \left(\langle C_i' \rangle_s - \langle C_i' \rangle_t \right)}_{\text{commonly neglected}} \tag{5.83}$$

The shear-stress tensor has the same Newtonian form as for a single-species gas (Eq. 5.63), where gradients of the mixture mass velocity appear in the expression and the coefficient of viscosity is a mixture-averaged value, denoted by μ_{mix}. Similar to that for a single species (Eq. 5.64), the heat flux vector has the same Fourier heat flux term now with a mixture-averaged coefficient of thermal conductivity, denoted by κ_{mix}. However, the heat flux vector has two additional energy transport terms arising from the species diffusion velocities, $\langle C_i' \rangle_s$. These diffusion velocities also appear directly in the species mass conservation equations (Eq. 5.76).

Since diffusion velocities are defined relative to the mixture mass velocity (Eq. 5.74), there is inherent coupling between species. The result of Chapman–Enskog analysis is that the set of diffusion velocities must be determined by solving the following system of equations, often referred to as the Stefan–Maxwell equations:

$$\sum_t \frac{n_s n_t}{n^2 \mathscr{D}_{st}} \left(\langle C_i' \rangle_t - \langle C_i' \rangle_s \right) = \mathbf{G_s} \tag{5.84}$$

where

$$\mathbf{G_s} = \frac{\partial}{\partial x_i}\left(\frac{n_s}{n}\right) + \underbrace{\left(\frac{n_s}{n} - \frac{\rho_s}{\rho}\right)\frac{\partial}{\partial x_i}(\ln p)}_{\text{commonly neglected}} + \underbrace{K_s^T \frac{\partial}{\partial x_i}(\ln T)}_{\text{commonly neglected}} \tag{5.85}$$

This system of equations determines the diffusion velocities $\langle C_i' \rangle_s$ to within a constant, which can be evaluated through conservation of mass (Eq. 5.75). In addition, the set of *binary diffusion coefficients* ($\mathscr{D}_{st}$) defined for each species pair (s, t), and the *coefficient of thermal diffusion* (D_s^T) defined for each species (s), now appear in Eqs. 5.83–5.85. The coefficients K_s^T are related to the species thermal diffusion coefficients (refer to Eq. 8.1-3 in Hirschfelder et al. (1954), for example). We have presented the above expressions for gas mixtures without derivation. Rigorous derivations of these expressions, including all coefficients, can be found in Chapman and Cowling (1952) and Hirschfelder et al. (1954).

The preceding expressions represent molecular conservation equations where the transport terms are evaluated in the near-equilibrium limit through Chapman–Enskog analysis and are therefore directly comparable to the Navier–Stokes equations. When compared to the most typical form of the compressible Navier–Stokes equations used for multispecies flow, we find that a number of terms are commonly neglected. In Eq. 5.85 for example, the forcing of diffusion velocities due to temperature gradients, with the coefficient K_s^T, related to thermal diffusion, is commonly neglected. Furthermore, forcing of diffusion velocities due to pressure gradients is also commonly neglected. As a result, as seen in Eqs. 5.84 and 5.85, diffusion velocities arise from gradients in species mole fractions (n_s/n) and the associated set of binary diffusion coefficients, $\mathscr{D}_{st}$. In addition, the final term in the heat flux vector (Eq. 5.83), also including the thermal diffusion coefficient, is not typically maintained in the Navier–Stokes model.

5.3.4 General Transport Properties of Polyatomic Mixtures

A goal of this chapter is to analyze consistency between the Boltzmann equation and the most commonly used forms of the Navier–Stokes equations in the near-equilibrium limit. Specifically, given a molecular model with a set of parameters describing molecular interactions, the equations presented in this chapter can be used to determine the set of continuum parameters appearing in the Navier–Stokes equations. In this manner, solutions of the Boltzmann and the Navier–Stokes equations are ensured to be consistent in the limit of near-equilibrium flow. To remain concise, we do not analyze the commonly neglected terms in the preceding equations. Instead, we complete the analysis of all remaining terms and coefficients in Eqs. 5.76–5.78 and Eqs. 5.82–5.85.

Specifically, the diffusion velocities $\langle C_i' \rangle_s$ arise from species mole fraction gradients with an associated set of binary diffusion coefficients $\mathscr{D}_{st}$. These diffusion velocities appear directly in the species mass conservation equations. The shear-stress tensor has the Newtonian form with gradients of the mixture mass velocity and a mixture averaged value of the coefficient of viscosity (μ_{mix}). Finally, the heat flux vector has the Fourier term proportional to the temperature gradient with a mixture averaged value of the coefficient of thermal conductivity (κ_{mix}) and a second term corresponding to the diffusive flux of species and associated enthalpy. We now further discuss each term and provide molecular-based expressions for the required transport coefficients.

Similar to the expressions for a single species gas (Eqs. 5.65 and 5.66), Chapman–Enskog analysis applied to a gas mixture results in integral expressions for the transport coefficients. Binary transport coefficients specific to each species pair (i, j) are first calculated and the mixture coefficients are obtained by well-defined averages of the binary coefficients. The binary viscosity coefficient and binary diffusion coefficient are given by

$$\mu_{ij} = \frac{5}{8}\frac{kT}{\Omega^{(2,2)}} \tag{5.86}$$

and

$$\mathscr{D}_{ij} = \frac{3}{16}\frac{kT}{nm_r\Omega^{(1,1)}} \tag{5.87}$$

where Ω is called the "collision integral."

Ω is simply an integral appearing in the transport coefficient expressions, and results from Chapman–Enskog analysis maintaining only the first Sonine polynomial expansion term (previously discussed in Section 5.3.2). The general form for the collision integral is

$$\Omega^{(l,s)} = \sqrt{\frac{kT}{2\pi m_r}}\int_o^{+\infty} e^{-\gamma^2}\gamma^{2s+3}Q^{(l)}\,d\gamma \tag{5.88}$$

where $\gamma^2 \equiv \frac{1}{2}m_r g^2/kT$, and m_r is the reduced mass, $m_r = (m_i m_j)/(m_i + m_j)$. Within the collision integral is an expression for the "collision cross section" Q, written generally as

$$Q^{(l)} = \int_0^{4\pi}(1 - \cos^l\chi)\sigma\, d\Omega \tag{5.89}$$

where χ is the scattering angle of a collision and $\sigma\, d\Omega$ is the differential cross-section appearing in the Boltzmann collision integral (refer to Eq. 1.69 and related discussion in Chapter 1). The physical meaning of the collision cross section is discussed at length in the next section. At this point, it is important to note that each collision cross section, collision integral, and binary transport coefficient (Eqs. 5.86 and 5.87) are specific to a collision *pair* involving species i and j, and the relative speed g is that of the collision pair as well.

The binary diffusion coefficients $\mathscr{D}_{ij}$ (Eq. 5.87) can be used directly in Eqs. 5.84 and 5.85 to evaluate the diffusion velocities required in the species conservation equations (Eq. 5.76).

As outlined in Hirschfelder et al. (1954), the mixture viscosity is determined through Chapman–Enskog analysis to be given by

$$\mu_{\text{mix}} = -\frac{\begin{vmatrix} H_{11} & H_{12} & H_{13} & \cdots & H_{1\nu} & \chi_1 \\ H_{12} & H_{22} & H_{23} & \cdots & H_{2\nu} & \chi_2 \\ H_{13} & H_{23} & H_{33} & \cdots & H_{3\nu} & \chi_3 \\ \cdot & \cdot & \cdot & & \cdot & \cdot \\ \cdot & \cdot & \cdot & \ddots & \cdot & \cdot \\ \cdot & \cdot & \cdot & & \cdot & \cdot \\ H_{1\nu} & H_{1\nu} & H_{1\nu} & \cdots & H_{\nu\nu} & \chi_\nu \\ \chi_1 & \chi_2 & \chi_3 & \cdots & \chi_\nu & 0 \end{vmatrix}}{\begin{vmatrix} H_{11} & H_{12} & H_{13} & \cdots & H_{1\nu} \\ H_{12} & H_{22} & H_{23} & \cdots & H_{2\nu} \\ H_{13} & H_{23} & H_{33} & \cdots & H_{3\nu} \\ \cdot & \cdot & \cdot & & \cdot \\ \cdot & \cdot & \cdot & \ddots & \cdot \\ \cdot & \cdot & \cdot & & \cdot \\ H_{1\nu} & H_{1\nu} & H_{1\nu} & \cdots & H_{\nu\nu} \end{vmatrix}} \tag{5.90}$$

where $s = 1, \ldots, \nu$ (ν is the total number of species), χ_s is the mole fraction, and

$$H_{ii} = \frac{\chi_i^2}{\mu_{ii}} + \sum_{k=1,k\neq i}^{\nu} \frac{2\chi_i\chi_k}{\mu_{ik}} \frac{m_i m_k}{(m_i + m_k)^2} \left[\frac{5}{3A_{ik}^*} + \frac{m_k}{m_i} \right] \tag{5.91}$$

$$H_{ij,i\neq j} = -\frac{2\chi_i\chi_j}{\mu_{ij}} \frac{m_i m_j}{(m_i + m_j)^2} \left[\frac{5}{3A_{ij}^*} - 1 \right] \tag{5.92}$$

where

$$A_{ij}^* = \frac{1}{2} \left[\frac{\Omega_{ij}^{(2,2)}}{\Omega_{ij}^{(1,1)}} \right] \tag{5.93}$$

where the binary viscosity coefficients (μ_{ij}) were given previously in Eq. 5.86.

Therefore, given an interatomic potential or collision cross sections ($Q^{(1)}$ and $Q^{(2)}$) for each species pair in a gas mixture, the binary coefficients μ_{ij} and $\mathscr{D}_{ij}$ can be calculated from Eqs. 5.86 and 5.87, which can be further used to determine the mixture viscosity μ_{mix} using Eq. 5.90 and the diffusion velocities $\langle C_i' \rangle_s$ by solving Eq. 5.84.

The thermal conductivity of a mixture of *monatomic* gases can be found using similar expressions as the mixture viscosity. Derivation of these expressions can be found in Chapter 8 of Hirschfelder et al. (1954); however, they cannot be used for polyatomic gases with internal energy and, therefore, we do not discuss them in detail in this text. A simplified model, called the Eucken correction, is commonly used in continuum models to account for the internal energy of the molecules and the influence on energy transport.

Specifically, Eucken proposed that the coefficient of thermal conductivity (κ) is directly proportional to the coefficient of viscosity (μ),

$$\kappa_{\text{mix}} = \mu_{\text{mix}} \left(\frac{5}{2} c_v^{\text{tr}} + c_v^{\text{rot}} + c_v^{\text{vib}} \right) \tag{5.94}$$

Here, $c_v^{\text{tr}} = \frac{3}{2}\frac{k}{2m_r}$, $c_v^{\text{rot}} = \frac{k}{2m_r}$ and, $c_v^{\text{vib}} = \frac{\zeta_{\text{vib}}}{2}\frac{k}{2m_r}$, where ζ_{vib} is the vibrational degrees of freedom available. Supporting theory and derivations of this approximate relation can be found in Chapter 7 of Hirschfelder et al. (1954) where the expression is shown to increase in accuracy in the limit of near-equilibrium flow, which is precisely the regime that we are analyzing in this section.

It is noted that high-temperature flows, where the vibrational energies of molecules become excited, often involve thermal nonequilibrium where the vibrational energy mode is not in equilibrium with the rotational and translational modes. In this situation, continuum models often add a separate energy equation (for the vibrational mode only) to the Navier–Stokes equations. This energy equation has a vibrational heat flux term with a separate vibrational thermal conductivity ($\kappa_{\text{mix-vib}}$). The mathematical link to the Boltzmann equation becomes less rigorous and certainly more complex in this case. Discussion of such physics is presented later in Chapters 6 and 7 in the context of establishing consistency between DSMC and CFD models.

The final consideration in linking the Boltzmann equation to the Navier–Stokes equations in the limit of near-equilibrium flow involves the difficulty in solving Eq. 5.84 for the diffusion velocities $\langle C_i' \rangle_s$. A rigorous treatment of species diffusion within continuum CFD simulations would require the solution of a system of equations within each computational cell and during each timestep of the simulation just to compute the diffusion velocities. This is computationally expensive and also creates difficulty for implicit algorithms where linearization of the governing equations is required. Although solving the Stefan–Maxwell equations within a CFD simulation is performed for certain applications (see Magin and Degrez 2004a, 2004b, for example), this full treatment of diffusion is rare.

As an alternative, an accurate and computationally efficient treatment of diffusion is the self-consistent effective binary diffusion (SCEBD) model (Ramshaw and Chang 1996). In this model, each diffusion velocity for a species s is determined by treating the gas as an *effective binary mixture* where species s diffuses relative to a single "composite" species that represents all of the other species. The diffusion velocity of the composite species is constructed to be an average of the species it represents, where the weighting factors have a specific form (Ramshaw and Chang 1996). The SCEBD model assumption reduces the Stefan–Maxwell system of equations (Eqs. 5.84 and 5.85) to the following explicit expression for each species diffusion flux:

$$J_s \equiv \rho_s \langle C_i' \rangle_s = -c M_{ws} D_s \mathbf{G_s} + y_s c \sum_k M_{wk} D_k \mathbf{G_k} \tag{5.95}$$

where $c = \sum_s \rho_s / M_{ws}$ is the total molar concentration, $y_s = \rho_s/\rho$ is the species mass fraction, and D_s is the effective binary diffusion coefficient for species s. D_s is evaluated as a weighted sum of binary diffusion coefficients ($\mathscr{D}_{st}$ in Eq. 5.87) for species s paired with each other species t,

$$D_s = \left(1 - \frac{w_s}{w}\right)\left(\sum_{t \neq s} \frac{\chi_t}{\mathscr{D}_{st}}\right)^{-1} \tag{5.96}$$

where χ_t is the mole fraction of species t, and w_s and w are species weighting factors given by

$$w_s = \frac{\rho_s}{\sqrt{M_{ws}}} \tag{5.97}$$

and

$$w = \sum_s w_s \tag{5.98}$$

Equations 5.95–5.98 represent a computationally efficient, yet accurate, way to compute the diffusive fluxes directly from the binary diffusion coefficients of species pairs ($\mathscr{D}_{st}$) without the need to solve the Stefan–Maxwell system of equations. Although the SCEBD model can readily be used with the full forcing function $\mathbf{G_s}$ (Eq. 5.85), as mentioned previously, the most common forms of the Navier–Stokes equations maintain only the forcing due to gradients in species mole fractions, and neglect the terms due to pressure and temperature gradients. The result of this assumption leads to the standard Fick's law for diffusion:

$$J_s \equiv \rho_s \langle C_i' \rangle_s = -\rho D_s \frac{\partial y_s}{\partial x_i} + y_s \sum_k \rho D_k \frac{\partial y_k}{\partial x_i} \tag{5.99}$$

Finally, it is also common to use a constant diffusion coefficient, D, for all species. When such a simplified model is used, it is common to determine D from the thermal conductivity of the gas,

$$D_s = D_{\text{mix}} = \text{Le}\frac{\kappa_{\text{mix}}}{\rho c_p} \tag{5.100}$$

where Le is the Lewis number, a constant that is typically set close to 1.4. This simple treatment of diffusion assumes that all species diffuse into all other species with equal efficiency and assumes a constant relationship between D_{mix} and κ_{mix}.

This completes the analysis of the transport properties required by the most commonly used formulations of the compressible Navier–Stokes equations for polyatomic gas mixtures. The equations provided in this section enable the determination of continuum transport properties from molecular interaction properties ensuring a high degree of consistency in the limit of near-equilibrium flow.

5.4 Evaluation of Collision Cross Sections and Transport Properties

In this section, we focus on how interatomic potentials can be used to determine cross sections and how these cross sections can be used to determine transport properties of the gas. We use simple interatomic potentials as examples, namely the hard-sphere, inverse-power law, and Lennard–Jones (LJ) expressions, discussed in Chapter 1. The material in this section is relevant to the formulation of cross section *models* for the DSMC method described in Chapter 6.

5.4.1 Collision Cross Sections

An important quantity used in molecular modeling and analysis of the Boltzmann equation is the *differential cross section*: $\sigma d\Omega$. The differential cross section represents the probability that a molecule involved in a collision scatters into a specific solid angle element ($d\Omega$), defined as

$$d\Omega \equiv \sin\chi \, d\chi \, d\epsilon \tag{5.101}$$

where χ and ϵ are angles defined in the collisional frame of reference, as presented previously in Fig. 1.14 of Chapter 1. The differential cross section has a compact notation that is convenient for mathematical analysis and is also a measurable quantity, such as in molecular beam experiments. The differential cross section appears directly in the collision integral of the Boltzmann equation (Eq. 1.77) and therefore appears in the integral expressions for transport coefficients resulting from Chapman–Enskog analysis (Eqs. 5.86–5.89). Recall, from Fig. 1.14 and related discussion, that the differential cross section is defined as $\sigma \, d\Omega \equiv b \, db \, d\epsilon$, where b is the distance of closest approach of the centers-of-mass of colliding molecules and is referred to as the "impact parameter." In general, cross sections can be expressed equivalently using either notation depending on preference. Therefore, the general expression for a collision cross section in Eq. 5.89 can be rewritten as

$$Q^{(l)} = \int_0^{4\pi} \left(1 - \cos^l \chi\right) \sigma \, d\Omega = 2\pi \int_0^{\infty} \left(1 - \cos^l \chi\right) b \, db \tag{5.102}$$

where the scattering angle is a function of the impact parameter, relative speed, and interatomic potential, $\chi = \chi(b, g, \psi)$.

As evident from Eqs. 5.86–5.89, there are two different cross sections that appear in the expressions for viscosity and diffusion coefficients, namely $Q^{(2)}$ for viscosity and $Q^{(1)}$ for diffusion. In many texts and articles in the literature, these cross sections are referred to as the "viscosity cross section" (σ_μ) and the "momentum cross section" (σ_M), respectively. Therefore,

$$\sigma_\mu = \sigma_\mu(g) \equiv Q^{(2)} = 2\pi \int_0^{\infty} \sin^2 \chi \, b \, db \tag{5.103}$$

and

$$\sigma_{\mathrm{M}} = \sigma_{\mathrm{M}}(g) \equiv Q^{(1)} = 2\pi \int_0^\infty (1 - \cos \chi) b\, db \tag{5.104}$$

It is important to recognize that σ_μ and σ_{M} are similar (but not equal) to the total collision cross section, σ_{T}, defined in Chapter 1 (Eq. 1.69). Also, since cross sections are evaluated for a specific relative speed, they are a function of g. The relation between the various cross sections and their dependence on relative speed is analyzed later in this section.

Finally, combining the preceding expressions for the collision cross sections with Eqs. 5.86–5.88, the binary coefficients of viscosity and diffusion can be written as

$$\mu_{ij} = \frac{\frac{5}{8}\sqrt{2\pi m_r kT}}{\left(\frac{m_r}{2kT}\right)^4 \int_0^\infty g^7 \sigma_\mu(g) e^{-m_r g^2/2kT} dg} \tag{5.105}$$

and

$$\mathscr{D}_{ij} = \frac{\frac{3}{16}\sqrt{2\pi kT/m_r}}{\left(\frac{m_r}{2kT}\right)^3 n \int_0^\infty g^5 \sigma_M(g) e^{-m_r g^2/2kT} dg} \tag{5.106}$$

These expressions for the collision cross sections and binary transport coefficients will be used frequently in the remainder of the textbook as they represent the three levels of physical modeling discussed in the introduction to this chapter (Fig. 5.1). Specifically, the potential energy surface (PES) ψ that governs interatomic forces is what determines the scattering angle χ of individual collisions. ψ is the model input for molecular dynamics calculations. The collision cross sections (Eqs. 5.103 and 5.104) are simply the integrated result of all impact parameters b that lead to a finite scattering angle χ. Since in a dilute gas impact parameters and the pre-collision orientations of molecules are completely random, it is not necessary to model every possible collision arrangement and only the integrated result (i.e., the cross section) is required. For this reason, the collision cross section is a physically meaningful and convenient quantity in many molecular models of dilute gases. It appears in the collision integral in the Boltzmann equation and serves as the main model input for the DSMC method.

In the limit of near-equilibrium velocity distribution functions (i.e., the Chapman–Enskog assumption leading to the Navier–Stokes equations), the collision cross sections are further integrated over the corresponding near-equilibrium distribution of relative speeds (g). This results in the "collision integral," which is a function of temperature that appears in the denominator of the transport coefficient expressions.

As evident from Eqs. 5.105 and 5.106, without the denominator, the transport coefficients would be proportional to $\sqrt{T}$. This is a result of the molecular transport being proportional to the mean thermal speed, which is proportional to $\sqrt{T}$. However, the denominator represents the average cross section of molecule pairs locally within the gas. Therefore, as the average cross section is reduced, the efficiency of transport increases since molecules are able to transport their mass, momentum, and energy further before experiencing a collision. We now investigate some specific examples of molecular interactions leading to specific values of the collision cross sections and transport property coefficients.

5.4.2 Hard-Sphere Interactions

For a binary collision, the scattering angle of a given collision can be related to the potential energy function (ψ) as follows

$$\chi(g,b) = \pi - 2b\int_{r_m}^{+\infty} \frac{dr/r^2}{\sqrt{1-\left(\frac{b}{r}\right)^2 - \frac{\psi(r)}{1/2 m_r g^2}}} \tag{5.107}$$

where r_m is the distance of closest approach during the collision, and it's value is equal to the largest root of the denominator in the integrand in Eq. 5.107. A derivation of Eq. 5.107, starting from the equations of motion, can be found on pages 45–51 of Hirschfelder et al. (1954).

Let us consider the case of hard-sphere molecules. For a hard-sphere, the interatomic potential function is

$$\psi(r) = \begin{cases} 0 & \text{if } r > d \\ +\infty & \text{if } r \leq d \end{cases} \tag{5.108}$$

In this case, $r_m = d$, since $\lim_{(r\to d^+)} \psi(r) = 0$, and $\psi(r) = 0$ maximizes the root in the denominator of Eq. 5.107. Furthermore, since for $r = d + \delta d \to \chi = 0$, the integral limits can be rewritten as

$$\chi(g,b) = \pi - 2b\int_{d}^{0} \frac{-d(1/r)}{\sqrt{1-\left(\frac{b}{r}\right)^2}} \tag{5.109}$$

which can be further reduced to

$$\chi(g,b) = \pi + 2\int_{0}^{b/d} \frac{-dy}{\sqrt{1-y^2}} \tag{5.110}$$

where $y = b/r$. This has the solution

$$\chi(g,b) = \begin{cases} 2\cos^{-1}(b/d) & \text{if } b < d \\ 0 & \text{if } b \geq d \end{cases} \tag{5.111}$$

and, as expected, it is evident that χ is not a function of g for hard-sphere collisions. This scattering angle result can now be used to determine the viscosity cross section using Eq. 5.103. Substitution yields

$$\begin{aligned}\sigma_\mu &= 2\pi \int_0^d 4\sin^2\left(\frac{\chi}{2}\right)\cos^2\left(\frac{\chi}{2}\right) b\,db \\ &= 2\pi \int_0^d 4\left(1-\left(\frac{b}{d}\right)^2\right)\left(\frac{b}{d}\right)^2 b\,db = \frac{2}{3}\pi d^2 \end{aligned} \tag{5.112}$$

Here, it is interesting to note that for hard-sphere scattering, and in fact for a number of models based on hard-sphere scattering presented later in Chapter 6, that the viscosity cross section is related to the total cross section ($\sigma_T = \pi d^2$), by a constant: $\sigma_\mu = (2/3)\sigma_T$. Furthermore, analogous to the result in Eq. 5.112, the result for the momentum cross section is simply $\sigma_M = \sigma_T$. Therefore in many cases, the total cross section σ_T, appearing directly in the collision integral of the Boltzmann equation (Eq. 1.69), is directly related to the viscosity and momentum cross sections (σ_μ and σ_M) that appear in the transport property expressions.

For the hard-sphere result from Eq. 5.112, the collision integral (Eq. 5.88) becomes

$$\Omega^{(2,2)} = \sqrt{\frac{kT}{2\pi m_r}} \int_0^{+\infty} e^{-\gamma^2}\gamma^7 \left(\frac{2}{3}\pi d^2\right) d\gamma = 3\sqrt{\frac{kT}{2\pi m_r}}\left(\frac{2}{3}\pi d^2\right) \tag{5.113}$$

and the viscosity for hard-sphere molecules is therefore

$$\mu^{HS} = \frac{5}{16}\frac{1}{\pi d^2}\sqrt{2\pi m_r kT} \tag{5.114}$$

In summary, using a hard-sphere interatomic potential results in a collision cross section that is independent of relative velocity (g), and therefore a collision integral and viscosity coefficient inversely proportional to the hard-sphere diameter squared (d^2) and proportional to $\sqrt{T}$. This temperature dependence is not accurate for real gases and is an artifact of the hard-sphere assumption.

5.4.3 Inverse Power-Law Interactions

Consider an inverse power-law interatomic potential function

$$\psi(r) = \frac{a}{r^{\eta-1}} = \frac{a}{r^\alpha}, \quad \alpha \equiv \eta - 1 \tag{5.115}$$

which results in an interatomic force

$$F(r) = -\frac{d}{dr}\left(\frac{a}{r^{\eta-1}}\right) \propto -\frac{1}{r^\eta} \tag{5.116}$$

For this case, the scattering angle (Eq. 5.107), becomes

$$\chi(g,b) = \pi - 2\int_{r_m}^{+\infty} \frac{-d(b/r)}{\sqrt{1-\left(\frac{b}{r}\right)^2 - \left(\frac{ar^{-\alpha}}{1/2 m_r g^2}\right)}} \tag{5.117}$$

This expression can be simplified to the following:

$$\chi(g, b) = \pi - 2\int_0^{y_m} \frac{dy}{\sqrt{1 - y^2 - \frac{1}{\alpha}\left(\frac{y}{\beta}\right)^{\alpha}}}, \tag{5.118}$$

by invoking the following definitions:

$$y = \frac{b}{r}, \qquad y_m = y(\beta) = \frac{b}{r_m}, \text{ where } \quad \beta = b\left(\frac{1/2m_r g^2}{\alpha a}\right)^{1/\alpha} \tag{5.119}$$

such that $1 - y_m^2 - \frac{1}{\alpha}(\frac{y_m}{\beta})^{\alpha} = 0$. Thus, the scattering angle is now formulated in terms of β for a given value of α. To integrate the scattering angle into the cross section, Eq. 5.103 must be written in terms of β. Since

$$\beta = b\left(\frac{1/2m_r g^2}{\alpha a}\right)^{\frac{1}{\alpha}} \rightarrow d\beta = db\left(\frac{1/2m_r g^2}{\alpha a}\right)^{\frac{1}{\alpha}} \tag{5.120}$$

and further multiplying by b and rewriting in terms of β, we have

$$b\,d\beta = b\,db\left(\frac{1/2m_r g^2}{\alpha a}\right)^{\frac{1}{\alpha}} \rightarrow b\,db = \beta\,d\beta\left(\frac{\alpha a}{1/2m_r g^2}\right)^{\frac{2}{\alpha}} \tag{5.121}$$

Therefore, the viscosity cross section can be written as

$$\sigma_\mu = 2\pi\left(\frac{\alpha a}{1/2m_r g^2}\right)^{\frac{2}{\alpha}} \int_0^{+\infty} \left[1 - \cos^2\chi(\beta)\right]\beta\,d\beta = 2\pi\left(\frac{\alpha a}{1/2m_r g^2}\right)^{\frac{2}{\alpha}} A^{(2)}(\alpha) \tag{5.122}$$

where $A^{(2)}(\alpha)$ can be evaluated numerically. Substitution of this viscosity cross section into the collision integral, Eq. 5.88, yields

$$\Omega^{(2,2)} = \sqrt{\frac{2\pi kT}{m_r}} A^{(2)}(\alpha)\left(\frac{\alpha a}{kT}\right)^{\frac{2}{\alpha}} \int_o^{+\infty} \gamma^{(7-4/\alpha)} e^{-\gamma^2} d\gamma \tag{5.123}$$

The integrand is a standard gamma function, and therefore,

$$\Omega^{(2,2)} = \sqrt{\frac{2\pi kT}{m_r}} A^{(2)}(\alpha)\left(\frac{\alpha a}{kT}\right)^{\frac{2}{\alpha}} \frac{1}{2}\Gamma(4 - \frac{2}{\alpha}), \tag{5.124}$$

and the viscosity is obtained by substitution into Eq. 5.86 to yield

$$\mu^{\mathrm{IPL}} = \frac{5}{8}\sqrt{\frac{2km_r}{\pi}}\left(\frac{k}{\alpha a}\right)^{\frac{2}{\alpha}} \left[\frac{1}{A^{(2)}(\alpha)\Gamma(4 - \frac{2}{\alpha})}\right]\left[T^{\frac{1}{2}+\frac{2}{\alpha}}\right] \tag{5.125}$$

Thus, an inverse power-law interaction leads to a viscosity with a temperature dependence determined by the power-law exponent α. The special case where $\alpha = 4$ (and therefore $\eta = 5$) is called a "Maxwell" molecule, for which the viscosity varies linearly with T. Such a linear dependence on temperature is typically not accurate for most gases, for example, nitrogen and oxygen have a viscosity with proportionality closer to $T^{0.7}$.

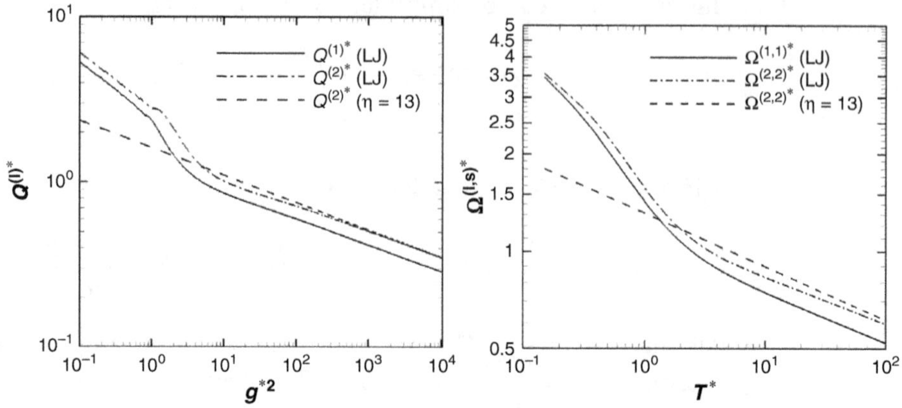

(a) Nondimensional collision cross sections. (b) Nondimensional collision integrals.

Figure 5.3 Viscosity and momentum cross sections and collision integrals for Lennard–Jones and inverse power law potential energy functions. Results are normalized by the Hard-Sphere result given in section 5.4.2. Specifically, $Q^{(l)*} \equiv Q^{(l)}/Q^{(l)}_{HS}$ and $\Omega^{(l,s)*} \equiv \Omega^{(l,s)}/\Omega^{(l,s)}_{HS}$, where $g^{*2} = m_r g^2/2\epsilon$ and $T^* = kT/\epsilon$.

5.4.4 General Interatomic Potentials

For a general interatomic potential $\psi(r)$, one can numerically integrate the integrals in Eqs. 5.103–5.107. An example of a simple PES, which is often used for the atomic interactions in noble gases, is the Lennard–Jones potential:

$$\psi(r_{ij}) = 4\epsilon\left[\left(\frac{s_0}{r_{ij}}\right)^{12} - \left(\frac{s_0}{r_{ij}}\right)^{6}\right]. \tag{5.126}$$

This simple function captures the weak attraction between two atoms at a finite separation and the strong repulsion experienced as the separation distance goes to zero. Specifically, referring to Fig. 1.2 of Chapter 1, ϵ corresponds to the energy minimum and s_0 corresponds to the separation distance at this minimum energy. One can directly substitute ψ into the equation for the scattering angle (Eq. 5.107), and numerically integrate the integrals in Eqs. 5.103–5.107. Using a standard fourth-order accurate Simpson quadrature rule to compute the integrals, the results for the collision integrals corresponding to the LJ PES are shown in Fig. 5.3.

As seen in the log-log plot shown in Fig. 5.3(a), as the relative collision speed (g) increases the collision cross section is reduced significantly. Molecule pairs with high relative speeds spend less time interacting on the potential energy surface. As a result, for a given impact parameter b, the deflection angle χ is reduced as g increases. The strong dependence of collision cross section on relative collision speed is an important physical effect that must be accounted for to determine correct transport properties and their dependence on the gas temperature.

Table 5.1 Atomic Parameters

	ε/k [K]	s_0 [Å]	m [amu]
Ar	124.0	3.418	39.9
He	10.23	2.576	4.0
Xe	229.0	4.060	131.3

k is Boltzmann's constant.

In Fig. 5.3(a), for $g \to +\infty$, the slope of $Q^{(2)}(g)$ approaches that of the inverse repulsion law $F(r_{ij}) \sim 1/r_{ij}^{13}$. This indicates that for high relative speeds, the collision cross section has a power-law dependence on g. However, for lower temperatures, where collisions are less energetic (small g), the power-law approximation becomes less valid.

Recalling that $\mu \sim \sqrt{T}/\Omega^{(2,2)}$, the collision integral is numerically integrated using Simpson's rule and the results are plotted in Fig. 5.3(b). For large T, $\Omega^{(2,2)} \sim T^{-\vartheta}$, with ϑ approaching 1/6, which corresponds to $\eta = 2/\vartheta + 1 = 13$ for the the inverse repulsion law, as expected, giving a temperature exponent of 2/3. In the low T range ($T < 1500$ K), $\log(\Omega^{(2,2)}) \neq -\vartheta \log(T)$, and moreover the curve becomes steeper, i.e., a larger *average* ϑ is necessary. The inverse power-law potential accurately reproduces the LJ potential at high temperatures, provided that η is set appropriately, whereas it becomes less physically valid at low temperatures.

Analogous results for the momentum cross section ($Q^{(1)}$) and the corresponding collision integral $\Omega^{(1,1)}$, are plotted in Fig. 5.3(a) and (b), respectively. The trends are identical and together these integral results highlight the influence of relative collision speed on the collision cross sections, which leads to a strong temperature dependence for the collision intergrals and thus for the transport coefficients. In Chapter 6, where the DSMC method is presented, we will introduce cross section *models* that are constructed to capture the most salient physics of molecular interactions without requiring the integration of collision dynamics on a PES.

Finally, we can use the expressions derived in this chapter to compute mixture quantities directly from atomic interaction parameters. Consider mixtures of noble gases, helium, xenon, and argon. Interactions between these atoms are well modeled by the LJ potential. The potential parameters for each element are contained in Table 5.1. For cross interactions, the following mixing rules were used:

$$\varepsilon_{ij} = \sqrt{\varepsilon_i \varepsilon_j}, \tag{5.127}$$

$$s_{0i,j} = \frac{s_{0i} + s_{0j}}{2}. \tag{5.128}$$

Although there is no theoretical justification, mixture viscosities computed with such combining relations are found to be in good agreement with the

Table 5.2 Mixture Viscosities Computed from LJ Interatomic Potential Parameters

Case	μ_1 [10^{-5} kg m^{-1} s^{-1}]	T [K]
Xe(1.5%)-He	2.14	300
Xe(3.0%)-He	2.25	300
Xe(6.0%)-He	2.41	300
Xe(9.0%)-He	2.51	300
Ar(11.5%)-He	1.462	162
Ar(24.7%)-He	1.500	162
Ar(44.0%)-He	1.477	162

available experimental data for a number of species (Hirschfelder et al. 1954). A more rigorous way is to evaluate ε_{ij} and $s_{0i,j}$ from first principles.

The binary coefficient μ_{ij} of each species pair can be calculated from Eq. 5.86 and further used to determine the mixture viscosity μ_{mix} using Eq. 5.90. As an example, the coefficient of viscosity for various mixtures is numerically integrated and the results are presented in Table 5.2.

An excellent review of available data and collision integral results required to calculate the mixture transport properties over a wide temperature range has been presented by Wright, Bose, Palmer, and Levin (2005) for air species, and by Wright, Hwang and Schwenke (2007) for the atmospheres of Mars and Venus.

The collision cross sections discussed in this chapter were assumed not to be functions of the internal structure of the molecule. However, the scattering angle could certainly be a function of the rotational and vibrational energy of colliding molecules, and thus the cross sections could be functions of relative velocity and internal energy. This is typically not the case except when molecules are highly stretched at very high temperatures (high energies) for which data from experiments and first-principles calculations is currently limited or nonexistent.

Also, the complexity of a PES can vary widely depending on the atomic interactions of interest and can range from having a few fitting parameters (such as ϵ and s_0 in Eq. 5.126) to more complex hypersurfaces having hundreds or thousands of fitting parameters (Paukku, Yang, Varga and Truhlar 2013). The determination of the potential energy between atoms and the fitting of a general potential surface lies in the field of computational chemistry and will not be described in this text. In this section, we described how simple PESs can be used to determine collision cross sections (used in DSMC collision models) and ultimately determine gas transport properties (used in CFD models). For high energy collisions involving (quantized) vibrational energy excitation and chemical reactions, more sophisticated PESs are required involving additional considerations that are discussed in Chapter 7.

5.5 Summary

In this chapter, the mathematical connection between the Boltzmann equation and the continuum Navier–Stokes equations was described.

Beginning with the Boltzmann equation, moments corresponding to molecular mass, momentum, and energy were taken, resulting in a set of molecular conservation equations. These equations included general averages over the local velocity distribution function and were therefore accurate for any degree of nonequilibrium. The molecular conservation equations where shown to have the same form as the continuum Navier–Stokes equations, however, they included transport terms that had no closed-form expressions for an arbitrary velocity distribution.

In the limit of near-equilibrium, through Chapman–Enskog analysis, the precise velocity distribution function that reduces the conservation equations to the Navier–Stokes equations including Newtonian, Fourier, and Fick transport laws was derived. This Chapman–Enskog distribution function was found to be a first-order perturbation to a Maxwell–Boltzmann equilibrium distribution, where the deviation from equilibrium was quantified by the gradient-length Knudsen number. In the limit of near-equilibrium (small Kn), the Chapman–Enskog distribution function is an accurate representation of the gas at the molecular level and therefore, in this limit, the Navier–Stokes equations are accurate as well.

The transport terms were evaluated in the near-equilibrium limit using the Chapman–Enskog distribution function. For the flow of a single-species gas, the transport terms were shown to be identical to those in the compressible Navier–Stokes equations. The transport property expressions for polyatomic gas mixtures are significantly more complex than commonly used forms of the multispecies Navier–Stokes equations. Transport terms that are commonly neglected in continuum models were highlighted, while expressions for all commonly included transport terms were presented and analyzed.

The analysis led to equations where, given an interatomic potential that dictates forces between atoms, one can compute the collision cross sections that appear in the Boltzmann equation and are required for DSMC simulations. The cross section expressions can then be used to evaluate further the transport properties that appear in the Navier–Stokes equations and are required for CFD simulations. While the equations corresponding to a monatomic simple gas were rigorously derived from the Boltzmann equation, for practical reasons, a number of simplifications are generally required for polyatomic gas mixtures. Such simplifications commonly employed in continuum models include the Eucken correction for thermal conductivity and the self-consistent effective binary diffusion (SCEBD) model.

Finally, it is important to stress that although much of the material in this chapter focused on the consistency between molecular and continuum descriptions in the near-equilibrium limit, the purpose of this textbook is to provide the theory and methods required to model gas flows in *any* degree of nonequilibrium (spanning from free molecular to continuum flow). Indeed, the DSMC method simulates the Boltzmann equation, including all relevant physics, and thus makes none of the assumptions and simplifications discussed in this chapter. Despite the mathematical complexity of the Boltzmann equation that has historically restricted the scope of solutions, the combination of the DSMC method and modern computational power now enable the numerical solution of three-dimensional nonequilibrium flows over complex geometries. The reminder of the textbook is devoted to describing the numerical algorithms that can be used to solve the equations presented in Chapters 1–5 and thereby obtain accurate solutions for a wide range of nonequilibrium flows.

6 Direct Simulation Monte Carlo

6.1 Introduction

The direct simulation Monte Carlo (DSMC) method was created by Graeme Bird, a professor of aeronautical engineering at the University of Sydney, Australia. The first publication, appearing in Physics of Fluids in 1963, demonstrated the DSMC method as an alternative to molecular dynamics (MD) for a hard-sphere gas (Bird 1963). Since that time, Bird has written two textbooks on the DSMC method (Bird 1994, 2013) and an entire field of research has emerged with hundreds of publications on the DSMC method and its application to scientific and engineering problems. The DSMC method has become a powerful and widespread technique for the simulation of nonequilibrium gases where the molecular nature of the gas must be accounted for.

To begin this chapter, it is important to identify the range of flow conditions and engineering applications for which the DSMC method is most appropriate. As described by Bird (1994), dilute gas conditions can be separated into various regimes through nondimensional quantities.

First, the transition between a dilute gas and a dense gas occurs when the size (d) of the molecules themselves (the extent of their interatomic forces) becomes comparable to the mean separation distance (δ) between molecules in the gas. As the molecular size becomes comparable to the molecular separation distance, the mean-free-path and mean-collision-time scales vanish. Indeed, the molecules in a condensed phase (a solid or liquid) are in a constant collisional state, experiencing continual forces due to neighboring molecules. Although the molecules in a dense gas may have nonzero mean-free-path and mean-collision-time scales, these scales begin to approach the Angstrom and femtosecond scales used in molecular dynamics simulations. Furthermore, in dense gases, multimolecule interactions are frequent and must be accounted for. In contrast, in a dilute gas, in which the mean molecular spacing is large compared to molecular size, collisions are predominately binary in nature. Such binary collisions occur over time scales that are much shorter than the mean collision time and can be thought to occur

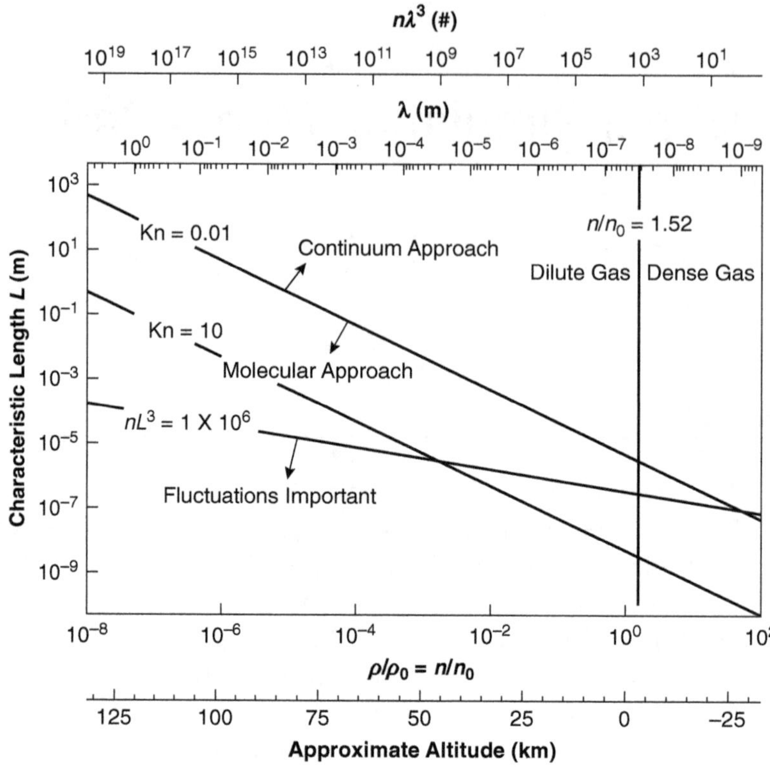

Figure 6.1 Gas flow regimes and implications for physical models.

instantaneously. As described in Chapter 1, this leads to the ideal gas equation of state.

For air at standard conditions (1 atm of pressure and 0 degrees Celsius), *Loschmidt's number*, $n_0 = 2.68684 \times 10^{19}$, represents the number of molecules within one cubic centimeter of gas. In this case, the average volume available to a molecule is $1/n_0$ and thus the mean molecular spacing is $\delta = n_0^{-1/3}$. One can approximate the molecular size by the hard-sphere diameter corresponding to the viscosity of air at standard conditions. Using Eq. 5.114, this value is $d = 4.15 \times 10^{-10}$ m. This leads to a ratio of $\delta/d = 8$, which is just large enough that air at standard conditions is very accurately modeled as a dilute gas. However, for ratios of $\delta/d \leq 7$, corresponding to approximately $n/n_0 \geq 1.52$ (the vertical line in Fig. 6.1), dilute gas assumptions begin to become inaccurate.

Second, if the characteristic length scale (L) becomes small enough that relatively few molecules are present within a volume of interest, then real fluctuations (spatial and temporal fluctuations) in gas properties may become significant and require consideration for practical, engineering, purposes. In Fig. 6.1, the flow regime where local fluctuations become significant is the region where $nL^3 < 10^6$. Below this line in Fig. 6.1, the characteristic volume (L^3) would therefore contain fewer than one million molecules. With fewer than one million molecules within the volume of interest, macroscopic properties such as bulk velocity, density, and temperature, may noticeably fluctuate from their expected steady-state values.

One may consider a single cubic mean free path as the smallest relevant volume in a dilute gas. The hard-sphere mean-free-path (λ), determined from Eq. 1.144 with $d = 4.15 \times 10^{-10}$ m, is inversely proportional to n and is also included in Fig. 6.1. Furthermore, the corresponding number of molecules per cubic mean free path ($n\lambda^3$) is included on the upper axis. Finally, the approximate altitude values corresponding to the plotted range of $n/n_0 = \rho/\rho_0$ are included on the lower axis. Here, the same exponential relation between density and altitude (h) used for Problem 1.1 in Chapter 1 is assumed. For altitudes above sea level, there are many millions of molecules present within each cubic mean free path, and thus real fluctuations in gas properties are completely negligible. However, near sea-level conditions, where the mean free path is approximately 68×10^{-9} m, there may be only thousands of molecules present per cubic mean free path. For gas flows involving micro electromechanical systems (MEMS), where length scales of interest are small, gas fluctuations can become significant and may need to be accounted for in engineering analysis.

Third, as discussed in detail in Chapter 1, the Knudsen number ($\text{Kn} \equiv \lambda/L$) determines whether a dilute gas flow is expected to exhibit nonequilibrium behavior necessitating a molecular modeling approach, or near-equilibrium behavior where methods from continuum gas dynamics provide an accurate model. In Fig. 6.1, this boundary is marked by the line where $\text{Kn} = 0.01$. In such a flow, a molecule would undergo approximately 100 collisions as it traverses the length scale of interest, L, and thus the gas in this vicinity would be expected to be in a nonequilibrium state. Clearly, nonequilibrium flow is expected for low-density conditions or for very small objects at standard density conditions. For $\text{Kn} > 10$, the flow is approximately free molecular where very few collisions occur within the volume of interest. The region where $0.01 < \text{Kn} < 10$ is often referred to as the *transition* regime where collisions cannot be neglected; however, the continuum fluid equations become inaccurate.

Figure 6.1 contains a great deal of information that can quickly be used to determine the flow regime and magnitudes of molecular parameters, corresponding to a given engineering problem.

Much of the development of the DSMC method has focused on high-altitude flight, where it is an accurate and efficient computational method for practical, three-dimensional engineering flows. As the flow becomes continuum, the mean-free-path and mean-collision-time scales, that must be resolved within a DSMC simulation, require prohibitively large computational resources. Of course, it is precisely in this regime where efficient continuum methods from the field of computational fluid dynamics (CFD) become accurate, and a molecular approach is no longer required. Often, defining Kn using the size of the vehicle (or object) for L and the free-stream conditions for the value of λ, gives a good indication of whether the flow requires molecular or continuum modeling approaches. As seen in Fig. 6.1, even as the gas density approaches standard conditions, if the length scale of the flow is sufficiently small, then continuum modeling becomes inaccurate and molecular modeling is required. There are a number of engineering systems associated with this gas flow regime, for example, flow inside a hard disk drive and flow through microchannels, pumps, valves, and nozzles. While high-altitude flows are usually associated with high-velocity flow, these small-scale, higher-density applications involve mainly low-velocity flow. Simulating low-velocity flow with the DSMC method can be computationally challenging, since the thermal velocity of the molecules is much higher than the bulk velocity of the gas, which, for these applications, is often less than 1 m/s. As a result, to obtain average macroscopic properties with statistical scatter below acceptable tolerances, larger numbers of molecular samples must be recorded within a DSMC simulation compared to higher velocity flow simulations. Despite this challenge, on current computer architectures, the DSMC method is quite capable of providing accurate solutions for these low-speed flows. For very low speed flows, variance reduction algorithms have been developed by Baker and Hadjiconstantinou (2005) and Homolle and Hadjiconstantinou (2007).

Another challenge, that arises in many flows, is that there may be small-length-scale flow features embedded within continuum flow around a larger object. There could be fine geometry details for which pressure forces, shear stress, and heat transfer may be important. An example in aerospace is the flow over sharp leading edges. Despite a continuum flow about a vehicle, the gas in the vicinity of a sharp leading edge may be in a state of strong nonequilibrium. In this case it may be appropriate to set L as the leading edge radius. As a result, accurate prediction of the gas flow and the transfer of momentum and energy to the sharp leading edge may require a molecular modeling approach. Furthermore, especially for high-speed flows, sharp gradients in gas flow properties may be present. Examples include shock waves, rapid expansions in the wake of a vehicle, and thin boundary layers next to a vehicle surface. In such cases, it is more appropriate to set L as the *gradient length* and the gradient-length Knudsen number becomes

$$\mathrm{Kn}_{GL-Q} \equiv \lambda \left| \frac{\nabla Q}{Q} \right| \tag{6.1}$$

where Q represents a macroscopic flow quantity such as density, bulk velocity, or temperature. Recall, the nonequilibrium perturbation terms in the Chapman–Enskog distribution function are proportional to this quantity (Eq. 5.52). A suggested cutoff value indicating that a localized region may be in a nonequiibrium state is $\mathrm{Kn}_{GL-Q} \geq 0.05$ (Boyd et al. 1995; Wang and Boyd 2003). This can be interpreted as indicating that a macroscopic flow property is undergoing a 5% change over a distance of λ, in which case it is intuitive that during this rapid change in state, the gas may locally be in a state of nonequilibrium. To accurately and efficiently simulate flows that are mainly continuum but have important localized regions of nonequilibrium, much research has been devoted to hybrid particle-continuum numerical methods (Schwartzentruber and Boyd 2006; Schwartzentruber et al. 2007, 2008a, 2008b, 2008c; Deschenes and Boyd 2011). Such hybrid particle-continuum numerical methods are not yet used for engineering analysis; rather development of these methods is currently an area of active research, and beyond the scope of this textbook.

Finally, it is interesting to consider the flow regime (seen in the bottom right-hand corner of Fig. 6.1) where the boundaries between dilute and dense gases, nonequilibrium and continuum gases, and fluctuating and nonfluctuating gases, all overlap. For such flows, fluctuations in the gas are significant and the ideal gas equation of state may no longer be accurate due to the high density and therefore significant multimolecule intermolecular forces. Such flows may be accurately simulated by either a molecular approach (such as DSMC) or a continuum approach (such as CFD), although both approaches should model fluctuations accurately. Recent research (Donev, Garcia, and Alder 2008; Donev, Bell, Garcia, and Alder 2010) has shown that in this regime the DSMC method is able to naturally model real fluctuations in the gas, that DSMC algorithms can be altered to model a different equation of state, and finally that DSMC can be combined with fluctuating hydrodynamics CFD methods, resulting in a hybrid code capability. Such research sheds much light on this particular flow regime; however, it is beyond the scope of this textbook.

This textbook focuses on the DSMC method applied to flows spanning conditions from continuum to free-molecular flow in dilute gases where real fluctuations are insignificant. As described in Part I of this textbook, the relevant governing equation is the Boltzmann equation. The DSMC stochastic particle method emulates the physics of the Boltzmann equation. For the case of a monatomic simple gas, the DSMC method has been shown to provide a solution to the Boltzmann equation as the number of simulated particles approaches infinity and the timestep and collision-cell sizes approach zero (Wagner 1992). There are a number of partial differential equation (PDE)–based numerical techniques to solve various forms of the Boltzmann equation. The main challenges associated with PDE–based methods include the high dimensionality of the Boltzmann equation, solving the collision integral including all relevant collision physics, and applying boundary conditions. While such methods are the subject of continued research, the DSMC method has been an established engineering tool for decades, able to provide accurate predictions for nonequilibrium flows. The stochastic particle nature of the DSMC method is able to efficiently handle the multidimensionality associated with nonequilibirum flows. In addition, the DSMC particle-based method enables considerable flexibility in modeling advanced physics (such as internal energy excitation, chemical reactions, gas surface interactions, extension to plasma flows, and even fluctuations in the gas state), that often cannot be formulated within the Boltzmann equation. Furthermore, the inherent nature of the DSMC method enables a wide variety of collision models, ranging from phenomenological to quantum-mechanics based, to be incorporated in a general and modular manner within a DSMC code. Over the past 50 years, many physical models have been developed for applications ranging from microflows, to astrophysics, to vapor deposition and condensation, to hypersonic flight.

Chapters 6 and 7 of this text describe how the theoretical material from Chapters 1 to 5 form the basis of modern DSMC physical models, and how the resulting equations and numerical algorithms can be implemented into a simulation code able to produce engineering predictions for nonequilibrium flows.

6.2 DSMC Basics

6.2.1 Fundamentals

The DSMC method leverages three physical characteristics of a dilute gas:

(1) Molecules move in free flight without interaction for time scales on the order of the local mean collision time.
(2) The impact parameters and initial orientations of colliding molecules are random.
(3) There are an enormous number of molecules per cubic mean free path and only a small fraction need be simulated to obtain an accurate molecular description of the flow.

These three techniques/assumptions are highly accurate for dilute gases, and, combined, they enable the DSMC method to simulate macroscopic nonequilibrium flow fields produced by flow around complex geometries.

A visual depiction of these dilute gas characteristics is portrayed in Fig. 6.2(a), (b), and (c). As described in Part I of this text, the defining characteristic of a nonequilibrium flow is the local departure (within a small gas volume) from Maxwell–Boltzmann equilibrium distribution functions for molecular velocity and internal energy. As described in Chapter 5, CFD methods inherently assume only small departures from equilibrium distribution functions, whereas the DSMC method is able to predict the distribution functions (within each small gas volume) with no assumptions or restrictions on the shape of the distribution. An example of a non–Maxwellian distribution for the x-velocity of molecules, within a cubic mean free path, is shown on the right-hand side of Figs. 6.2(a), (b), and (c).

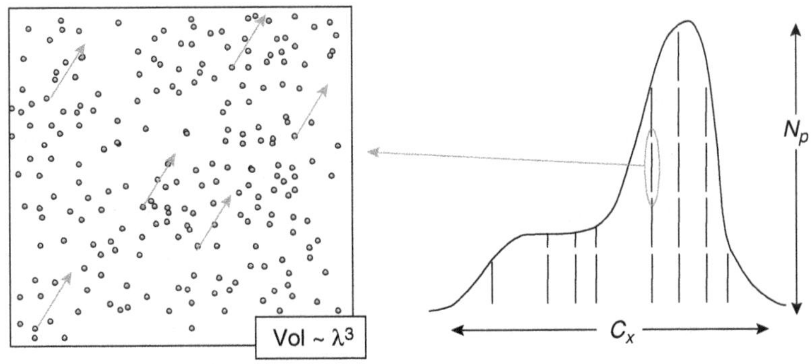

(a) Number of real molecules occupying a narrow range of molecular velocities (a specific positive *x*-velocity).

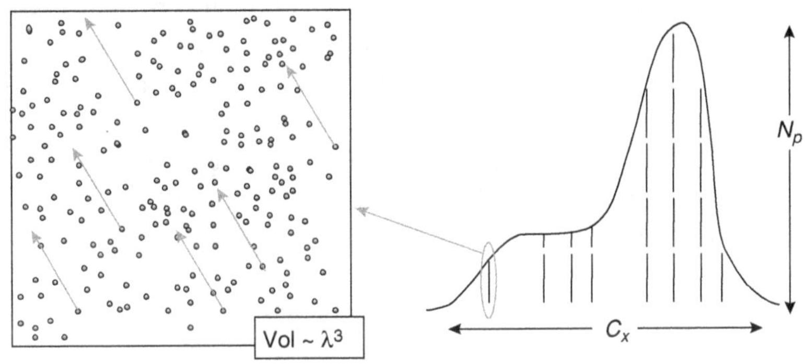

(b) Number of real molecules occupying a narrow range of molecular velocities (a specific negative *x*-velocity).

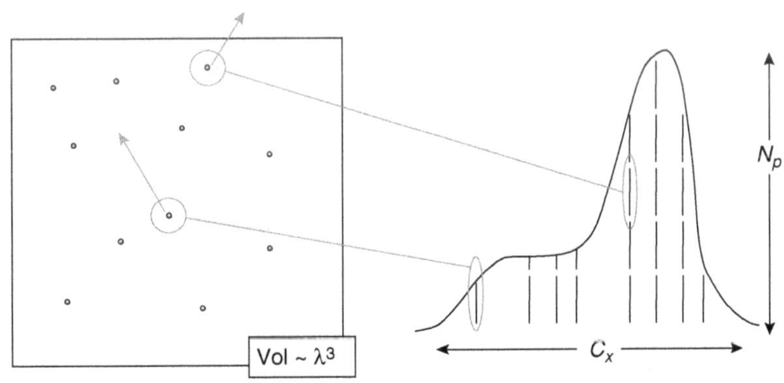

(c) Number of DSMC *simulated particles* occupying a narrow range of molecular velocities. An allowable collision pair within a DSMC cell is circled.

Figure 6.2 Schematic of the underlying characteristics of the DSMC method.

First, as previously discussed, there can be many millions of molecules contained within each cubic mean free path. This implies that each molecule experiences only one collision (on average) as it moves across this volume. Outside of such collisions, the separation distance between molecules is much greater than the extent of intermolecular forces and therefore molecules move in straight lines along their center of mass velocity vectors. Indeed, molecules are very small and there is a great deal of empty space. The first technique/assumption of the DSMC method is that simulated molecules can be moved in straight lines for a fraction of their mean collision time without any loss in accuracy. In addition to being physically accurate, the numerical accuracy of this technique is also apparent from pure MD simulations of dilute gases (Valentini and Schwartzentruber, 2009a).

Second, and also apparent from pure MD simulations of dilute gases, the impact parameters of colliding molecules in a dilute gas are completely random. Except for highly specialized circumstances, there is no inherent bias that all molecules begin to interact with a certain separation distance (b), or at a specific angle (ϵ), or at a specific point in their rotational orbit or vibrational period, or any other impact parameter for that matter. The second technique/assumption of the DSMC method is that these impact parameters are random and do not need to be deterministically simulated as they are with MD.

Finally, based purely on statistical arguments, the third technique/assumption of the DSMC method is that it is not necessary to simulate the properties of all real molecules contained within each cubic mean free path. Rather, highly precise distribution functions, and thus a full molecular description of the flow, can be obtained by modeling only a small fraction of the real number of molecules.

Each technique is related to the other two, and the combination of these three techniques enables the DSMC method to accurately simulate macroscopic nonequilibrium flow fields. Specifically, within a small volume of gas, many molecules will have velocity vectors that lie within the same narrow range. This situation is depicted in Fig. 6.2(a) for a narrow range of positive x-velocities, and in Fig. 6.2(b) for a narrow range of negative x-velocities. Let us now consider some approximate magnitudes to frame the discussion of Fig. 6.2. Instead of tracking four million molecules with the same x-velocity shown in Fig. 6.2(a) and also tracking one million molecules with the same x-velocity shown in Fig. 6.2(b) (where these magnitudes dictate the shape of the velocity distribution function), the same function could be obtained by considering only four molecules of the first velocity and one molecule of the second velocity (as depicted by the line segments comprising the distribution functions in Figs. 6.2(a) and 6.2(b)). This is precisely what the DSMC method does. As depicted in Fig. 6.2(c), each *simulated particle* is representative of a large number of real molecules contained within that small volume of gas. For the current discussion, each simulated particle would represent one million real molecules.

A key concept of the DSMC method is that a simulated particle does not represent a distribution of real molecules. Rather, a simulated particle represents a large number of *identical* real molecules. This enables collisions between pairs of simulated particles to be modeled with precisely the same physical considerations as collisions between pairs of real molecules.

Since the DSMC method does not simulate every real molecule in the system, the deterministic nature of molecular movement and collisions (as simulated with MD, for example) is lost. A DSMC simulation cannot determine precisely which real molecules actually collide and what impact parameters characterize the initial conditions of each collision. This loss of determinism is also a result of moving simulated particles in straight lines for a fraction of their mean collision time. In effect, although the distribution functions are accurately resolved at spatial and temporal scales of the mean free path and mean collision time, the precise locations of the real molecules comprising those distributions are no longer known below these scales. That is, the positions of simulated particles *within* a DSMC cell (significantly below the mean free path scale) is irrelevant to the method. This situation is depicted in Fig. 6.2(c), where in fact it is perfectly appropriate to select two simulated particles to collide despite their velocity vectors pointing away from each other. Although this may seem unphysical, recall that each simulated particle represents a large number of identical real molecules sweeping through the volume (whose precise positions are spread throughout the volume as shown in Figs. 6.2(a) and (b)), and therefore one would expect some number of these molecule pairs (with these specific velocities) to collide. Another way of rationalizing the DSMC collision pair circled in Fig. 6.2(c) is that the position of simulated DSMC particles is not physically precise at scales below the mean free path for at least two reasons. First, the positions of simulated particles injected through simulation boundaries are not known/specified at scales below the mean free path, and second, these simulated particles are then moved according to the local *mean* collision time through the computational domain.

In this manner, collisions between simulated particles within each DSMC cell are performed stochastically. During each timestep, simulated particles are randomly paired up within each cell, and subsequently tested for a collision. Thus all possible collision pairs, consisting of all classes of molecular velocities, internal energies, and chemical species, are sampled in a Monte Carlo fashion from the actual distributions of particles within each cell. Indeed, one popular method of solving the high-dimensional collision integral, appearing in the Boltzmann equation, is through Monte Carlo integration. This is one main reason why the DSMC method provides a very efficient technique for simulating the Boltzmann equation.

Since simulated particles are randomly paired for collisions within each cell, DSMC collision cells must be sized at or below the local mean free path. If cells were significantly larger than the mean free path, molecules separated by large distances could be randomly selected to collide. Such collisions would transfer mass, momentum, and energy over unphysically large distances. The same error would be produced if particles were moved in straight lines for timesteps much larger than the mean collision time. The resulting errors in the computed flow field would be analogous to the numerical dissipation error that results from using a coarse mesh or too-large timestep for a CFD calculation. Essentially, sharp flow gradients would not be accurately resolved and heat and momentum transfer to surfaces would also be incorrect.

It is noted that there are proposed acceleration techniques for the DSMC method that model some level of subcell resolution (analogous to higher-order reconstruction techniques in PDE-based continuum methods). Such techniques aim to use larger cells and larger timesteps while still maintaining solution accuracy. Other acceleration techniques use variable timesteps and varying particle weights, combined with cloning and deleting of particles, in order to minimize the number of simulation particles while maintaining solution accuracy. A wide variety of techniques have been proposed in the literature (examples include Kannenberg and Boyd 2000; Burt, Josyula, and Boyd 2011, 2012; Burt and Josyula 2014; and Galitzine and Boyd 2015) and their accuracy and use is often specific to a given problem. Most of these techniques build directly on core DSMC algorithms and do not require significant modifications to a standard DSMC code. It is therefore important to clearly describe the core algorithms of the DSMC method first, which is the purpose of Chapters 6 and 7.

In summary, the central DSMC algorithm can be summarized as follows:

(1) Generate particles at inflow boundaries.
(2) Move all particles in straight lines along their molecular velocity vectors for a timestep less than the local mean collision time.
   - Apply boundary conditions to particles that collide with a surface.
   - Remove particles that exit the simulation domain.
(3) Perform collisions stochastically within each cell, assuming impact parameters for collision pairs to be random.
   - Collision cell sizes should be smaller than the local mean free path.
(4) Sample particle properties in each cell.
(5) Return to (1).

In this manner, simulated particles move through a computational domain where they collide with each other and with surfaces according to prescribed boundary conditions. As depicted schematically in Fig. 6.3, molecular properties can be sampled in each cell to compute velocity and internal energy distribution functions, as well as average quantities (macroscopic quantities) such as density, bulk velocity, temperature, and pressure. If there are many particles in each cell then such average quantities and even distribution functions can be resolved with high precision at each timestep of the simulation. This may be desired for the simulation of unsteady flows using DSMC. However, most applications of the DSMC method involve steady state flows. In this case, once the simulation has reached steady state, the molecular properties in each cell can be continually sampled over many timesteps (time-averaged during steady state). This allows a small number of particles per

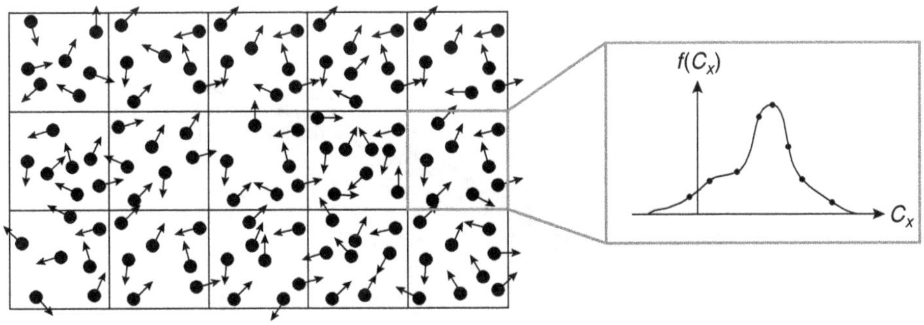

Figure 6.3 Schematic of DSMC simulation particles within collision cells and sampled distribution function.

cell (at any given instant) to be sampled over many timesteps during steady state, yielding highly precise average quantities and even distribution functions. It is noted that unsteady flow solutions may also be obtained using a small number of particles per cell, provided a large number of simulations are performed and the results are ensemble averaged. As will be discussed in Example 6.1, it is generally desirable for computational efficiency to minimize the number of simulated particles per cell, where this minimum number of particles is determined through further statistical considerations.

In summary, the DSMC method takes advantage of the inherent properties of dilute gases by using simulated particles that each represent a large number of identical real molecules, moving particles with timesteps on the order of the mean collision time, and stochastically selecting collision pairs and initial orientations within volumes (computational cells) on the order of the mean free path. These are rigorous simplifications based on sound physical principles. Present DSMC methods then go one step further and use probabilistic rules to determine the local collision rate and collision outcomes, thus introducing *collision models*. After a brief discussion of particle movement and sorting, the remainder of this chapter is devoted to detailed discussion of collision models for viscosity, thermal conductivity, diffusivity, and treatment of internal energy. Chapter 7 then presents models for high-temperature thermochemistry involving rotational and vibrational excitation and chemical reactions.

6.2.2 Particle Movement and Sorting

In this text, only a basic description of particle movement and sorting within cells is presented. Moving millions of particles through a domain, detecting collisions with complex surface geometry, and sorting within an arbitrary grid, can certainly be a difficult, tedious, computer programming task. However, the basic concepts used by even the most complex DSMC codes are actually straightforward.

Two examples of computational grids, adapted to the local mean free path, for hypersonic flow over a planetary probe geometry are shown in Fig. 6.4.

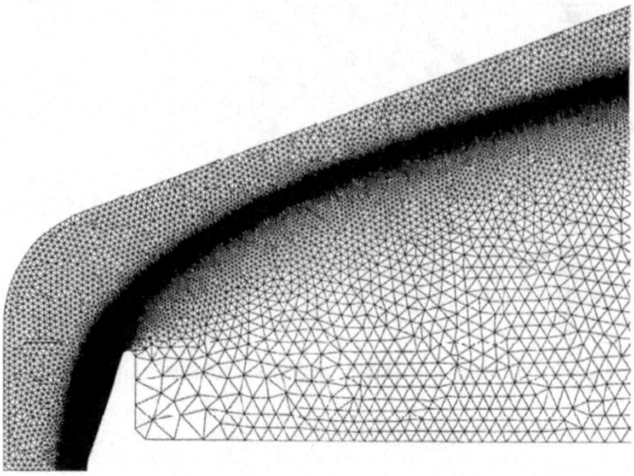

(a) Unstructured body-fitted grid.

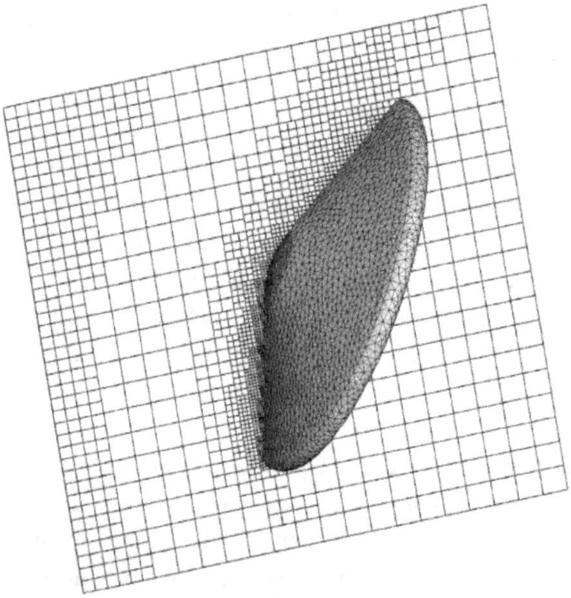

(b) Cartesian-based grid employing a "cut cell" technique.

Figure 6.4 Examples of flow field grids used in DSMC implementations.

The first grid (Fig. 6.4(a)) is an unstructured triangular grid used by the MONACO DSMC code (Dietrich and Boyd 1996), which is able to employ any general unstructured grid topology in two or three dimensions. Such grids can naturally be "body-fitted" grids, where the geometry of a solid body or domain perimeter can be exactly defined by cell faces (edges). The second grid (Fig. 6.4(b)) is a multilevel Cartesian grid used by the Molecular Gas Dynamic Simulator (MGDS) code (Gao, Zhang, and Schwartzentruber 2011; Nompelis and Schwartzentruber 2013). This particular grid is a three-level embedded grid, with arbitrary refinement at each level. Cartesian grids can also employ binary refinement, for example quadtree (2D) or octree (3D) organization, where cells are always refined (or de-refined) by a factor of 2 in each coordinate direction. For flow involving no significant density or temperature gradients (i.e., constant mean free path), uniform Cartesian grids may be employed. Unlike body-fitted grids, Cartesian-based grids require "cut cell" algorithms to simulate any body or domain geometry that is not Cartesian. There are two main aspects to such cut cell algorithms. The first aspect involves properly detecting particle surface collisions during the free-flight movement of simulation particles. The second aspect involves computation of the "cut volumes" resulting from the intersection of body or domain surfaces with the Cartesian flow field grid. As will be described in the next section, the flow volume of each cell (V_{DSMC}) must be known to determine the correct collision rate. Both the intersection of particles' velocity vectors with arbitrary surfaces and the calculation of cut volumes are purely geometrical problems that can be solved exactly (within machine precision). A variety of algorithms, from a variety of computational fields, are available for such operations (e.g., see Zhang and Schwartzentruber 2012) and they are not described in this text.

The fact that simulation particles must be locally sorted within cells is an aspect of DSMC that merits further discussion.

One approach is to simply move particles for their full timestep, determine their new coordinates, and then map these coordinates to a specific flow field cell. This is depicted in Fig. 6.5(a) for a particle with velocity $\overline{C}$ moving through a uniform Cartesian grid. In this case, the movement is trivial; however, if this type of movement procedure is used on a more complex grid, mapping a set of coordinates to a specific cell within an arbitrary unstructured 3D grid can be computationally expensive. Mapping a set of coordinates within a Cartesian based grid is more computationally efficient, especially if the grid is uniform or has quadtree or octree organization. If the new particle position lies within a solid body or outside the computational domain, then appropriate boundary conditions must be applied. Finally, since most DSMC implementations use domain decomposition for execution on multiple processors, mapping of coordinates to a specific cell may not be trivial if the cell geometry and boundary geometry is partitioned and, therefore, not globally known by each processor.

A second approach is to continually sort particles into cells *during* movement; referred to as "ray-tracing." Ray-tracing is a general and efficient method of tracking particle movement that is widely used in the computer graphics industry and is also used by many DSMC codes. The procedure is depicted schematically in Fig. 6.5(b) for movement within an unstructured 2D triangular grid. However, the same procedure can be used for any grid, including Cartesian-based grids, as shown by the second particle in Fig. 6.5(a). One approach to ray-trace a particle is to compute the time-to-hit each face of the current cell. As depicted in Fig. 6.5(b) this can be computed as $\Delta t_{\mathrm{hit}-f} = \Delta x_f/(\overline{C} \cdot \hat{n}_f)$, where $\hat{n}_f$ is the face unit normal

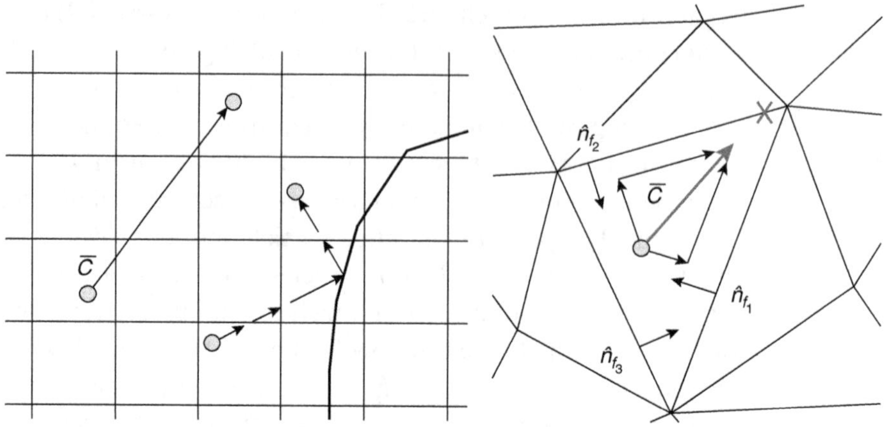

(a) Cartesian movement procedure and "cut cell" treatment. (b) Ray-tracing movement procedure.

Figure 6.5 Particle tracking procedures relevant to the DSMC method.

vector and Δx_f is the normal distance between the particle and the face. The particle is then advanced for the minimum positive time, $\Delta t_{\text{move}} = \min(\Delta t_{\text{hit}-f})$. Since, for most grids, cell/face connectivity is known, after this partial move the particle can be immediately reassigned to the correct neighboring cell. The identical movement process is then repeated for the remaining time ($\Delta t = \Delta t - \Delta t_{\text{move}}$), within the new cell and according to the new cell faces. Of course if the minimum time-to-hit a face is greater than the desired simulation timestep ($\Delta t_{\text{move}} > \Delta t$), then the particle can be moved for the full simulation timestep (Δt) and remain within the current cell. This simple ray-trace procedure is generally applicable to any grid topology.

In addition, if complex triangulated surface geometry is "cut" from a flow field cell (such as in Fig. 6.4(b)), then these surface elements can simply be treated as additional cell faces (i.e., triangular or quadrilateral, planar elements) that are included when calculating $\Delta t_{\text{move}} = \min(\Delta t_{\text{hit}-f})$. In general, if the planar elements are smaller than the cell, these boundary faces may form a complex set of planes within the cell. Since the preceding equation for $\Delta t_{\text{hit}-f}$ calculates only the time-to-hit the *plane* of the face, a further calculation to ensure the particle trajectory actually intersects the face (not just the plane of the face) is required (Zhang and Schwartzentruber 2012).

Finally, since domain decomposition boundaries are tied to cell boundaries, ray-tracing movement naturally detects when a particle is moving between parallel partitions. Thus, the same ray-trace algorithm can be used to move any particle within any cell (including cut cells containing surface geometry) within any type of computational grid. It is noted that for ray-tracing on Cartesian-based grids, the calculations involving surface normal vectors can be made more computationally efficient (fewer operations), since cell faces are always aligned in Cartesian directions.

At this point, it is important to understand that after all particles are moved and sorted locally into cells, all remaining physical models operate only on the list of particles contained within each cell. Thus, the algorithms presented in the remainder of this chapter are to be applied to the particles within a single cell, independently of any other cell in the computation.

6.2.3 Collision Rate

In general, for each computational cell, during each simulation timestep, the number of collisions between classes of molecules, the number of inelastic and reacting collisions, and the outcomes of all such collisions must be determined.

The number of collisions that occur within a DSMC cell during a single timestep (the simulated collision rate) is directly linked to the local transport properties of the gas. In the limit of equilibrium flow, a DSMC simulation should reproduce the equilibrium collision rate of a real gas and therefore reproduce the real gas viscosity. However, for nonequilibrium, a DSMC simulation should ensure that the correct collision rate is applied to molecule pairs of all molecule classes. While the equilibrium collision rate is often known from experimental viscosity measurements, the details of nonequilibrium collision rates often require information from theory of atomistic interactions (a potential energy surface). In fact, for dilute gases, a consistent theory bridges from the interatomic potential between atoms, to the collision cross sections used in DSMC, to the equilibrium transport properties of the gas. This theory was described previously in Chapter 5.

In a gas composed of hard-sphere molecules with diameter d, the probability that two molecules, both located within a volume of gas V, collide during time Δt is proportional to the volume swept out by their interaction normalized by the volume of gas under consideration,

$$P \propto \pi d^2 g \Delta t / V \tag{6.2}$$

where g is the relative velocity of the molecules' center of mass. If the volume V contains N molecules, this probability (Eq. 6.2), applies to all $N(N-1)/2$ possible molecule pairs. Since each DSMC particle represents a large number (W_p) of *identical* real molecules (i.e., all having the same velocity and molecular properties), then the following probability should be applied to all $N_p(N_p-1)/2$ particle pairs within the cell:

$$P_{\text{DSMC}} = (\pi d^2 W_p) g \Delta t_{\text{DSMC}} / V_{\text{DSMC}} \tag{6.3}$$

In this manner, the total number of collisions to be performed within a DSMC cell during a single timestep for a hard-sphere gas is

$$N_{\text{coll}} = \frac{1}{2} N_p (N_p - 1) \frac{(\pi d^2 W_p) \langle g \rangle \Delta t_{\text{DSMC}}}{V_{\text{DSMC}}} \tag{6.4}$$

where $\langle g \rangle$ represents the average relative velocity of all particle pairs in the DSMC cell. If the particles within the cell have a Maxwellian velocity distribution, then the average relative velocity of particle pairs is $\langle g \rangle = 4\sqrt{kT/\pi m}$ (see Eq. B.15 in Appendix B), and therefore the collision rate per particle for hard-spheres in the DSMC cell is,

$$\nu_{\mathrm{DSMC}}^{\mathrm{HS}} = \frac{N_{\mathrm{coll}}}{N_p \Delta t_{\mathrm{DSMC}}} = 2\, W_p\, d^2 \frac{N_p - 1}{V_{\mathrm{DSMC}}} \sqrt{\frac{\pi k T}{m}} \tag{6.5}$$

For a simple gas, in the limit as $W_p \to 1$ (and thus $N_p - 1 \to N - 1 \approx N$, for large N), this is identical to the analytical expression for the hard-sphere collision rate in an equilibrium gas (Z/n in Eq. 1.138). However, it is important to note that since the probability of collision (Eq. 6.3) is applied to each particle pair individually, the DSMC method correctly predicts different collision rates for different particle-pair classes. For example, for hard-sphere particles, pairs with higher relative velocities collide more frequently compared to pairs with lower relative velocities. Thus, in addition to reproducing the correct equilibrium collision rate (Eq. 6.5), DSMC also predicts the correct *nonequilibrium* collision rate and is accurate for an arbitrary (non-Maxwellian) velocity distribution function, found within the interior of a shock wave, for example.

As demonstrated in Chapter 5, the cross section of a molecule pair is not constant ($\sigma \neq \pi d^2$) and, as seen in Fig. 5.3(a), is generally a strong function of the relative velocity ($\sigma = \sigma(d, g)$). Since the average relative velocity of molecule pairs in a gas is proportional to the gas temperature, this significantly alters the temperature dependence of the local collision rate (and therefore transport properties) compared to the hard-sphere result, and must be accounted for. Since DSMC evaluates the probability of collision for each molecule pair individually, it is trivial to extend the hard-sphere procedure to more realistic interactions, simply by allowing the cross section for a molecule pair to include a dependence on relative velocity. The probability applied to each molecule pair now becomes

$$P_{\mathrm{DSMC}} = \sigma(d, g) g\, W_p \Delta t_{\mathrm{DSMC}} / V_{\mathrm{DSMC}} \tag{6.6}$$

and the number of collisions performed within a DSMC cell becomes

$$N_{\mathrm{coll}} = \frac{1}{2} N_p (N_p - 1) \frac{\langle \sigma(d, g) g \rangle\, W_p \Delta t_{\mathrm{DSMC}}}{V_{\mathrm{DSMC}}} \tag{6.7}$$

where $\langle \sigma(d, g) g \rangle$ represents the average of all molecule pairs in the cell. The functional form of $\sigma(d, g)$ can be set arbitrarily within a DSMC simulation, enabling significant flexibility in physical modeling. However, there are a number of specific functional forms, outlined in Section 6.3, that are both physically accurate and numerically efficient. Once a functional form for $\sigma(d, g)$ is specified, the DSMC algorithm for selecting molecule pairs to actually collide within a DSMC cell is very simple, and requires only a few additional considerations related to numerical efficiency. The collision rate algorithm described here is the No-Time-Counter (NTC) algorithm of Bird Bird (1994), which has proven to be a very accurate and efficient algorithm for DSMC.

First, an obvious problem is evident for small numbers of particles per cell (N_p). For example, in a DSMC simulation of a gas at equilibrium that contains only $N_p = 10$, application of Eq. 6.7, will result in an underestimation in the simulated collision rate of 10%, compared to a simulation with $W_p = 1$ (i.e., $N_p = N$), during a single timestep. However, for steady state DSMC simulations, which are sampled over many timesteps, it is the time-averaged collision rate that is important. Indeed, N_p is a fluctuating property within a DSMC simulation that follows a Poisson distribution, for which it can be shown that $\langle N_p(N_p - 1)\rangle = \langle N_p \langle N_p \rangle\rangle$ (e.g., refer to Garcia (2000), pp. 356 and 359). Thus, if averaged over many timesteps during steady state, the application of Eq. 6.7 produces identical results as the the application of

$$N_{\text{coll}} = \frac{1}{2} N_p \langle N_p \rangle \frac{\langle \sigma(d,g)g \rangle\, W_p \Delta t_{\text{DSMC}}}{V_{\text{DSMC}}} \tag{6.8}$$

which is, in fact, proportional to $(N_p W_p)^2 = N^2$ even for small N_p. It is noted that the application of Eq. 6.8 requires the storage and update of an average value of N_p (i.e., $\langle N_p \rangle$) be maintained in each DSMC cell. Thus although the use of Eqs. 6.7 and 6.8 produce identical results, Eq. 6.7 may be preferred owing to computational efficiency.

Experience has shown that for most flows, $N_p > 10$ is sufficient to simulate the local collision rate accurately. For multi-species and chemically reacting flows, higher values of N_p may be required to resolve collision rates for trace species. For the simulation of unsteady flows, the use of either Eq. 6.7 or Eq. 6.8 becomes problematic for small values of N_p since the numerical fluctuations in $W_p N_p$ may become comparable in magnitude to the actual variation in N due to the unsteady nature of the flow. However, typically for unsteady flows, large values of N_p are required to resolve macroscopic properties at each timestep. For large values of N_p the application of either Eq. 6.7 or Eq. 6.8 will be accurate.

Second, the probability applied to each of the $N_p(N_p - 1)$ pairs is typically small. It is common to find that fewer than 20% of the particles within a DSMC cell collide during a single timestep. This fraction can certainly vary substantially; however, it clearly should not exceed 100%. Instead of applying a small probability to $O(N_p^2)$ particle pairs, one can obtain an identical result by applying a larger probability to a smaller number of particle pairs. First, a maximum for the number of particle pairs expected to collide is determined as

$$N_{\text{coll-max}} = \frac{1}{2} N_p (N_p - 1) \frac{[\langle \sigma(d,g)g \rangle]_{\max}\, W_p \Delta t_{\text{DSMC}}}{V_{\text{DSMC}}} \tag{6.9}$$

where $[\langle\sigma(d,g)g\rangle]_{\max}$ is an estimate of the maximum expected value of $\sigma(d,g)g$ within the cell. Since only an integer number of pairs can be collided in a DSMC cell within a single timestep, the number of particle pairs to be tested is set as

$$N_{\text{pairs-tested}} = \text{floor}\,(N_{\text{coll-max}} + 0.5) \tag{6.10}$$

Next, the number must be limited by the actual number of pairs present within the cell. Assuming that each particle can collide only once per timestep, this number must be limited as

$$1 \le N_{\text{pairs-tested}} \le \text{floor}\,(N_p/2) \tag{6.11}$$

Thus, $N_{\text{coll-max}}$ is the number of collision pairs that exactly corresponds to a constant value of $[\langle\sigma(d,g)g\rangle]_{\max}$ applied to all particle pairs (an estimated upper limit), whereas $N_{\text{pairs-tested}}$ is the corresponding number of collision pairs that will actually be tested in the DSMC cell, owing to the discrete nature of the simulation. The ratio of the two values,

$$F_{\text{correction}} = N_{\text{coll-max}}/N_{\text{pairs-tested}} \tag{6.12}$$

must be accounted for at each timestep.

The true collision rate is then simulated by randomly selecting $N_{\text{pairs-tested}}$ particle pairs (where a particle cannot be included in multiple pairs), and accepting each pair for an actual collision with probability

$$P_{\text{DSMC}} = F_{\text{correction}} \times \frac{\sigma(d,g)g}{[\langle\sigma(d,g)g\rangle]_{\max}} \tag{6.13}$$

Thus the product $N_{\text{pairs-tested}} \times P_{\text{DSMC}}$ (the simulated collision rate within a DSMC cell for a single DSMC timestep) reduces exactly to the expression in Eq. 6.7. It is important to note that the quantity $[\langle\sigma(d,g)g\rangle]_{\max}$ cancels from numerator and denominator and thus its exact value does not influence the simulated collision rate at all; rather its value only alters the numerical efficiency of the algorithm (the number of pairs tested). In fact, the most efficient algorithm would have $P_{\text{DSMC}} \approx 1$. In practice, the value of $[\langle\sigma(d,g)g\rangle]_{\max}$ is updated within each cell as the simulation proceeds.

It is certainly possible that a computed value of P_{DSMC} is greater than unity. This implies that the particles in the pair should have collided more than once during the current timestep. In this case, DSMC is unable to simulate the correct collision rate during that timestep and within that cell. Such occurrences should be minimized/eliminated, and the number of such occurrences should be output to the user so that the user is aware of the degree of inaccuracy.

There are a few reasons why a value of $P_{\text{DSMC}} > 1$ may be calculated. One reason is that the estimated value of $[\langle\sigma(d,g)g\rangle]_{\max}$ may be too small. This would lead to a small number of particle pairs being tested for collision, when in fact the actual collision rate (proportional to $\sigma(d,g)g$) dictates that more particle pairs should collide. In a steady state DSMC simulation, it is acceptable if this occurs during the initial transient period. Since the value of $[\langle\sigma(d,g)g\rangle]_{\max}$ is continually updated in each cell, the number of such occurrences should decrease towards zero during the transient, and remain at a very low level during steady state sampling. For unsteady flows, the local collision rate within a cell may be continually changing, making the choice of $[\langle\sigma(d,g)g\rangle]_{\max}$ difficult. In this case, choosing a conservatively large estimate for its value may be necessary with an associated reduction in computational efficiency. Another typical reason why $P_{\mathrm{DSMC}} > 1$ may be calculated is that the timestep is too large. In this case, even if $[\langle\sigma(d,g)g\rangle]_{\max}$ is appropriately set, a large timestep will result in $P_{\mathrm{DSMC}} > 1$, through the $F_{\mathrm{correction}}$ term in Eq. 6.13. As seen in Eq. 6.9, the desired number of collisions to be tested is directly proportional to Δt_{DSMC}. Since the DSMC cell only has $N_p/2$ pairs, a large timestep may imply that more than $N_p/2$ pairs should collide during that timestep within that cell. The simulation, therefore, cannot reproduce the true collision rate since the particles should have undergone multiple collisions during that timestep. As long as the timestep is chosen to be less than the local mean collision time, such occurrences should be infrequent.

Finally, an elegant feature of the NTC method (Eqs. 6.9 through 6.13) is that it remains unaltered for simulations of multispecies gas mixtures. As an example, consider a gas mixture composed of species A and species B particles. In this case, N_p still refers to the total number of particles within the cell, but now N_{A} of these particles are species A and N_{B} of these particles are species B, such that

$$N_p = N_{\mathrm{A}} + N_{\mathrm{B}} \tag{6.14}$$

There are still $N_p(N_p - 1)/2$ possible pairings between particles; however, it can readily be shown that the number of pairs specific to each species combination can be written as

$$\underbrace{\frac{N_p(N_p-1)}{2}}_{\text{total pairs}} = \underbrace{\frac{N_{\mathrm{A}}(N_{\mathrm{A}}-1)}{2}}_{\text{A–A pairs}} + \underbrace{\frac{N_{\mathrm{B}}(N_{\mathrm{B}}-1)}{2}}_{\text{B–B pairs}} + \underbrace{N_{\mathrm{A}}N_{\mathrm{B}}}_{\text{A–B pairs}}\,. \tag{6.15}$$

Recall that, for an equilibrium gas, Eq. 6.4 reduced to the theoretical collision rate for like-species in Eq. 6.5, which was identical to the result for Z/n derived earlier in Eq. 1.138. In the same manner, application of the NTC method to the two component mixture (Eq. 6.15) results in the theoretical collision rate for both like-species and unlike-species given by Eq. 1.138 (Z/n where $\delta_{\mathrm{AA}} = \delta_{\mathrm{BB}} = 1$ and $\delta_{\mathrm{AB}} = 0$). Furthermore, since the statistical behavior of N_{A} and N_{B} are the same as for N_p, the NTC method applied to a multicomponent mixture at steady state will also be accurate for small values of N_{A} and/or N_{B} as discussed previously in reference to Eqs. 6.7 and 6.8. Finally, analogous to the case of a single species, instead of testing all possible particle pairs for a collision, it is accurate and more computationally efficient to test only a maximum-expected number of pairs. In a multispecies simulation, this subset of pairs is still based on the total number of particles within the cell (Eqs. 6.9 through 6.12). Since the pairs are formed randomly, the subsets of A–A, B–B, and A–B pairs will be statistically the same as found in the cell as a whole. It is noted that the maximum-expected cross section, $[\langle\sigma(d, g)g\rangle]_{\text{max}}$, is now the maximum found for any species pair within the cell. Thus, the value is still conservative and, as previously discussed below Eq. 6.13, has no influence on the simulated collision rate. Therefore, the NTC method (Eqs. 6.9 through 6.13) can be applied to gas mixtures with no special treatment or modification. Collisions between the various species pairs will be selected at the correct rate based on the cross section $\sigma(d, g)$ associated with each species pair.

In summary, the preceding simple algorithms enable the DSMC method to accurately simulate equilibrium and nonequilibirum collision rates and ultimately determine which particles within a given DSMC cell should collide during each timestep. The main model required in these algorithms is the cross section of a particle pair and the algorithms are general to gas mixtures. In the upcoming sections, specific functional forms for $\sigma(d, g)$ and scattering angles for post-collision properties will be presented, along with their link to gas transport properties.

6.2.4 Cell and Particle Properties

As a reference for the remainder of the chapter, it is useful to summarize the basic properties (data) associated with simulation particles and collision cells. Since each simulation particle represents an identical number of real molecules, the properties of a simulated particle are the same as those of a real molecule.

Particle properties:

- i (species type)
- x, y, z (position in space)
- C_x, C_y, C_z (center of mass velocity components)
- ϵ_{rot} (rotational energy)
- ϵ_{vib} (vibrational energy)

Species and collision pair properties:

Data is also required for each species i, where the species data required by a simulation is often dependent on the physical models employed. Examples of species data include, the molecular weight (M_w), available rotational and vibrational degrees of freedom ($\zeta_{\text{rot}}, \zeta_{\text{vib}}$), and characteristic temperatures for rotation, vibration, and dissociation processes ($\theta_{\text{rot}}, \theta_{\text{vib}}, \theta_d$). For example, these parameters were routinely used in Chapter 4.

Furthermore, since DSMC models involve collision *pairs*, a number of model parameters, specific to each species pair (i, j), are typically required. Such collision pair data are required for viscosity, diffusion, and thermal conductivity models, as well as for internal energy excitation and chemical reaction models.

It is also useful to associate certain data with each computational cell. A list of typical data stored in each cell includes the following:

Cell properties:

- $type$ (cell type: flow interior, bordering a boundary, etc.)
- $x_0, y_0, z_0, x_E, y_E, z_E$ (bounding vertices for Cartesian grids)
- $(x_v, y_v, z_v)_f$ (each vertex v of cell face f for unstructured grids)
- $(\hat{n}_x, \hat{n}_y, \hat{n}_z)_f$ (normal vector of cell face f for unstructured grids)
- V (cell volume)
- W_p (local particle weight)
- Δt (local timestep)
- $[\langle\sigma(d, g)g\rangle]_{\max}$ (used to evaluate the maximum expected collision rate)
- $(\sum N_p, \sum C_x, \sum C_y, \sum C_z, \sum C_x^2, \sum C_y^2, \sum C_z^2, \sum \epsilon_{\mathrm{rot}}, \sum \epsilon_{\mathrm{vib}})_i$ [cumulative sums of particle properties (samples), for each species (i), used to determine macroscopic flow quantities (refer to Appendix D)]

DSMC calculations can require millions of computational cells and many millions of simulated particles. As a result, DSMC calculations require the storage and bookkeeping of a large amount of data. Complex data structures are often employed for data organization and can provide the flexibility required for a general DSMC implementation. Particle data are the dominant contributor to memory, however, the global number of particles, as well as the local number within each cell, can vary greatly within a DSMC simulation. Thus, statically allocated arrays of particles must either be conservatively large (unused memory) or resized often (computationally and memory intensive). Likewise, since the mesh resolution in DSMC (i.e., the local mean free path) depends on the solution itself, ideally, the number of cells would change during a simulation and data structures should account for this possibility as well. The choice of data structures that best achieves desired flexibility and computational performance is application dependent. DSMC code developers are recommended to consult the recent literature for various approaches to specific classes of applications.

At this point, we can now proceed to describe the physical models that are used to collide pairs of simulated particles within each DSMC cell. The models generally involve the use of both cell and particle data (listed earlier), as well as model parameters specific to each possible species pairing. Ultimately, collision models update the properties of simulated particles involved in collisions in a physically realistic manner. Models that lead to the correct simulation of gas viscosity, thermal conductivity, and diffusion are discussed next.

6.3 Models for Viscosity, Diffusivity, and Thermal Conductivity

The majority of DSMC collision models are specific to the collision pair, for example, species i colliding with species j. In fact, all of the equations and most of the model parameters that appear in this section are specific to a species pairing (i,j). Instead of using subscripts i,j on the model parameters, they are simply implied, and example problems are included to demonstrate their use.

6.3.1 The Variable Hard-Sphere Model

The most widely used DSMC collision cross section model for elastic collisions (no internal energy transfer) is the variable hard-sphere (VHS) model (Bird 1994). The VHS model allows the total cross section ($\sigma_{\mathrm{T}}^{\mathrm{VHS}}$) of an interaction to have a dependence on the relative velocity of the collision pair. As discussed in Chapter 5, the integrated cross section of an interaction has a strong dependence on relative collision velocity and must be considered to predict the correct temperature dependence of the viscosity coefficient, clearly established from experimental measurements. Specifically, the form of $\sigma(d, g)$ corresponding to the VHS model, that could be used in Eq. 6.13, uses a variable diameter,

$$d = d_{\mathrm{ref}} \left(\frac{g_{\mathrm{ref}}}{g} \right)^{\nu} \tag{6.16}$$

and therefore

$$\sigma_{\mathrm{T}}^{\mathrm{VHS}} = \pi d_{\mathrm{ref}}^2 \left(\frac{g_{\mathrm{ref}}}{g} \right)^{2\nu} \tag{6.17}$$

This expression can be viewed as a curve fit to the integrated cross sections obtained from a PES (shown previously in Fig. 5.3(a)). The value of the power-law exponent (ν) can be set to match the slope, and the reference values (d_{ref} and g_{ref}) can be set to "anchor" the magnitude of the curve fit at a specific value of g. As evident from Fig. 5.3(a), a power-law fit is quite accurate over a wide range of relative velocities.

However, accurate PESs and cross section data are unavailable for many gas species of interest. Experimental viscosity data, on the other hand, are widely available for most gases. Without detailed (and validated) cross section data available, it is logical to choose collision model parameters based on viscosity data. Indeed, this is how the majority of DSMC collision models have been developed.

For example, in a simple hard-sphere gas ($\nu = 0$), by rearranging Eq. 5.114, the HS diameter that would achieve a simulated viscosity of μ at a specific temperature T, would be

$$d^2 = \frac{5}{16}\sqrt{\frac{mkT}{\pi}}\left(\frac{1}{\mu}\right) \tag{6.18}$$

Thus, for a uniform temperature flow, a DSMC simulation employing the hard-sphere collision model would accurately simulate the correct gas viscosity, provided d was set using Eq. 6.18. It is important to note, however, that although the correct equilibrium viscosity is simulated (the overall collision rate in the gas), the nonequilibrium collision rate (the collision rate between different parts of the velocity distribution function) is not based on any underlying data, and therefore is not strictly validated.

To derive expressions for the viscosity and diffusion coefficients corresponding to the VHS cross section, it is first noted that the relations between the total cross section and the viscosity and momentum cross sections, derived in Chapter 5 (Eq. 5.112), still hold. Specifically, since σ is not a function of χ for the VHS model, Eqs. 5.103 and 5.104 lead to the result,

$$\sigma_\mu^{\mathrm{VHS}} = \frac{2}{3}\sigma_{\mathrm{T}}^{\mathrm{VHS}} \text{ and } \sigma_{\mathrm{M}}^{\mathrm{VHS}} = \sigma_{\mathrm{T}}^{\mathrm{VHS}} \tag{6.19}$$

On substituting this VHS viscosity cross section (given by Eqs. 6.17 and 6.19) into the equation for the viscosity coefficient (Eq. 5.105), the simulated viscosity corresponding to the VHS model under equilibrium conditions is analytically obtained as

$$\mu^{\mathrm{VHS}} = \frac{\frac{15}{8}\sqrt{2\pi m_r k}\left(\frac{2k}{m_r}\right)^{\nu} T^{1/2+\nu}}{\Gamma(4-\nu)\pi d_{\mathrm{ref}}^2 g_{\mathrm{ref}}^{2\nu}} \tag{6.20}$$

Since the reference values are constants, the VHS model simulates a gas viscosity that has a power law dependence on the gas temperature. In addition to prescribing a reference diameter (d_{ref}) as required by the hard-sphere model, the VHS model requires a reference value for the relative velocity (g_{ref}). As proposed by Bird (1994), a reasonable reference value would be the mean value of g encountered in the collisions occurring in an equilibrium gas at a reference temperature T_{ref}. Although a different reference value could be chosen, this value is physically reasonable and also simplifies the model expressions, since the mean value of g can be written in terms of the gas temperature T at equilibrium. The theory required to derive average quantities found in collisions within a gas is contained in Appendix B. The relevant result for g_{ref} is derived in Eq. B.22, and is given by

$$g_{\mathrm{ref}}^{2\nu} \equiv \left\langle g^{2\nu}\right\rangle_{\mathrm{collisions}}^{\mathrm{VHS}} = \left(\frac{2kT_{\mathrm{ref}}}{m_r}\right)^{\nu}\frac{1}{\Gamma(2-\nu)} \tag{6.21}$$

That is, the mean value of $g^{2\nu}$ averaged over all *collisions* in a VHS gas (with power-law exponent ν) in equilibrium at temperature T_{ref} is given by Eq. 6.21. Here $\Gamma()$ is the Gamma function described in Appendix E. Substituting this *definition* for $g_{\text{ref}}^{2\nu}$ into Eq. 6.20 leads to the final expression for the viscosity simulated by the VHS model:

$$\mu^{\text{VHS}} = \mu_{\text{ref}}^{\text{VHS}} \left(\frac{T}{T_{\text{ref}}} \right)^{\omega} \tag{6.22}$$

where

$$\mu_{\text{ref}}^{\text{VHS}} = \frac{15\sqrt{2\pi m_r k T_{\text{ref}}}}{2(5-2\omega)(7-2\omega)\pi d_{\text{ref}}^2} \tag{6.23}$$

and

$$\omega \equiv \nu + 1/2 = 2/\alpha + 1/2, \tag{6.24}$$

with α being the exponent for the inverse power-law model (Eq. 5.115).

The diffusion coefficient simulated by the VHS collision model can be written in closed form by substituting the momentum cross section for the VHS model (σ_M^{VHS} given in Eqs. 6.17 and 6.19) into Eq. 5.106 (the first approximation for the diffusion coefficient from Chapman–Enskog analysis). The result is

$$D^{\text{VHS}} = \frac{\frac{3}{8}\sqrt{\pi}\left(\frac{2kT}{m_r}\right)^{1/2+\nu}}{\Gamma(3-\nu)\, n\, \pi d_{\text{ref}}^2\, g_{\text{ref}}^{2\nu}} \tag{6.25}$$

Using the definition in Eq. 6.21, this reduces to

$$D^{\text{VHS}} = D_{\text{ref}}^{\text{VHS}} \left(\frac{T}{T_{\text{ref}}} \right)^{\omega} \tag{6.26}$$

where

$$D_{\text{ref}}^{\text{VHS}} = \frac{3\sqrt{2\pi k T_{\text{ref}}/m_r}}{4(5-2\omega)n\pi d_{\text{ref}}^2}. \tag{6.27}$$

Note that for the case of HS cross sections ($\omega = 1/2$), that $\mu_{\text{ref}}^{\text{VHS}}$ reduces to the HS viscosity (Eq. 5.114), and the diffusion coefficient further reduces to

$$D^{\text{HS}} = \frac{3\sqrt{2\pi k T/m_r}}{16 n\pi d_{\text{ref}}^2} \tag{6.28}$$

In addition to the preceding relations for transport coefficients, one can also determine quantities such as the mean collision time (τ_{coll}) and the mean free path (λ) corresponding to the VHS model. The equations for these quantities along with those for other common average quantities can be found in Appendix D for general multispecies gas mixtures.

Finally, the Schmidt number (Sc)

$$\text{Sc} \equiv \frac{\mu}{\rho D} \tag{6.29}$$

is a nondimensional number often used in continuum modeling to relate momentum diffusivity (viscosity) to mass diffusivity. Using the preceding analysis, the corresponding value for the VHS is model is

$$\mathrm{Sc}^{\mathrm{VHS}} = \frac{5}{7 - 2\omega} \tag{6.30}$$

It is important to note that the value of ω is chosen to match the temperature dependence of μ or D, and therefore, the value of Sc is fixed. Indeed, as evident from the preceding equations, the VHS model can match a desired coefficient of viscosity or diffusivity, but not both independently.

At this point, the necessary equations have been derived to perform some DSMC simulations. Specifically, the cross section model given by Eqs. 6.17 and 6.21, can be used to select a number of collision pairs within a cell at each timestep (using Eqs. 6.9–6.11), and then for each pair, performing the collision with the probability given in Eq. 6.13. Random scattering angles are used to assign post-collision particle velocities as described in section C.1.1 of Appendix C.

Example 6.1 Collision Rate in a Simple Gas at Equilibrium

In this example, a uniform simple gas is simulated. Some statistical aspects of the DSMC method become apparent and the simulated collision rate is verified with theoretical expressions.

A small volume of argon gas at rest with a density of $\rho = 8 \times 10^{-5}$ kg/m^3 and a temperature of $T = 500$ K is simulated with DSMC using the NTC collision rate algorithm. The VHS collision model is used where $d_{\mathrm{ref}} = 3.915 \times 10^{-10}$ m, $T_{\mathrm{ref}} = 273$ K, and $\omega = 0.81$. The computational domain consists of a box with dimensions of $5 \times 4 \times 2.5$ mm in x, y, and z coordinate directions, respectively. The domain is divided into 24 collisions cells ($4 \times 3 \times 2$ in x, y, and z directions, respectively).

At the beginning of the simulation, the box is filled with simulation particles corresponding to the gas conditions. Specifically, $N_p = 20$ or $N_p = 20,000$ particles are randomly positioned within each cell and the particle velocities are initially assigned from a Maxwell–Boltzmann distribution (refer to section A.1.3 in Appendix A) corresponding to zero bulk velocity and temperature T. The particle weight (W_p) is set to achieve the desired density ρ. A timestep of $\Delta t_{\mathrm{DSMC}} = 1.427 \times 10^{-7}$ seconds is used and particles that collide with the box walls undergo specular reflection. In this manner, the volume is initialized with a uniform, equilibrium gas which should remain in equilibrium throughout the simulation.

During each timestep, only a fraction of the particles in each cell are selected to undergo collisions using the NTC algorithm (Eqs. 6.9–6.11 and Eq. 6.13). In this example we define the "collision rate fraction" as the number of collisions performed within the box divided by the number of particles within the box. Simulation results for the case where $N_p = 20$ (and therefore 480 particles within the box) are plotted in Fig. 6.6(a). Here, the discrete and

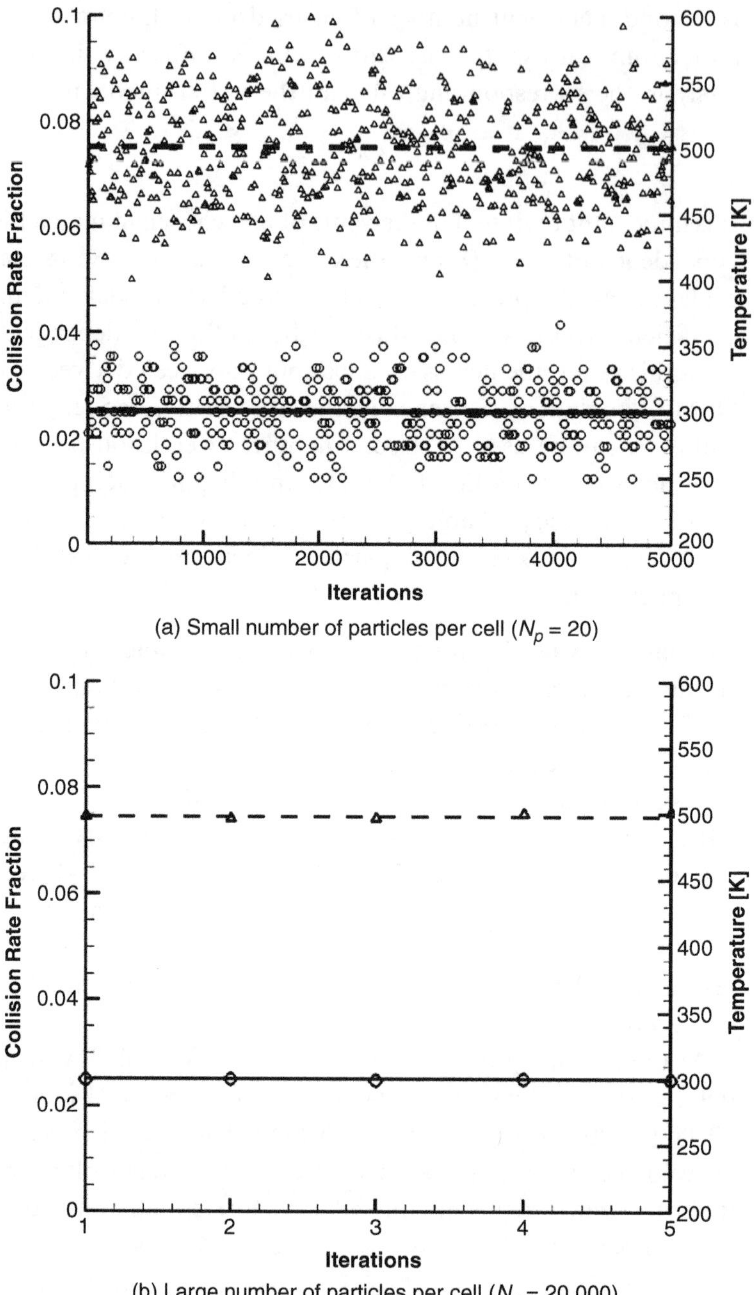

Figure 6.6 Instantaneous collision rate and temperature computed for a uniform, equilibrium, argon gas. Symbols represent quantities calculated at each timestep and lines represent time averaged quantities. Circles and solid line refer to the collision rate fraction. Triangles and dashed line refer to the gas temperature.

statistical nature of the DSMC simulation is evident. The collision rate fraction, plotted at each iteration, is seen to range between 0.01 and 0.04, which corresponds to between 5 and 20 collisions within the box during any single timestep. However, since the gas is in a steady state, the important result is the average of the collision rate fraction "sampled" over many iterations. The average value sampled over 5000 iterations (shown by the solid black line in Fig. 6.6(a)) is equal to 0.0251. The *simulated* collision rate per particle is therefore

$$\nu_{\text{sim}} = \frac{\langle \text{Collision Rate Fraction} \rangle}{\Delta t_{\text{DSMC}}} = 1.759 \times 10^5 \text{ collisions/particle/sec} \tag{6.31}$$

An exact analytical expression for the collision rate corresponding to the VHS model can be obtained using the procedure for the HS model previously derived in Eqs. 1.133–1.138. Specifically for the VHS model, the number of collisions experienced by species A with species B per particle of species A per unit time is

$$\nu_{\text{AB}}^{\text{VHS}} = \frac{n_{\text{B}} \pi d_{\text{ref}}^2}{1 + \delta_{\text{AB}}} \sqrt{\frac{8kT_{\text{ref}}}{\pi m_r}} \left(\frac{T}{T_{\text{ref}}} \right)^{1-\omega} \tag{6.32}$$

For this example, $A = B$, the reduced mass becomes $m_r = m_{\text{Ar}}/2$, and using the specified argon gas properties the collision rate is $\nu_{\text{Ar}}^{\text{VHS}} = 1.752 \times 10^5$ collisions per particle per second. Therefore, the simulated collision rate (ν_{sim}) accurately reproduces the analytical collision rate ($\nu_{\text{Ar}}^{\text{VHS}}$), within approximately 0.4% for this case.

The instantaneous temperature at each timestep (calculated with Eq. 5.17 using all particles in the box) is also plotted in Fig. 6.6(a). The temperature calculated at any single timestep is seen to range significantly from 400 to 600 K. However, the time-averaged "sampled" value for temperature is very close to 500 K as expected.

The results in Fig. 6.6(a) are typical of most steady state DSMC simulations. The collision rate and macroscopic gas properties (such as density, bulk velocity, and temperature) evaluated within each cell at each timestep exhibit large statistical fluctuations. To reduce statistical scatter below a desired tolerance, particle properties must be "sampled" over many iterations to obtain macroscopic properties and molecular distribution functions (refer to Appendix D for more details).

Simulation results for the case where $N_p = 20,000$ (and therefore 480,000 particles within the box) are plotted in Fig. 6.6(b). Since there are now 1000 times more particles in each cell, in order to obtain the same number of samples in the averaged collision rate, only five iterations are required for time averaging. As seen in Fig. 6.6(b), the instantaneous collision rate and temperature, evaluated at each timestep, now has very little statistical scatter due to the large value of N_p. In this case, the average collision rate fraction is found to be 0.0250 which gives a simulated collision rate of $\nu_{\text{sim}} = 1.752 \times 10^5$ collisions/particle/sec, in excellent agreement with the analytical result ($\nu_{\text{Ar}}^{\text{VHS}}$).

This example demonstrates typical statistical scatter present in a DSMC simulation and how particle properties must be either sampled over many timesteps during steady state, or sampled over large N_p. Typically, to limit memory requirements and for overall computational efficiency, it is desirable to minimize N_p. However, as evident in the collision rate results in Fig. 6.6(a), the discrete nature of the simulation sets a lower limit on the value of N_p. For example, if $N_p = 4$, either no particle pairs, one pair, or both pairs collide during a given timestep. With such a large percentage variation in the collision rate, even an average taken over many timesteps may not converge to the correct collision rate. If trace species are present or if rare collision events are important, such as certain energy transitions or chemical reactions, then N_p must be raised appropriately.

Finally, it is important to note that the results presented in this example could be obtained by using only a single collision cell (with constant N_p) and neglecting particle movement and boundary interactions entirely, since a uniform gas is simulated. Such a zero-dimensional approach is useful for rapidly testing collision algorithms. However, including particle movement within separate collision cells introduces fluctuations in N_p and is a more stringent test.

Example 6.2 Normal Shock Wave in Argon

In this example, a normal shock wave in argon gas is simulated using the VHS collision model.

The pre-shock conditions (state 1) are that of a high Mach number ($M_1 = 9$) flow of argon at a temperature of $T_1 = 300$ K and a density of $\rho_1 = 1.069 \times 10^{-4}$ kg/m^3. These conditions match experiments performed by Alsmeyer (1976). Using the jump equations across a normal shock wave, the post-shock conditions (state 2) are $M_2 = 0.456$, $T_2 = 7856$ K, and $\rho_2 = 4.123 \times 10^{-4}$ kg/m^3. The length of the simulation domain is 8 cm, which is approximately equal to $100\lambda_1$ and cells are uniformly sized to be $0.25\lambda_1$.

The computational domain was initialized with particles drawn from a Maxwell–Boltzmann velocity distribution function (VDF) corresponding to state 1 for $x < 4$ cm and corresponding to state 2 for $x > 4$ cm, with appropriate number densities. The particle weight, W_p, was set to obtain approximately $N_p = 1000$ particles in each cell in the freestream region. Before each timestep, any particles residing in either the first 10 cells or the final 10 cells of the domain were deleted and regenerated from Maxwell–Boltzmann distributions corresponding to state 1 and state 2 respectively. This technique enforces pre- and post-shock boundary conditions. The basic DSMC algorithm is iterated with timesteps that are a fraction of the freestream mean collision time, τ_1. The steady-state shock wave profile requires approximately $20\tau_1$ to develop. After this initial transient period, the solution is sampled during a further simulation time of $5\tau_1$.

It is noted that this simulation is highly resolved, and that accurate solutions can be obtained with fewer particles and larger cell sizes. However, for normal shock wave simulations using this technique, it is desirable to use a

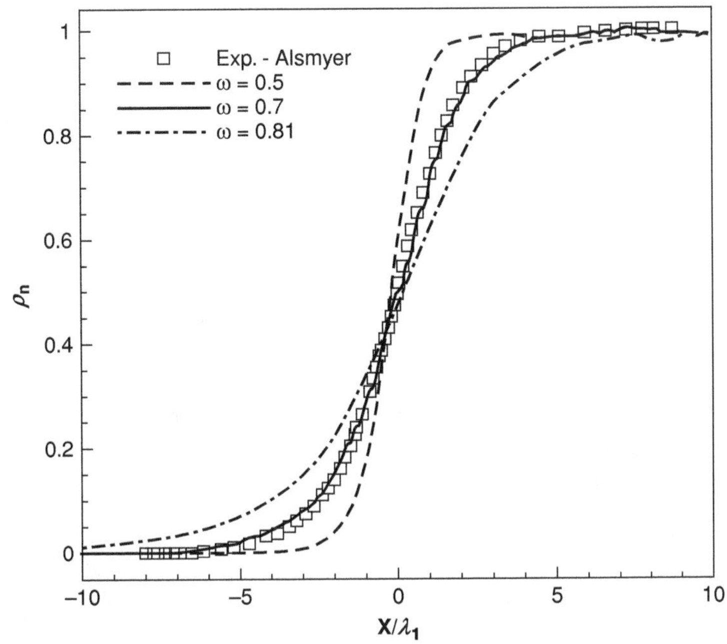

Figure 6.7 Normalized density profiles in a Mach 9 argon shock wave.

large number of particles per cell and sample the solution over as few iterations as possible, to avoid any movement of the shock wave (during sampling) caused by statistical fluctuations. This is discussed in more detail in Bird (1994), in addition to other strategies for simulating normal shock waves. The technique described previously, however, is accurate and simple to implement for many shock wave conditions.

The collision parameters used for the simulation are, $d_{\text{ref}} = 3.974$ Å, $T_{\text{ref}} = 273$ K, and three different values of the viscosity law exponent, $\omega = 0.5, 0.7, 0.81$ are used. Shock wave profiles are typically plotted in normalized variables,

$$q_n(x) \equiv \frac{q(x) - q_1}{q_2 - q_1}$$

where q is a flow property. The normalized density profiles resulting from the three simulations are shown in Fig. 6.7, where the VHS model using $\omega = 0.7$ best matches the experimental data of Alsmeyer. More comparisons can be found in Valentini and Schwartzentruber (2009b). Note that the simulation using $\omega = 0.5$ corresponds to a hard-sphere collision model and predicts a shock wave that is much thinner than the experimental result. This is expected, since the VHS model more accurately captures the reduction in cross section for higher relative collision velocities (experienced in this temperature range). The reduction in cross section enables both pre- and post-shock molecules to be transported further (before colliding), therefore resulting in a thicker shock compared to the hard-sphere result.

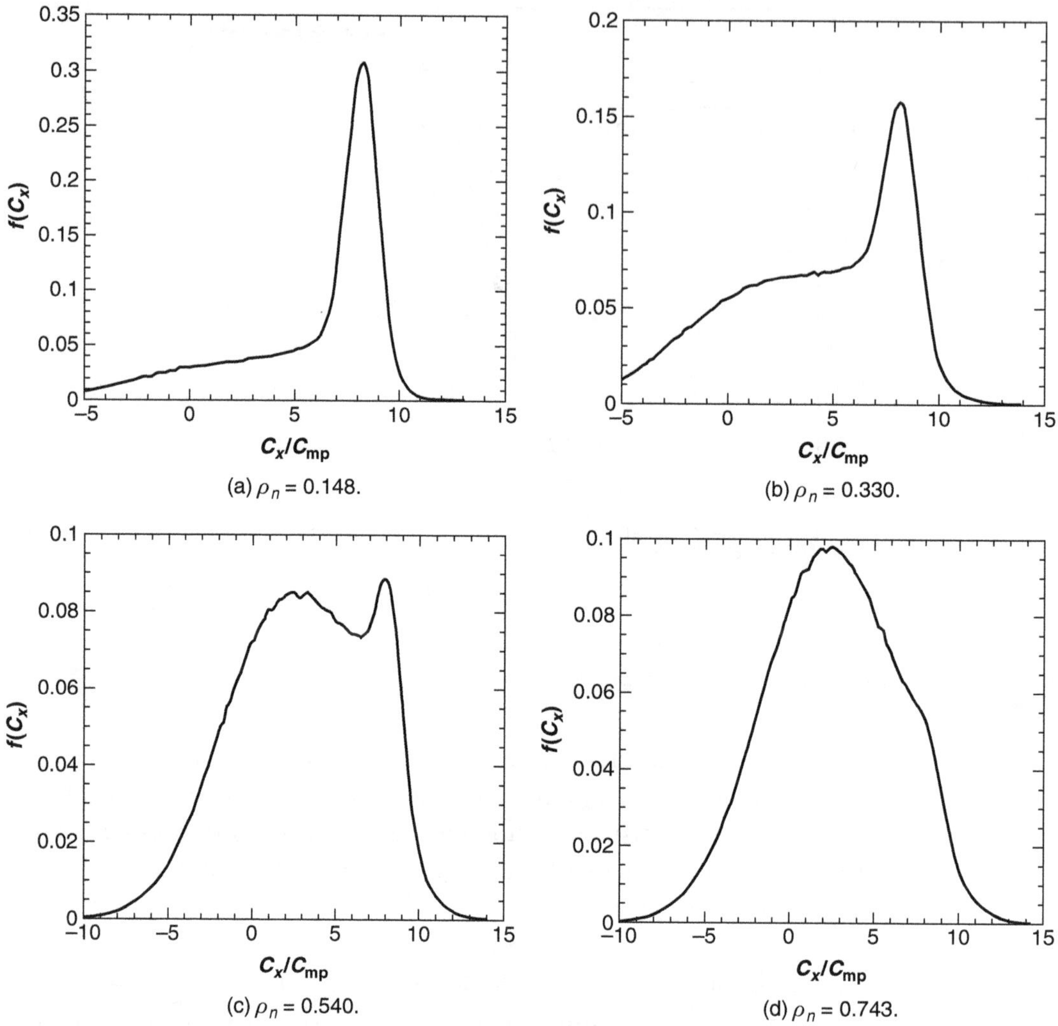

Figure 6.8 Distribution functions for x-velocity within a normal shock wave in argon.

The x-velocity distribution functions at four locations within the shock wave (simulated using $\omega = 0.7$) are shown in Fig. 6.8. In the upstream region of the shock, the VDF (Fig. 6.8(a)) is seen to have a narrow peak at high velocity, corresponding to the cold but high-speed freestream flow. However, a low-velocity tail is beginning to appear that results from a finite number of post-shock molecules propagating upstream. The VDFs in the center of the shock (Figs. 6.8(b) and 6.8(c)), exhibit a “bimodal” mixture of the pre- and post-shock VDFs. Finally, in the downstream portion of the shock (Fig. 6.8(d)), the VDF is approaching a Maxwell–Boltzmann distribution corresponding to state 2.

Example 6.3 Normal Shock Wave in a Mixture of Helium and Xenon
In this example, a normal shock wave is simulated for a gas mixture composed of two species with very different molecular weights, helium and xenon, for

Table 6.1 VHS Model Parameters

Collision Partners	ω	d_{ref} [Å]	T_{ref} (K)
He–He	0.66	2.33	273
Xe–Xe	0.85	5.74	273
He–Xe	0.755	4.035 (3.65 modified)	273

which diffusive transport must be considered. The flow conditions match experiments performed by Gmurczyk, Tarczynski, and Walenta (1979) and further analysis of this problem can be found in Valentini, Tump, Zhang, and Schwartzentruber (2013).

Specifically, the pre-shock conditions (state 1) are that of a moderate Mach number ($M_1 = 3.61$) mixture of helium (98.5% by mole) and Xenon (1.5% by mole) at a temperature of $T_1 = 300$ K and a mixture density of $\rho_1 = 8.0 \times 10^{-5}$ kg/m^3. Using the jump equations across a normal shock wave, the post-shock conditions (state 2) are $M_2 = 0.5$, $T_2 = 1480$ K, and $\rho_2 = 2.6 \times 10^{-4}$ kg/m^3.

As in the previous example problem, the length of the simulation domain is 8 cm. Owing to the disparate masses of He and Xe, and also due to the small mole fraction of Xe, this shock wave simulation is more computationally expensive than the previous example problem for argon. The particle weight, W_p, was set to obtain approximately $N_p = 4000$ particles in each cell in the freestream region. This results in, on average, 3940 He particles and 60 Xe particles in each cell in the freestream. The same boundary condition and iteration procedures are performed as described in the previous example problem. Since the freestream mean free path (λ_1) and mean collision time (τ_1) are that of the *mixture*, they are not representative of each species individually. Based on mixture properties, a transient period of $1000\tau_1$ was allowed, after which the solution was sampled during a further simulation time of $100\tau_1$.

As described in Chapter 5, an interatomic potential can be used to determine viscosity and diffusion coefficients through Eqs. 5.86 and 5.87 which are functions of the collision integrals. The collision integrals resulting from the LJ 12-6 potential were plotted previously in Fig. 5.3(b) and the viscosity and diffusion coefficients resulting from Eqs. 5.86 and 5.87 are now plotted in Fig. 6.10. The LJ potential parameters for He–He, Xe–Xe, and Xe–He pairs, were listed previously in Table 5.1. As described in Valentini et al. (2013), these LJ parameters result in mixture viscosity and diffusion coefficients that agree well with experimental measurements. Furthermore, pure molecular dynamics (MD) simulations of this normal shock wave were performed Valentini et al. (2013), and are reproduced in Fig. 6.9. For this example, these MD simulation results, using the LJ 12-6 potential, are taken as an accurate baseline solution to which DSMC results using the VHS model are now compared.

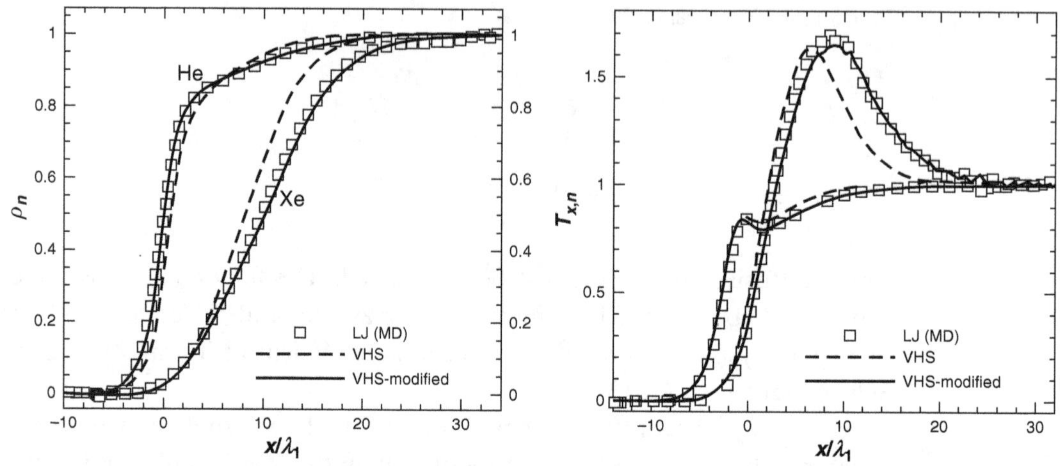

(a) Density profiles. (b) Temperature (x-component only) profiles.

Figure 6.9 Normalized He and Xe profiles for a Mach 3.61 normal shock wave (1.5% Xe and 98.5% He) predicted by DSMC and pure MD simulation.

The VHS model parameters are listed in Table 6.1. Specifically, the resulting VHS viscosity coefficients $\mu_{\text{He-He}}$ and $\mu_{\text{Xe-Xe}}$ match the viscosity of Helium and Xenon at room temperature, and also match well the trend with temperature. The ω parameter for Xe–He collision pairs is set as the average of the He–He and Xe–Xe values, however, two different values are used for d_{ref}. The first value for d_{ref} is simply the average of the He–He and Xe–Xe values. The second value for d_{ref} is chosen to better match the diffusion coefficient predicted by the LJ potential. Figure 6.10 plots the viscosity and diffusion coefficients for the cross-species pair (He–Xe) determined from the LJ potential, and both sets of VHS parameters. When the VHS parameters for the cross-species pair is taken as the average of the single-species pairs, there is a noticeable discrepancy in both viscosity and diffusion coefficients compared to the LJ result. However, if the value of d_{ref} (specific to He–Xe) is reduced from 4.035 to 3.65, the modified VHS value for the diffusion coefficient comes into close agreement with the LJ result over the full temperature range experienced in the shock wave. Note that this modification to match the diffusion coefficient ($\mathscr{D}_{\text{He-Xe}}(\text{T})$) does not necessarily increase the agreement for the viscosity coefficient ($\mu_{\text{He-Xe}}$).

DSMC solutions using both VHS models are shown in Fig. 6.9 and compared to the pure MD solution using the LJ potential. In Fig. 6.9(a), the numerical solutions predict a separation in the species densities within the shock. The lighter He atoms "experience" the shock first since their velocities are significantly affected by both collisions with Xe and with other He atoms. Whereas, the heavier Xe atoms are affected significantly only by collisions with other Xe atoms. Such species separation within shock waves has been experimentally documented (for example see Gmurczyk et al. (1979)) and the LJ solution in Fig. 6.9(a) has been verified to agree closely with the experimental result for these shock conditions (Valentini et al., 2013).

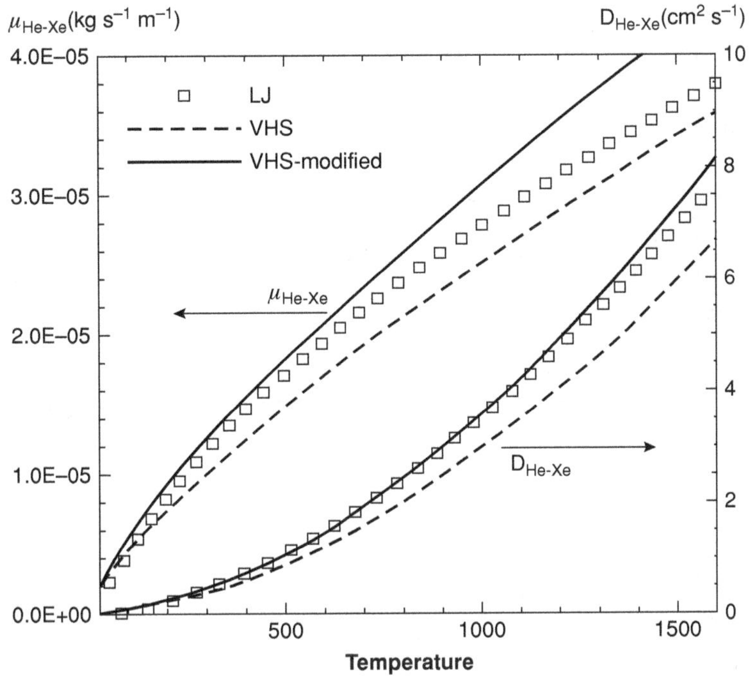

Figure 6.10 Binary viscosity and diffusion coefficients, for He–Xe interactions, corresponding to various models.

As seen in Fig. 6.9(a), the profiles from the modified VHS model (where d_{ref} for He–Xe collisions is set to match the diffusion coefficient) are in much better agreement with the pure MD result. This modification has the effect of increasing the diffusion coefficient and therefore leads to a larger species separation within the shock as predicted by pure MD. To demonstrate improved agreement beyond the density profiles, Fig. 6.9(b) shows that the VHS solution with modified parameters is in much better agreement with MD for the x-translational temperature (T_x) profiles. Since T_x is a measure of the standard deviation of the x-velocity distribution function, it is significantly affected by the bimodal nature of the VDF, and is a more sensitive parameter than the density.

Thus, by matching the cross-species VHS parameters with a known binary diffusion coefficient, $\mathscr{D}_{\text{He-Xe}}(T)$ (in this case the LJ result), the VHS model can very accurately simulate this stringent test case involving disparate species masses and concentrations within a strongly nonequilibrium flow involving a wide temperature range. In general, this strategy of choosing VHS parameters for like-species from viscosity data and for unlike-species from diffusion data is quite accurate for a wide range of nonequilibrium flows. It is important to note that while the VHS parameters in Table 6.1 are quite accurate in the temperature range considered, they may be inaccurate at temperatures above or below this range. Researchers using DSMC should carefully determine VHS parameters based on the best available data for viscosity or diffusion coefficients from the appropriate temperature range, on a case-specific basis.

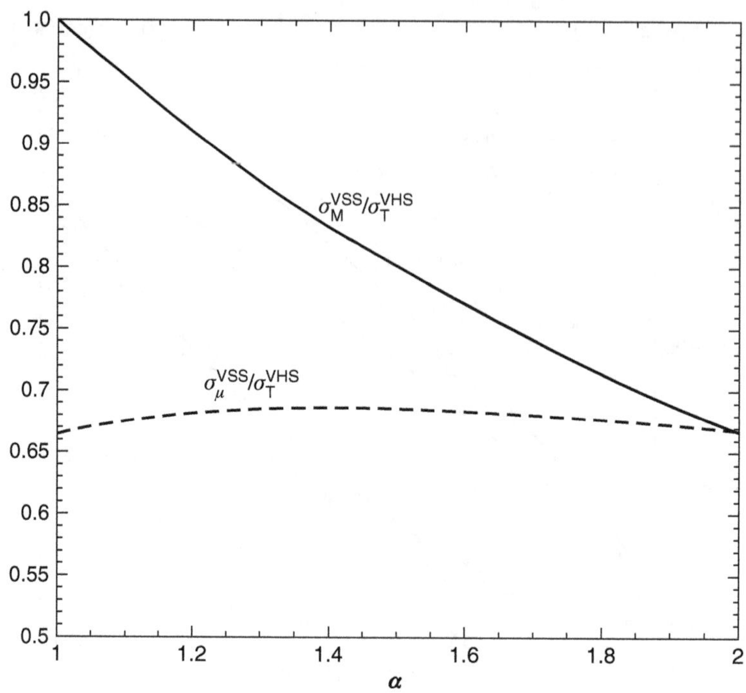

Figure 6.11 Viscosity and momentum cross sections corresponding to the VSS model, in function of α.

6.3.2 The Variable Soft-Sphere Model

As discussed above, the VHS model parameters for a specific species-pair (i,j) can be set to obtain a desired viscosity coefficient μ_{ij} or a diffusion coefficient D_{ij}, but not both independently. This is evident from Eqs. 6.22 and 6.26. The variable soft sphere model was proposed by Koura and Matsumoto (1991, 1992), which includes a more realistic scattering model compared to the isotropic, hard-sphere scattering used in the VHS model. As a result, the VSS model is able to accurately reproduce the correct ratio between the viscosity and momentum cross sections. The VSS model employs the same total cross section model as the VHS model (Eq. 6.17); however, the scattering angle is determined by

$$\chi = 2\cos^{-1}\left[\left(\frac{b}{d}\right)^{(1/\alpha)}\right] \tag{6.33}$$

Thus, for the VSS model, using Eqs. 5.103 and 5.104, the viscosity and momentum cross sections become

$$\sigma_\mu^{VSS} = \frac{4\alpha}{(\alpha+1)(\alpha+2)}\sigma_T^{VHS} \text{ and } \sigma_M^{VSS} = \frac{2}{(\alpha+1)}\sigma_T^{VHS} \tag{6.34}$$

where σ_T^{VHS} is given in Eq. 6.17. The variation of viscosity and momentum cross sections with the parameter α is shown in Fig. 6.11 for the typical range $1 < \alpha < 2$. Clearly, the viscosity cross section is weakly dependent on α, whereas the momentum cross section is strongly dependent on α.

On substitution into the equations for the first coefficient of viscosity and diffusion (Eqs. 5.105 and 5.106), the resulting expressions differ from the VHS viscosity and diffusion coefficients only by a constant. The expressions are

$$\mu^{\rm VSS} = \frac{\frac{5}{16}(\alpha+1)(\alpha+2)\sqrt{2\pi m_r k}\left(\frac{2k}{m_r}\right)^{\nu} T^{1/2+\nu}}{\alpha\Gamma(4-\nu)\pi d_{\rm ref}^2 g_{\rm ref}^{2\nu}} \tag{6.35}$$

and

$$D^{\rm VSS} = \frac{\frac{3}{16}(\alpha+1)\sqrt{\pi}\left(\frac{2kT}{m_r}\right)^{1/2+\nu}}{\Gamma(3-\nu)\, n\, \pi d_{\rm ref}^2\, g_{\rm ref}^{2\nu}} \tag{6.36}$$

It can be shown that the mean value of $g_{\rm ref}^{2\nu}$ found in collisions in a VSS gas is identical to that in a VHS gas, and therefore the same definition (Eq. 6.21) is employed here. The resulting expressions for viscosity and diffusion coefficients are still given by Eqs. 6.22 and 6.26, however, with different reference constants of:

$$\mu_{\rm ref}^{\rm VSS} = \frac{5(\alpha+1)(\alpha+2)\sqrt{2\pi m_r k T_{\rm ref}}}{4\alpha(5-2\omega)(7-2\omega)\pi d_{\rm ref}^2} = \frac{(\alpha+1)(\alpha+2)}{6\alpha}\mu_{\rm ref}^{\rm VHS} \tag{6.37}$$

and

$$D_{\rm ref}^{\rm VSS} = \frac{3(\alpha+1)\sqrt{2\pi k T_{\rm ref}/m_r}}{8(5-2\omega)n\pi d_{\rm ref}^2} = \frac{(\alpha+1)}{2}D_{\rm ref}^{\rm VHS} \tag{6.38}$$

The corresponding Schmidt number becomes

$$S_{\rm c}^{\rm VSS} = \frac{5}{3(7-2\omega)}\left(\frac{\alpha+2}{\alpha}\right) \tag{6.39}$$

As an example, for typical values of $\omega = 0.75$ and $\alpha = 1.5$, Sc $= 0.71$, which is in the appropriate range compared to experimental measurements in many dilute gases. Note that when $\alpha = 1$, these expressions reduce to the VHS expressions, and if in addition $\omega = 1/2$, they reduce to the hard-sphere expressions.

At this point, it is again important to stress that all of the preceding equations are specific to a collision pair (species i and j). Generally speaking, for pairs of like-species, the viscosity coefficient (μ_{ii}) is routinely available from experimental measurements, however the self-diffusion coefficient (D_{ii}) requires experiments using different isotopes of the same species, and is less widely available. For pairs of unlike-species, the diffusion coefficient (D_{ij}) can be interpreted from experimental measurements in two-component mixtures, however, interpreting the viscosity (μ_{ij}) is difficult since contributions from collisions between like- and unlike-species are difficult to separate. For this reason, the strategy employed in Example 6.3, using the

VHS model parametrized with viscosity for like-species and diffusivity for unlike-species, is quite reasonable and accurate for most cases. On the other hand, if mass and momentum transport information for all like-species and unlike-species is available, then the VSS model can be parameterized to better fit the data.

To parameterize the VSS model, in addition to determining values for ω and d_{ref}, one must also determine a value for α. The value of ω is chosen to match either the power-law dependence of the cross sections due to relative velocity and/or the power-law dependence of the transport coefficients on temperature. Then, since the viscosity cross section is typically a weak function of α (see Fig. 6.11), an initial estimate for alpha (typically $1 < \alpha < 2$) can be assumed, and a corresponding value of d_{ref} can be chosen to match either viscosity cross section data at a reference relative velocity, or viscosity coefficient data at a reference temperature. Then using this value of d_{ref}, an improved value of α could be determined to match momentum cross section data or diffusion coefficient data. If required, the process could be iterated and a least squares method could be employed.

It is important to note that the VSS viscosity and diffusion coefficients are still limited to the specific α dependence in the above equations. Furthermore, the VSS model assumes that α is a constant, when it is possible that its value depends on temperature. Finally, recall that both the VHS and VSS models assume a constant exponent (ω) for the power-law dependence of transport properties with temperature; a limitation that often prevents a single model parameterization from being accurate over a wide temperature range. Thus, similar to the VHS model, the VSS model must be carefully parameterized according to the transport information available for the temperature range of interest, and in some cases may provide more flexibility compared to the VHS model.

6.3.3 Generalized Hard-Sphere, Soft-Sphere, and LJ Models

The generalized hard-sphere (GHS) model was introduced by Hash and Hassan (1993), to enable a VHS-type model to include the attractive portion of the PES and thus a nonconstant power-law exponent for the transport properties.

In the GHS model, the total cross section is given by a sum of VHS-like cross section terms, each with a different power-law exponent:

$$\frac{\sigma_T}{s_0^2} = \sum \alpha_j \left(\frac{\epsilon_{\text{tr}}}{\epsilon}\right)^{-\psi_j} \tag{6.40}$$

where ϵ_{tr} is the translational energy associated with the collision pair,

$$\epsilon_{\text{tr}} = \frac{1}{2} m_r g^2 \tag{6.41}$$

and s_0^2, ϵ, α_j, and ψ_j are free model parameters that can be chosen based on aspects of a potential energy surface and/or transport property information.

On substitution into the equations for the first coefficient of viscosity and diffusion (Eqs. 5.105 and 5.106), the resulting expressions are

$$\mu^{\text{GHS}} = \frac{15\sqrt{2\pi m_r kT}}{8 s_0^2 \sum \alpha_j \Gamma(4-\psi_j)(kT/\epsilon)^{-\psi_j}} \tag{6.42}$$

and

$$D^{\text{GHS}} = \frac{3\sqrt{2\pi kT/m_r}}{8 s_0^2\, n \sum \alpha_j \Gamma(3-\psi_j)(kT/\epsilon)^{-\psi_j}} \tag{6.43}$$

Unlike for the VHS and VSS models, a reference diameter or reference cross section (i.e., reference relative velocity) cannot be defined for the GHS model and values for all of the GHS model parameters must be specified for each species pair. Although more parameters must be specified, in practice this allows more general viscosity and diffusion laws to be modeled (i.e., curve fit). As with the VHS and VSS models, the parameters may be related to (or fit to) features of the PES, viscosity and momentum cross section data, or viscosity and diffusion coefficient data.

Many interaction potentials contain two terms, one contributing to the repulsive force at short separation distances and a second term that models the weak attractive force at larger separation distances, for example,

$$\psi(r_{ij}) = \frac{\kappa}{r_{ij}^{\eta-1}} - \frac{\kappa'}{r_{ij}^{\eta'-1}} \tag{6.44}$$

However, there is no closed form expression for the viscosity or diffusion coefficient (recall the integrated results for the LJ 12-6 potential were shown previously in Fig. 5.3).

The viscosity and diffusion coefficient expressions that correspond to the GHS total cross section model with two terms are obtained directly from Eqs. 6.42 and 6.43, as

$$\mu^{\text{GHS}} = \frac{(5/8)\sqrt{2\pi m_r kT}/s_0^2}{(\alpha_1/3)\Gamma(4-\psi_1)(kT/\epsilon)^{-\psi_1} + (\alpha_2/3)\Gamma(4-\psi_2)(kT/\epsilon)^{-\psi_2}} \tag{6.45}$$

and

$$D^{\text{GHS}} = \frac{(3/8)\sqrt{2\pi kT/m_r}/(s_0^2\, n)}{\alpha_1\Gamma(3-\psi_1)(kT/\epsilon)^{-\psi_1} + \alpha_2\Gamma(3-\psi_2)(kT/\epsilon)^{-\psi_2}} \tag{6.46}$$

Furthermore, the Schmidt number could be calculated using Eq. 6.29. Since there is no closed-form expression for the viscosity coefficient for a general PES, the GHS parameters cannot be related to a PES in analytic closed form. However, with certain assumptions the exponent parameters in the GHS cross section expression (and corresponding viscosity and diffusion coefficient expressions) can be related to the exponents in the PES expression.

WEAK ATTRACTIVE INTERACTIONS (GHS-WEAK)

As shown in Chapman and Cowling (1952), if $\kappa' \ll \kappa$ (the case where the attractive force is much weaker than the repulsive force), the viscosity coefficient can be approximated by

$$\mu = \mu_{\text{repulsive}} \left(1 + \frac{S}{T^{(\eta-\eta')(\eta-1)}}\right)^{-1} \tag{6.47}$$

where $\mu_{\text{repulsive}}$ is the viscosity coefficient neglecting the attractive portion of the PES, η and η' are the exponents appearing in the potential function (Eq. 6.44), and S is a constant.

Starting with the two-term viscosity coefficient for the GHS model (Eq. 6.45), a similar functional form as Eq. 6.47 can be obtained by rearranging as

$$\mu = \frac{15\sqrt{2\pi m_r kT}(kT/\epsilon)^{\psi_1}}{8\alpha_1\Gamma(4-\psi_1)s_0^2}\left(1 + \frac{\alpha_2\Gamma(4-\psi_2)}{\alpha_1\Gamma(4-\psi_1)}(kT/\epsilon)^{\psi_1-\psi_2}\right)^{-1} \tag{6.48}$$

In this manner, the temperature exponent in the first term (ψ_1) can be related to the temperature exponent in the viscosity coefficient obtained when only the strong repulsive portion of the PES is considered. This was the case for both VHS and VSS models and so by comparing with Eq. 6.22

$$\psi_1 = 2/(\eta - 1) \tag{6.49}$$

Likewise, the temperature exponent in the second term of Eq. 6.48 can be compared to the second term in Eq. 6.47 to obtain

$$\psi_2 = (2 + \eta - \eta')/(\eta - 1) \tag{6.50}$$

Thus, under the assumption of a weak attractive interaction, the GHS parameters corresponding to the Lennard–Jones 6-12 potential (where $\eta = 13$ and $\eta' = 7$) would be $\psi = 1/6$ and $\psi' = 2/3$.

STRONG ATTRACTIVE INTERACTIONS (GHS-STRONG)

For the case where the attractive interaction is *not* assumed to be weak, Kunc, Hash, and Hassan (1995) derive a different expression for ψ_2. By analyzing exact viscosity integrals for the two-term PES in Eq. 6.44 (such as the Lennard–Jones results shown in Fig. 5.3), it can be shown that, at high temperatures, the viscosity coefficient has the following dependence:

$$\mu \propto T^{\frac{1}{2}+\frac{2}{\eta-1}} \tag{6.51}$$

whereas, at low temperatures, the attractive portion of the PES is dominant, and the viscosity coefficient has the dependence:

$$\mu \propto T^{\frac{1}{2}+\frac{2}{\eta'-1}} \tag{6.52}$$

As developed in Kunc et al. (1995), a potentially better choice for the exponents appearing in the GHS cross section expression are therefore

$$\psi_1 = 2/(\eta - 1) \quad \text{and} \quad \psi_2 = 2/(\eta' - 1) \tag{6.53}$$

For either the weak or strong GHS models, the parameters ϵ and s_0 are typically set equal to the energy minimum of the PES and the separation distance corresponding to this minimum, respectively (refer to Eq. 5.126 and accompanying discussion). With ψ_1, ψ_2, ϵ, and s_0, chosen, the remaining free parameters (α_1 and α_2) can then be determined by a least-squares fit to transport property data.

However, to gain some generality, the remaining parameters could also be fit using results of the collision integrals (instead of the transport coefficients). Using the definition that relates the transport coefficients to the collision integrals (Eqs. 5.86 and 5.87), combined with the GHS expressions for the transport properties (Eqs. 6.42 and 6.43), one can write the collision integrals corresponding to the GHS model:

$$\Omega^{(2,2)^*} = \frac{1}{6\pi}\left[\alpha_1\Gamma(4-\psi_1)\left(\frac{\epsilon}{kT}\right)^{\psi_1} + \alpha_2\Gamma(4-\psi_2)\left(\frac{\epsilon}{kT}\right)^{\psi_2}\right] \tag{6.54}$$

and

$$\Omega^{(1,1)^*} = \frac{1}{2\pi}\left[\alpha_1\Gamma(3-\psi_1)\left(\frac{\epsilon}{kT}\right)^{\psi_1} + \alpha_2\Gamma(3-\psi_2)\left(\frac{\epsilon}{kT}\right)^{\psi_2}\right] \tag{6.55}$$

where

$$\Omega^{(l,s)^*} \equiv \frac{\Omega^{(l,s)}}{\Omega^{(l,s)}_{\text{HS}}} \tag{6.56}$$

with the numerator and denominator given previously in Eqs. 5.88 and 5.113, respectively. Exact collision integrals have been computed using the Lennard–Jones potential (analogous to those plotted in Fig. 5.3(b)) and the results are tabulated in Chapman and Cowling (1952). Thus, for LJ-type potentials, the GHS parameters ψ_1, ψ_2, ϵ, and s_0 can be chosen consistent with the PES and the remaining parameters (α_1 and α_2) can be determined from a least-squares fit to the exact collision integrals. By fitting to the collision integral results, instead of the transport properties, the parameters determined are general to any molecule species pair (like or unlike) as long as their interaction is given by the two-term PES in Eq. 6.44.

Finally, although the GHS model was originally motivated by inclusion of the attractive portion of typical PES, with model parameters set based on the PES, in practice, the additional free parameters introduced by the GHS model simply enable more general cross section expressions to be formulated. This, in turn, gives further control over viscosity and diffusion properties,

Table 6.2 GHS-Weak Model Parameters for Argon

ϵ/k (K)	s_0 [Å]	ψ_1	ψ_2	α_1	α_2
124	3.418	1/6	2/3	3.85	3.10

as well as control over the temperature dependence of transport properties. If necessitated by the application, the GHS model could therefore be used to model more general transport property relations than the VHS and VSS models.

Example 6.4 "Cold" Normal Shock Wave in Argon

In this example, we simulate a normal shock wave in argon where the freestream temperature is very low. As a result, the attractive force between atoms may be important in the upstream shock region, the shock involves a large variation in temperature, and the shock interior is strongly nonequilibrium; a relevant test case for DSMC using the GHS collision model.

Specifically, the pre-shock conditions (state 1) are that of high Mach number ($M_1 = 10$) argon flow at temperature of $T_1 = 20$ K and a density of $\rho_1 = 7.6 \times 10^{-5}$ kg/m^3. Using the jump equations across a normal shock wave, the post-shock conditions (state 2) are $M_2 = 0.45$, $T_2 = 642$ K, and $\rho_2 = 2.95 \times 10^{-4}$ kg/m^3. Experiments by Alsmeyer (1976) measured shock wave structure under very similar conditions, but at Mach 7.183. A separate study investigated these conditions using pure MD (LJ 12-6) and DSMC with the VHS model (Valentini and Schwartzentruber 2009b), where no noticeable difference was found between MD and DSMC-VHS solutions. The current Mach 10 conditions induce a larger temperature variation and, in this example, we compare DSMC solutions using GHS and VHS models to see if modeling the attractive portion of the interatomic potential produces noticeable effects.

Since the flow conditions are similar to those used for Example 6.2, the DSMC simulation setup is identical to that used in Example 6.2, and is not repeated here. The VHS model parameters used in this example are also identical those used in Example 6.2, since they were found to give excellent agreement with experiment for the Mach 9 shock wave (where $T_1 = 300$ K). Specifically, the VHS parameters are $\omega = 0.7$, $d_{\text{ref}} = 3.974 \times 10^{-10}$ m, and $T_{\text{ref}} = 273$ K. The GHS parameters are listed in Table 6.2, where the GHS-weak assumption is employed. For this example, the GHS parameters were fit to reproduce the LJ 12-6 PES viscosity result over the range $20\text{ K} < T < 600\text{ K}$ (the LJ collision integral was plotted in Fig. 5.3(b)).

The viscosity coefficients resulting from the VHS and GHS models are plotted in Fig. 6.12, where they are compared with the result from the LJ PES. All three models produce a viscosity at 273 K in close agreement with the experimentally accepted value of $\mu = 2.1 \times 10^{-5}$ kg/m/sec. However, since the VHS model is restricted to a single power-law exponent (ω), it is not able

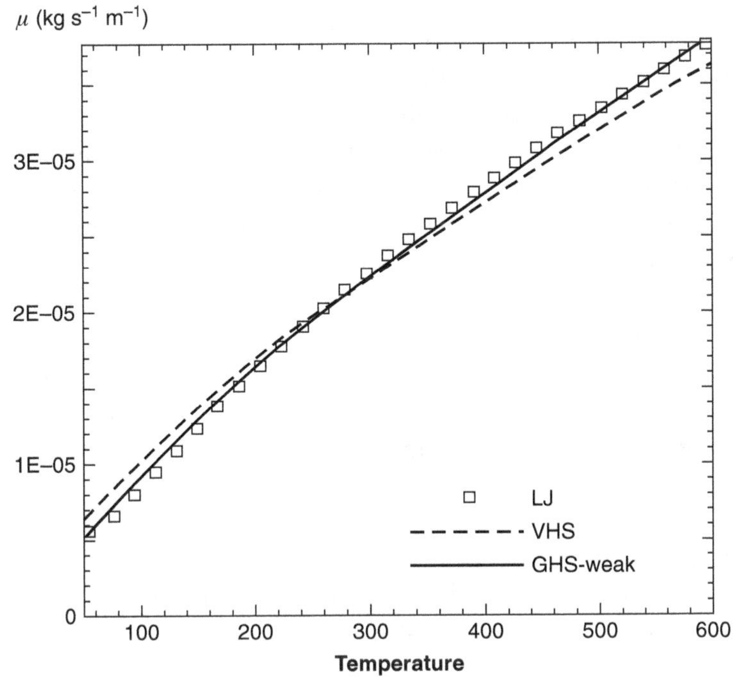

Figure 6.12 Viscosity coefficients for argon corresponding to various models.

to precisely match the temperature variation of the LJ result, whereas, the GHS model is.

As seen in Fig. 6.13, there is very little difference between VHS and GHS predictions for the density and temperature profiles through the shock.

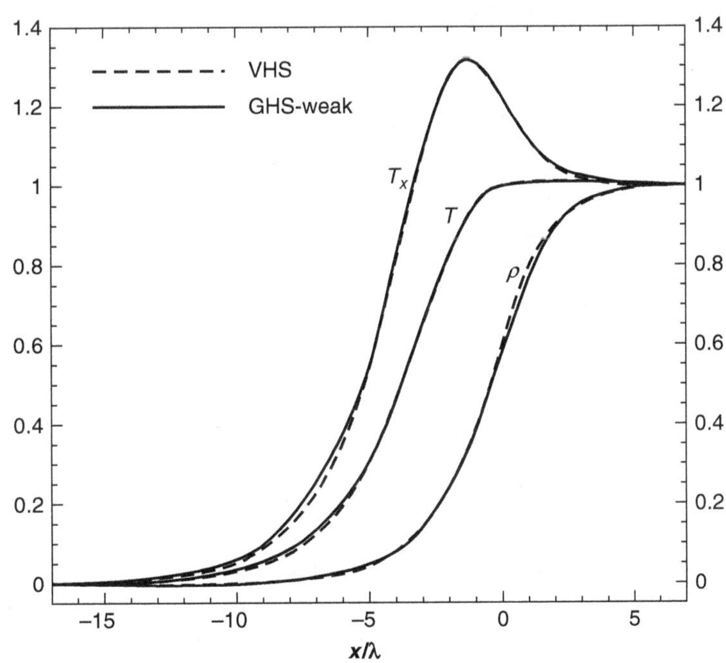

Figure 6.13 Density and temperature profiles within the shock wave.

Similar close agreement was observed, for both flow profiles and velocity distribution functions, when DSMC-VHS simulations were compared with pure MD simulations along with the experimental data from Alsmeyer (Valentini and Schwartzentruber, 2009b).

Therefore, for this nonequilibrium problem involving a wide temperature range where both the long-range attractive and short-range repulsive interactions may be important, it is still the repulsive interaction that dominates. As long as the VHS model is properly parameterized, there is only a small difference compared to the more general GHS model. This is due to the fact that only a small region of the shock wave (the upstream portion) involves low temperatures (low collision energies). One would expect larger differences between GHS and VHS solutions for flows that involve a wide temperature range *and* a substantial portion of the flow field being at low temperature.

GSS AND LJ MODELS

The GHS model was further extended to include soft-sphere scattering [the generalized soft-sphere (GSS) model] by Fan (2002), analogous to the VSS model described previously. The GSS model introduces an additional parameter α (compared to the GHS model) that is used to compute a nonisotropic scattering angle χ (using Eq. 6.33). This value of α can then be obtained by a least-squares fit to viscosity and diffusion coefficient data for each species pair interaction, in the same manner described previously for the VSS model. Full details of the model and a number of parameterizations for common gases can be found in Fan (2002).

Matsumoto and Koura (1991) proposed a DSMC simulation technique, where, for each collision within a DSMC simulation, the scattering angle χ was directly calculated by integration using the LJ PES (via Eq. 5.107). Although accurate, this approach is computationally expensive compared to VSS and GSS models, which use the model in Eq. 6.33. More recently, Venkattraman and Alexeenko (2012) developed a LJ collision model for DSMC. In this model, the scattering angle values (χ), calculated by integrating Eq. 5.107, are curve fit in a manner general to any LJ PES. Evaluating the curve fit for each DSMC collision is much more efficient than integrating Eq. 5.107 for each collision, and accuracy of the full LJ PES is maintained.

6.3.4 Thermal Conductivity

Through Chapman–Enskog analysis (refer to Chapter 5) for a monatomic gas, the first approximation of the coefficient of thermal conductivity (κ) was found to be directly proportional to the first approximation to the coefficient of viscosity (μ). However, for a gas with internal energy modes, the relation between thermal conductivity and viscosity becomes extremely complicated. In this case, the Eucken relation is typically employed to determine the coefficient of thermal conductivity:

$$\kappa = \mu\left(\frac{5}{2}c_{\mathrm{v}}^{\mathrm{tr}} + c_{\mathrm{v}}^{\mathrm{rot}} + c_{\mathrm{v}}^{\mathrm{vib}}\right) \tag{6.57}$$

Here, $c_{\mathrm{v}}^{\mathrm{tr}} = \frac{3}{2}\frac{k}{2m_r}$, $c_{\mathrm{v}}^{\mathrm{rot}} = \frac{k}{2m_r}$, and $c_{\mathrm{v}}^{\mathrm{vib}} = \frac{\zeta_{\mathrm{vib}}}{2}\frac{k}{2m_r}$, where ζ_{vib} is the vibrational degrees of freedom available. Again, this equation is specific to a species pair, where if the viscosity coefficient corresponding to the DSMC model is determined, then so is the thermal conductivity coefficient. To determine the thermal conductivity of a gas mixture, Eq. 5.94 can be used.

6.3.5 Model Parametrization

In summary, the hard-sphere cross section model fails to capture the strong dependence on relative collision velocity. The VHS model captures this dependence (accurately and efficiently) by introducing a power-law relation between cross section and relative collision velocity, while still employing simple isotropic scattering. This power-law is physically accurate for collision energies representative of temperatures above 273 K, and in many cases, still leads to accurate results at lower temperatures. In the near-equilibrium limit, the VHS model produces the physically correct temperature dependence of the transport coefficients, whereas the HS model does not. The VHS model can not be parameterized to both viscosity and diffusivity data independently for each species pair (arbitrary Schmidt number). If this is desired, the VSS model enables more flexibility to do so by modeling the scattering angle of collisions. The GHS and GSS models further extend the VHS and VSS models by including multiple power-law terms within the cross section expression. This enables long-range attractive forces (important at low collision energies) to be accurately modeled in addition to the short-range repulsive forces that dominate above 273 K. It is important to note, however, that for the majority of DSMC applications, models for scattering angles and attractive forces have very little influence on solution accuracy. As a result, the original VHS model of Bird continues to be a highly accurate and computationally efficient DSMC collision model.

As detailed in this section, a variety of DSMC collision models of varying complexity are available. Each of these collision models can be parameterized by fitting to aspects of a PES, or by fitting to differential or total cross section experimental data, or by fitting to experimental transport property data in the near-equilibrium limit (viscosity and diffusion coefficients). As more free parameters are included in successive models, the precision of such fitting may be improved. When deciding what model to use for a given application, several considerations are important.

First, sufficient experimental data to determine all model parameters may not be available for many flow conditions. For example, although like-species viscosity and unlike-species diffusivity data is widely available, unlike-species viscosity and like-species diffusivity data is often not available. For this reason, as demonstrated in Example 6.3, the practice of using the VHS model by fitting like-species parameters to viscosity data and unlike-species parameters to diffusivity data can be a highly accurate and prudent strategy for many flows.

Second, experimental data may only be available within a limited range of conditions, for example, within a limited temperature range. Using DSMC model parameters outside of the range in which they were parameterized and verified can easily lead to inaccurate results.

In principle, parameterizing DSMC models using PESs could provide all required information across a wide range of conditions. However, it is important to note that many simplified PESs in the literature have, themselves, been parameterized based on near-equilibrium transport properties. Thus, using GHS or GSS models fit to aspects of such PESs may not necessarily be more accurate than if they were fit to transport property data. A promising source of data for the construction and parametrization of new DSMC collision models are ab initio-based PESs constructed using methods from computational chemistry.

Finally, the DSMC models detailed in this section are phenomenological and their link to near-equilibrium transport properties was stressed. However, it is paramount to emphasize that once parameterized, DSMC models operate on individual collisions within arbitrary velocity distribution functions. Thus, for nonequilibrium flows (i.e., non-Maxwell–Boltzmann VDFs), where such transport relations become inaccurate (as discussed in Chapters 1 and 5), DSMC models embody realistic collision physics and can therefore model nonequilibrium gas states accurately. This is evident from extensive comparisons between DSMC solutions and experimental measurements of shock wave structure. At the same time, in the near-equilibrium limit, these DSMC models reduce analytically, through Chapman–Enskog theory, to the conventional transport laws widely used in continuum fluid dynamics.

The remainder of this chapter focuses on DSMC collision models involving internal energy transfer.

6.4 Internal Energy Transfer Modeling in DSMC

6.4.1 Continuum and Molecular Models

DSMC calculations of diatomic and polyatomic gases require storing and updating the internal energy of simulation particles. As discussed in Chapter 4, internal energy excitation and relaxation processes occur at finite rates. In continuum analyses, this is typically modeled using the Jeans and Landau–Teller equations (presented earlier in Eqs. 4.35 and 4.34) for translational-rotational and translational-vibrational relaxation, respectively. This section

where C is a constant. The maximum value of this probability occurs for

$$\frac{\epsilon_{tr}}{\epsilon_{coll}} = \frac{3/2 - \omega}{\zeta_{rot} + 1/2 - \omega} \tag{6.74}$$

giving

$$P_{max} = C(3/2 - \omega)^{3/2-\omega}(\zeta_{rot} - 1)^{\zeta_{rot}-1}\left[(\zeta_{rot} + 1/2 - \omega)\epsilon_{coll}\right]^{\zeta_{rot}-1} \tag{6.75}$$

and finally, the ratio of probability to the maximum probability can be written as

$$\frac{P}{P_{max}} = \left\{\frac{\zeta_{rot} + 1/2 - \omega}{3/2 - \omega}\left(\frac{\epsilon_{tr}}{\epsilon_{coll}}\right)\right\}^{3/2-\omega}\left\{\frac{\zeta_{rot} + 1/2 - \omega}{\zeta_{rot} - 1}\left(1 - \frac{\epsilon_{tr}}{\epsilon_{coll}}\right)\right\}^{\zeta_{rot}-1} \tag{6.76}$$

This equation can be incorporated into a simple acceptance–rejection procedure, the general procedure can be found in Appendix A. Specifically, for a given collision (with collision energy ϵ_{coll}) a random translational energy can be generated using a random number, $\epsilon_{tr}/\epsilon_{coll} = R_1$ $(0 \leq R_1 \leq 1)$. Equation 6.76 can be evaluated as, $P/P_{max} = a^a b^{-b} c^{-c} R_1^b (1 - R_1)^c$, where $a = \zeta_{rot} + 1/2 - \omega$, $b = 3/2 - \omega$, and $c = \zeta_{rot} - 1$. The resulting value can then be compared to a second random number $(0 \leq R_2 \leq 1)$, and if $P/P_{max} \geq R_2$ then the value of ϵ_{tr} is accepted as the post-collision translational energy of the collision pair. If the value of ϵ_{tr} is not accepted, then another R_1 value is randomly generated and the process is repeated until a value is accepted (called ϵ'_{tr}).

Since total collision energy is conserved, the post-collision internal energy is

$$\epsilon'_{rot} = \epsilon_{coll} - \epsilon'_{tr}, \tag{6.77}$$

which must be divided in some manner between the two particles involved in the collision. Given the total rotational energy associated with the pair of molecules (ϵ'_{rot}), Eq. 6.67 determines that the probability that one molecule has rotational energy $\epsilon'_{rot,1}$ is

$$P = D(\epsilon'_{rot,1})^{\zeta_{rot}/2-1}(\epsilon'_{rot} - \epsilon'_{rot,1})^{\zeta_{rot}/2-1} \tag{6.78}$$

where D is a constant. The maximum probability occurs when the internal energy is equally divided between the two colliding molecules, resulting in

$$\frac{P}{P_{max}} = 2^{\zeta_{rot}-2}\left(\frac{\epsilon'_{rot,1}}{\epsilon'_{rot}}\right)^{\zeta_{rot}/2-1}\left(1 - \frac{\epsilon'_{rot,1}}{\epsilon'_{rot}}\right)^{\zeta_{rot}/2-1} \tag{6.79}$$

Thus, a value of $\epsilon'_{rot,1}$ can be determined by applying an acceptance–rejection method to Eq. 6.79 (analogous to the procedure applied to Eq. 6.76). The remaining internal energy is then given to the other molecule involved in the collision, $\epsilon'_{rot,2} = \epsilon'_{rot} - \epsilon'_{rot,1}$, and the post-collision velocity vectors are determined by the appropriate scattering law using the value of ϵ'_{tr} determined above (refer to Appendix C). Note that for the case of diatomic molecules ($\zeta_{rot} = 2$), Eq. 6.79 shows that all values of $\epsilon'_{rot,1}$ are equally likely, and in that case an acceptance–rejection method is not even required.

In summary, particle pairs are first selected for an elastic collision using the NTC algorithm (Section 6.2.3) combined with a cross-section model (Section 6.3). A fraction of these collision pairs are further selected (with p_{rot}) for an inelastic collision that includes internal energy transfer. The post-collision energies are then sampled from equilibrium distributions based on the collision energy using the BL equations derived earlier.

LINK BETWEEN COLLISION NUMBER AND COLLISION PROBABILITY

The Borgnakke–Larsen method outlined above determines post-collision energies for those collisions involving internal energy transfer (inelastic collisions). This section focuses on determining *how many* inelastic collisions should be performed during a DSMC simulation. Once a pair of simulation particles have been selected for an elastic collision (using the NTC method with a cross section model such as VHS), each particle involved in the collision is further considered for internal energy exchange with an inelastic collision probability (Eq. 6.61).

Since these probabilities (p_{rot} or p_{vib}) are applied to the individual collision pairs selected within each DSMC cell, it is appropriate that they are formulated in terms of the properties of colliding particles. In DSMC, the distribution functions of colliding particles within each cell are generally nonequilibrium distributions. However, we also desire consistency with continuum models (Eq. 6.58) in the limit of equilibrium distribution functions, where the collision numbers are formulated in terms of temperature.

The general expression linking probabilities to collision numbers is through an integration over equilibrium distribution functions. Using translational-rotational energy exchange as an example, we have

$$\frac{1}{Z_{rot}(T_{tr}, T_{rot})} = \int_0^\infty \int_0^\infty \frac{p_{rot}(\epsilon_{tr}, \epsilon_{rot})}{C} f(\epsilon_{tr}; T_{tr}) f(\epsilon_{rot}; T_{rot})\, d\epsilon_{tr}\, d\epsilon_{rot} \tag{6.80}$$

where $f(\epsilon_i; T_i)$ (given previously in Eq. 6.64) is the equilibrium distribution function for energy mode i at temperature T_i, found within collisions. Note that in Eq. 6.80, we have included a connection factor C that will be determined in this section. This section establishes consistency between Z and p for the case where p, and therefore Z, are constants. A more general treatment is presented in Section 6.4.4.

In the case of constant Z and p, Eq. 6.80 reduces to

$$p_{rot} = \frac{C}{Z_{rot}} \tag{6.81}$$

Even though the collision number Z_{rot} represents the average number of collisions required to equilibrate the rotational energy mode (i.e., $\tau_{\text{rot}} = \tau_{\text{coll}} Z_{\text{rot}}$), the probability that should be used within a DSMC simulation may not be exactly $1/Z_{\text{rot}}$, as may seem intuitive. In order to understand why a connection factor is necessary and determine its value, we conduct an approximate mathematical analysis similar to that used by Lumpkin III, Haas and Boyd (1991), which equates the energy gain simulated by DSMC to that modeled by the Jeans equation.

Suppose a system is in a nonequilibrium state ($T_{\text{tr}} \neq T_{\text{rot}}$) associated with ζ_{tr} and ζ_{rot} degrees of freedom, and is relaxing toward equilibrium. At time t, assume the average relative translational and rotational energies of the system are $\langle \epsilon_{\text{tr}} \rangle(t)$ and $\langle \epsilon_{\text{rot}} \rangle(t)$, respectively. During a simulation timestep, Δt, the fraction of particles that will undergo a collision is $\Delta t/\tau_{\text{coll}}$, where τ_{coll} is the mean collision time. For each selected collision pair, the probability that they will undergo an inelastic collision during Δt is

$$P_{\text{inelastic}} = \frac{\Delta t}{\tau_{\text{coll}}} p_{\text{rot}} \tag{6.82}$$

The probability they will not undergo an inelastic collision during Δt is

$$P_{\text{elastic}} = 1 - \frac{\Delta t}{\tau_{\text{coll}}} p_{\text{rot}} \tag{6.83}$$

To assess consistency with the continuum Jeans equation, in addition to probabilities of rotational energy exchange, we must also determine the overall *amount* of rotational energy exchanged within DSMC. To do this, we first consider what the expected change in rotational energy is for a DSMC collision. Since we are using the BL energy redistribution method, described in the previous subsection, we know that post-collision energy ϵ'_{rot} is sampled from the distribution given in Eq. 6.72. Knowing this distribution enables us to determine the expectation value of the post-collision total rotational energy (denoted as $< \epsilon'_{\text{rot}} >$) for a given total collision energy (ϵ_{coll}). Specifically, by equating expressions for the average post-collision rotational energy, taken over all collision pairs,

$$\begin{aligned} &\int_0^\infty \int_0^{\varepsilon_{\text{coll}}} \varepsilon'_{\text{rot}} f(\varepsilon_{\text{coll}} - \varepsilon'_{\text{rot}}; T_{\text{coll}}) f(\varepsilon'_{\text{rot}}; T_{\text{coll}}) d\varepsilon'_{\text{rot}}\, d\varepsilon_{\text{coll}} \\ &= \int_0^\infty < \varepsilon'_{\text{rot}} > f(\varepsilon_{\text{coll}}, T_{\text{coll}})\, d\varepsilon_{\text{coll}} \end{aligned} \tag{6.84}$$

we arrive at the properly normalized expression for $< \epsilon'_{\text{rot}} >$:

$$< \varepsilon'_{\text{rot}} > = \frac{1}{f(\varepsilon_{\text{coll}}, T_{\text{coll}})} \int_0^{\varepsilon_{\text{coll}}} \varepsilon'_{\text{rot}} f(\varepsilon_{\text{coll}} - \varepsilon'_{\text{rot}}; T_{\text{coll}}) f(\varepsilon'_{\text{rot}}; T_{\text{coll}})\, d\varepsilon'_{\text{rot}} \tag{6.85}$$

It follows that

$$
\begin{aligned}
<\varepsilon'_{\rm rot}> &= \frac{1}{f(\varepsilon_{\rm coll}, T_{\rm coll})}\int_0^{\varepsilon_{\rm coll}} \frac{\varepsilon'_{\rm rot} e^{-\frac{-\varepsilon_{\rm coll}}{kT_{\rm coll}}}}{\Gamma(\frac{\zeta_{\rm tr}}{2})\Gamma(\frac{\zeta_{\rm rot}}{2})(kT_{\rm coll})^2}\left(\frac{\varepsilon_{\rm coll}-\varepsilon'_{\rm rot}}{kT_{\rm coll}}\right)^{\frac{\zeta_{\rm tr}}{2}-1} \\
&\times \left(\frac{\varepsilon'_{\rm rot}}{kT_{\rm coll}}\right)^{\frac{\zeta_{\rm rot}}{2}-1} d\varepsilon'_{\rm rot} \\
&= \frac{1}{f(\varepsilon_{\rm coll}, T_{\rm coll})}\frac{e^{-\frac{-\varepsilon_{\rm coll}}{kT_{\rm coll}}}}{\Gamma(\frac{\zeta_{\rm tr}}{2})\Gamma(\frac{\zeta_{\rm rot}}{2})}\left(\frac{\varepsilon_{\rm coll}}{kT_{\rm coll}}\right)^{\frac{\zeta_{\rm tr}+\zeta_{\rm rot}}{2}} \int_0^{\varepsilon_{\rm coll}} \left(1-\frac{\varepsilon'_{\rm rot}}{\varepsilon_{\rm coll}}\right)^{\frac{\zeta_{\rm tr}}{2}-1} \\
&\times \left(\frac{\varepsilon'_{\rm rot}}{\varepsilon_{\rm coll}}\right)^{\frac{\zeta_{\rm rot}}{2}} d\frac{\varepsilon'_{\rm rot}}{\varepsilon_{\rm coll}} \\
&= \frac{\Gamma(\frac{\zeta_{\rm tr}+\zeta_{\rm rot}}{2})}{\Gamma(\frac{\zeta_{\rm tr}}{2})\Gamma(\frac{\zeta_{\rm rot}}{2})}\varepsilon_{\rm coll}\int_0^1 (1-x)^{\frac{\zeta_{\rm tr}}{2}-1} x^{\frac{\zeta_{\rm rot}}{2}} dx \\
&= \frac{\Gamma(\frac{\zeta_{\rm tr}+\zeta_{\rm rot}}{2})}{\Gamma(\frac{\zeta_{\rm tr}}{2})\Gamma(\frac{\zeta_{\rm rot}}{2})}\frac{\Gamma(\frac{\zeta_{\rm tr}}{2})\Gamma(\frac{\zeta_{\rm rot}}{2}+1)}{\Gamma(\frac{\zeta_{\rm tr}+\zeta_{\rm rot}}{2}+1)}\varepsilon_{\rm coll} \\
&= \frac{\zeta_{\rm rot}}{\zeta_{\rm tr}+\zeta_{\rm rot}}\varepsilon_{\rm coll}
\end{aligned} \tag{6.86}
$$

Note that the solution in Eq. 6.86 involves the beta function

$$
B(p,q) = \int_0^1 (1-x)^{p-1} x^{q-1} dx = \Gamma(p)\Gamma(q)/\Gamma(p+q) \tag{6.87}
$$

The result $<\varepsilon'_{\rm rot}> = \zeta_{\rm rot}/(\zeta_{\rm tr}+\zeta_{\rm rot}) \times \varepsilon_{\rm coll}$ is physically meaningful, since the role of the BL model is to establish equilibrium between the translational and rotational energy modes.

We can now use this value of $<\epsilon'_{\rm rot}>$ to determine the amount of energy exchange during a DSMC timestep and directly compare the result with Jeans equation. Specifically, combining Eqs. 6.82 and 6.83, for a collision pair with energy $(\varepsilon_{\rm tr}, \varepsilon_{\rm rot})$ at time t, the expected value of rotational energy at time $t+\Delta t$ is then

$$
\begin{aligned}
\varepsilon_{\rm rot}(t+\Delta t) &= P_{\rm inelastic} <\varepsilon'_{\rm rot}> + P_{\rm elastic}\varepsilon_{\rm rot} \\
&= \frac{\Delta t}{\tau_{\rm coll}} p_{\rm rot} \frac{\zeta_{\rm rot}}{\zeta_{\rm tr}+\zeta_{\rm rot}}(\varepsilon_{\rm tr}+\varepsilon_{\rm rot}) + \left(1 - \frac{\Delta t}{\tau_{\rm coll}} p_{\rm rot}\right)\varepsilon_{\rm rot}
\end{aligned} \tag{6.88}
$$

The average rotational energy at time $t + \Delta t$, taken over all collisions, is then

$$\begin{aligned}
\langle\epsilon_{\rm rot}\rangle(t+\Delta t) &= \int_0^\infty\int_0^\infty \varepsilon_{\rm rot}(t+\Delta t) f(\varepsilon_{\rm tr}, T_{\rm tr}) f(\varepsilon_{\rm rot}, T_{\rm rot})\, d\varepsilon_{\rm tr}\, d\varepsilon_{\rm rot} \\
&= \int_0^\infty\int_0^\infty \varepsilon_{\rm rot} f(\varepsilon_{\rm tr}, T_{\rm tr}) f(\varepsilon_{\rm rot}, T_{\rm rot})\, d\varepsilon_{\rm tr} d\varepsilon_{\rm rot} \\
&\quad + \frac{\Delta t}{\tau_{\rm coll}} p_{\rm rot} \int_0^\infty\int_0^\infty \left(\frac{\zeta_{\rm rot}}{\zeta_{\rm tr}+\zeta_{\rm rot}}(\varepsilon_{\rm tr}+\varepsilon_{\rm rot}) - \varepsilon_{\rm rot}\right) \\
&\quad \times f(\varepsilon_{\rm tr}, T_{\rm tr}) f(\varepsilon_{\rm rot}, T_{\rm rot})\, d\varepsilon_{\rm tr}\, d\varepsilon_{\rm rot} \\
&= \langle\epsilon_{\rm rot}\rangle(t) + \frac{\Delta t}{\tau_{\rm coll}} p_{\rm rot} \int_0^\infty\int_0^\infty \frac{\zeta_{\rm tr}}{\zeta_{\rm tr}+\zeta_{\rm rot}} \left(\frac{\zeta_{\rm rot}}{\zeta_{\rm tr}}\varepsilon_{\rm tr} - \varepsilon_{\rm rot}\right) \\
&\quad \times f(\varepsilon_{\rm tr}, T_{\rm tr}) f(\varepsilon_{\rm rot}, T_{\rm rot})\, d\varepsilon_{\rm tr} d\varepsilon_{\rm rot}
\end{aligned} \tag{6.89}$$

Finally, with a first-order approximation and using Eq. 6.89, we have,

$$\begin{aligned}
\frac{d\langle\epsilon_{\rm rot}\rangle}{dt} &\approx \frac{\langle\epsilon_{\rm rot}\rangle(t+\Delta t) - \langle\epsilon_{\rm rot}\rangle(t)}{\Delta t} \\
&= \frac{1}{\tau_{\rm coll}} \frac{\zeta_{\rm tr}}{\zeta_{\rm tr}+\zeta_{\rm rot}} \left(\frac{\zeta_{\rm rot}}{\zeta_{\rm tr}}\langle\epsilon_{\rm tr}\rangle(t) - \langle\epsilon_{\rm rot}\rangle(t)\right) p_{\rm rot}
\end{aligned} \tag{6.90}$$

or equivalently,

$$\frac{dE_{\rm rot}}{dt} = \frac{1}{\tau_{\rm coll}} \frac{\zeta_{\rm tr}}{\zeta_{\rm tr}+\zeta_{\rm rot}} \left(E^*_{\rm rot}(t) - E_{\rm rot}(t)\right) p_{\rm rot} \tag{6.91}$$

Comparing Eq. 6.91 with the right-hand side of the Jeans equation (Eq. 6.58a), we then have

$$\frac{E^*_{\rm rot}(t) - E_{\rm rot}(t)}{Z_{\rm rot}\tau_{\rm coll}} = \frac{1}{\tau_{\rm coll}} \frac{\zeta_{\rm tr}}{\zeta_{\rm tr}+\zeta_{\rm rot}} \left(E^*_{\rm rot}(t) - E_{\rm rot}(t)\right) p_{\rm rot} \tag{6.92}$$

and therefore

$$p_{\rm rot} = \frac{\zeta_{\rm tr}+\zeta_{\rm rot}}{\zeta_{\rm tr}} \frac{1}{Z_{\rm rot}} = \frac{C}{Z_{\rm rot}} \tag{6.93}$$

This then gives the analytical connection factor C for the constant rotational collision probability model, and is the same as has been discussed by Lumpkin III et al. (1991) and Haas et al. (1994).

Therefore, for a DSMC simulation using the BL method to be consistent with the Jean's equation (using $Z_{\rm rot}$) in the near equilibrium limit, inelastic collisions must be performed with probability $p_{\rm rot}$ given in Eq. 6.93. The post-collision properties (translational and rotational energies) should then be sampled using the BL probability expressions (distributions) given in Eqs. 6.76 and 6.79.

However, there is another issue that requires further consideration; what are the precise values of $\zeta_{\rm tr}$ and $\zeta_{\rm rot}$ that should be used? It turns out that these should be set as the number of degrees of freedom that are actually *available to the simulated collision*. For example, if both particles' rotational energies are updated during the collision, then ζ_{rot} will have a different value than if only one molecule's rotational energy is updated. Deciding which energy modes can be updated during a collision is called the "selection procedure," and is described in the next section. Furthermore, a completely general description of the BL method for multispecies gases, for rotational and vibrational energy exchange, and when p_{rot} is not a constant is presented later in Section 6.4.4.

6.4.3 Inelastic Collision Pair Selection Procedures

During a real collision between two molecules with internal energy, generally speaking, the internal energy of both molecules will be altered. For example, this will be the case within a molecular dynamics simulation. However, DSMC is a stochastic simulation method (not deterministic), where the objective is to accurately model the evolution of molecular distribution functions and not model the details of every atom in the system. Therefore, within a DSMC collision, it may be perfectly acceptable and accurate to change the internal energy of only one of the molecules. The collision pair selection procedure is a subtle, yet crucial issue within a DSMC simulation. As will be explained in this section, care must be taken to ensure that all DSMC physical models (and associated equations) are consistent with the chosen selection procedure.

The effect of the collision selection procedure on the simulated relaxation process is most significant for gas mixtures, since some selection procedures inherently couple the relaxation probabilities and internal energy redistribution processes of the different species. For clarity, we summarize the three most widely used inelastic collision selection procedures as they apply to rotational relaxation. For inelastic collisions involving vibrational energy exchange, the procedures are identical, only the internal energy mode is altered.

(A) **Pair Selection** (Boyd 1990b; Lumpkin III et al. 1991): In this case, the collision pair is tested for an inelastic collision with a probability, and once the collision pair is selected, the energy of both particles in the pair is redistributed. Specifically, the total collision energy $\epsilon_{\text{coll}} = \epsilon_{\text{tr}} + \epsilon_{\text{rot,A}} + \epsilon_{\text{rot,B}}$ is redistributed between translational and rotational modes, as $\epsilon_{\text{coll}} = \epsilon'_{\text{tr}} + \epsilon'_{\text{rot}}$. The post-collision rotational energy ϵ'_{rot} is then distributed between the two collision partners as $\epsilon'_{\text{rot}} = \epsilon'_{\text{rot,A}} + \epsilon'_{\text{rot,B}}$. This was the procedure used to introduce the BL method in Section 6.4.2.

(B) **Particle Selection Permitting Double Relaxation** (Bird 1994): In this case, each particle in the collision pair is tested with a probability for inelastic collision individually. If the first particle is selected for an inelastic collision, the BL procedure is used to redistribute the collision energy

($\epsilon_{coll} = \epsilon_{tr} + \epsilon_{rot,A}$) between the translational energy of the *pair* and the rotational energy of only the *selected particle* ($\epsilon_{coll} = \epsilon'_{tr} + \epsilon'_{rot,A}$). Next, the second particle in the pair is tested with a probability for inelastic collision. If selected, the collision energy now includes the redistributed translational energy of the pair from the first collision and only the rotational energy of the second particle ($\epsilon'_{coll} = \epsilon'_{tr} + \epsilon_{rot,B}$). The BL procedure is employed again to redistribute this collision energy between a final post-collision translational energy and a rotational energy for the second particle ($\epsilon'_{coll} = \epsilon''_{tr} + \epsilon'_{rot,B}$). In this manner, if both particles are selected for an inelastic collision, there is some degree of coupling between their relaxation processes.

(C) **Particle Selection Prohibiting Double Relaxation** (Haas et al. 1994) In this case, the two particles in the collision pair are tested with a probability for inelastic collision individually. The collision energy is always the sum of the relative translational energy of the collision pair and the rotational energy of only the particle (*i*) being considered ($\epsilon_{coll} = \epsilon_{tr} + \epsilon_{rot,i}$). The BL procedure always redistributes post-collision energies between the translational energy of the pair and only the rotational energy of the particle considered ($\epsilon_{coll} = \epsilon'_{tr} + \epsilon'_{rot,i}$). Furthermore, if one particle is selected to undergo an inelastic collision, the other particle is not tested for an inelastic collision, and the relaxation process for the collision pair ends. Only if the first particle is not selected for an inelastic collision, is the identical procedure applied to the second particle in the pair.

Selection procedure (A) couples the relaxation probabilities and energy redistribution processes for collision pairs of different species, and thus couples the simulated relaxation process. Although not as direct as procedure (A), procedure (B) also couples the energy redistribution processes of species when both particles are selected for rotational relaxation (i.e., double relaxation). Furthermore, when both rotational and vibrational relaxation processes are considered, sequential testing for rotational followed by vibrational inelastic collisions, in selection procedures (A) and (B), will inherently couple the rotational and vibrational relaxation processes. Although such coupling may seem physically realistic, it is stressed that the DSMC collision models discussed here are phenomenological and are constructed to reproduce specified internal energy relaxation rates (collision numbers, Z) for a given energy mode and species interaction. Such consistency is highly desirable and this section focuses specifically on selection procedure (C).

The particle selection prohibiting double relaxation procedure was first introduced by Haas et al. (1994) and modified algorithms have been proposed by Gimelshein, Gimelshein, and Levin (2002) and Zhang and Schwartzentruber (2013). In the article by Zhang and Schwartzentruber (2013), it is proven analytically that all three algorithms produce the same simulated relaxation

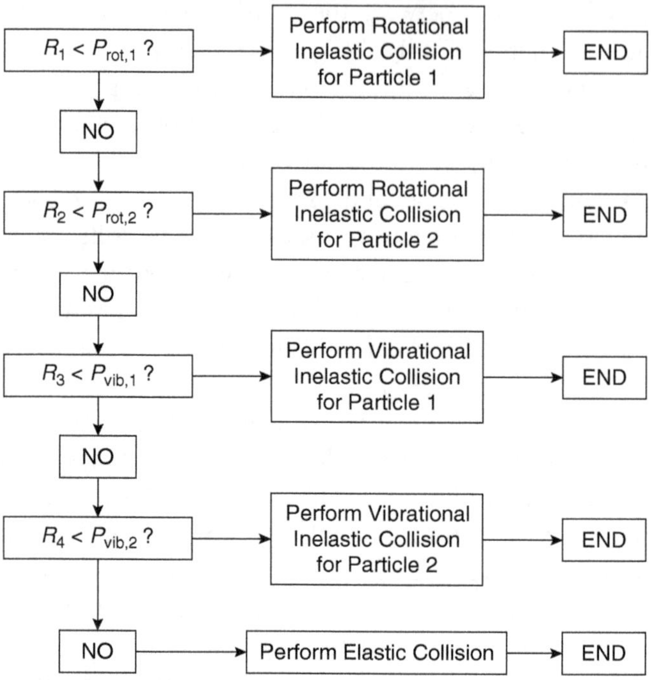

Figure 6.14 The process to select an inelastic collision using selection procedure (C). In the figure, $R_n(n = 1, 2, 3, 4)$ are uniform random numbers between 0 and 1, and $P_{\text{rot},i}$, $P_{\text{vib},i}$ $(i = 1 \text{ or } 2)$ are the rotational and vibrational inelastic collision probabilities used in DSMC for particle i.

rates, and therefore any of these three algorithms can be used to implement the particle selection prohibiting double relaxation procedure. This section summarizes the method of Zhang and Schwartzentruber (2013) first, followed by the method of Gimelshein et al. (2002).

SELECTION PROCEDURE OF ZHANG AND SCHWARTZENTRUBER

The logical steps followed by this selection procedure are depicted in Fig. 6.14. The specific notation used throughout the remainder of this section requires a careful description. As depicted in Fig. 6.14, before starting the collision procedure, the two particles in the pair must be assigned a number (either particle 1 or 2). We use a subscript, i, to denote the particle numbering of the pair ($i = 1$ or $i = 2$). We further note that i does not denote a specific particle type; thus particles $i = 1$ and $i = 2$ may be the same, or different, particle type (monatomic, diatomic, or polyatomic). In this manner, all parameters denoted by a subscript i (such as $p_{\text{rot},i}$, $Z_{\text{rot},i}$, $\zeta_{\text{rot},i}$, etc.) are specific to the relaxation process of particle i. For example, $\zeta_{\text{rot},i}$ and $\zeta_{\text{vib},i}$ are the rotational and vibrational degrees of freedom of only particle i, whereas ζ_{tr} represents the available translational degrees of freedom of the collision pair (and thus has no subscript). Furthermore, $Z_{\text{rot},i}$ and $Z_{\text{vib},i}$ are the rotational and vibrational inelastic collision numbers specific to the relaxation of particle i during a collision with the other particle in the pair. For example, if both particles are of the same type (A) then the collision numbers for the two particles would be equal (i.e., $Z_{\mathrm{rot},1} = Z_{\mathrm{rot},2} = Z_{\mathrm{rot,A|A}}$). However, if the two particles were of different types (A for $i = 1$, and B for $i = 2$, as an example), then the collision numbers would be $Z_{\mathrm{rot},1} = Z_{\mathrm{rot},A|B}$ and $Z_{\mathrm{rot},2} = Z_{\mathrm{rot,B|A}}$ where in general, $Z_{\mathrm{rot,A|B}}$ may be specified as not equal to $Z_{\mathrm{rot,B|A}}$.

Referring to Fig. 6.14, to begin, the two particles involved in the collision are randomly assigned as particle 1 and particle 2. Then each event in Fig. 6.14 is processed using a standard acceptance–rejection technique using the probabilities $P_{\mathrm{rot},1}$, $P_{\mathrm{rot},2}$, $P_{\mathrm{vib},1}$, $P_{\mathrm{vib},2}$ given by the following expressions:

$$A = 1, \qquad P_{\mathrm{rot},1} = A p_{\mathrm{rot},1} \tag{6.94a}$$

$$B = \frac{A}{1 - P_{\mathrm{rot},1}}, \qquad P_{\mathrm{rot},2} = B p_{\mathrm{rot},2} \tag{6.94b}$$

$$C = \frac{B}{1 - P_{\mathrm{rot},2}}, \qquad P_{\mathrm{vib},1} = C p_{\mathrm{vib},1} \tag{6.94c}$$

$$D = \frac{C}{1 - P_{\mathrm{vib},1}}, \qquad P_{\mathrm{vib},2} = D p_{\mathrm{vib},2} \tag{6.94d}$$

where

$$p_{\mathrm{rot},i} = \frac{\zeta_{\mathrm{tr}} + \zeta_{\mathrm{rot},i}}{\zeta_{\mathrm{tr}}} \frac{1}{Z_{\mathrm{rot},i}} \tag{6.95}$$

and

$$p_{\mathrm{vib},i} = \frac{\zeta_{\mathrm{tr}} + \Gamma_i}{\zeta_{\mathrm{tr}}} \frac{1}{Z_{\mathrm{vib},i}} \tag{6.96}$$

For vibration, if a continuous energy distribution is used, then $\Gamma_i = \zeta_{\mathrm{vib},i}$. Whereas for the simple harmonic oscillator (SHO) discrete energy level model, $\Gamma_i = \xi_{\mathrm{vib}}(T)^2 \exp(\theta_{\mathrm{vib}}/T)/2$, where $\xi_{\mathrm{vib}}(T) = \frac{2\theta_{\mathrm{vib}}/T}{\exp(\theta_{\mathrm{vib}}/T)-1}$ (see Eq. 3.132), T is the temperature, and is usually set as the cell averaged translational temperature, i.e., $T = T_{\mathrm{tr}}$, and θ_{vib} is the characteristic temperature of vibration (where all parameters are specific to particle i).

Finally, collision quantity dependent models do not have a direct relationship between the collision probability p and a collision number Z. For such models, p is now a function of some collision quantities (collision energies, for example) and should be used directly in Eq. 6.94. Examples of collision-quantity–based models are presented in Chapter 7.

SELECTION PROCEDURE OF GIMELSHEIN ET AL.

In the selection procedure by Gimelshein et al. (2002), the calculation of the simulated collision probability for inelastic collisions and the subsequent use of the acceptance–rejection technique are combined together. Specifically, a random number R_n between [0, 1] is selected and used within a series of inequalities. Inequalities of the form $R_n > A_1$ and $A_i < R_n < A_{i+1}$ ($i = 1, 2, 3$) are sequentially tested. If one inequality is true, then the corresponding inelastic collision relaxation is performed, and the procedure ends for the current collision pair. Only if one inequality does not hold, will the subsequent inequality be tested.

The expressions for A_i, as found in Fig. 1 of Gimelshein et al. (2002), are as follows:

$$A_1 = p_{\text{rot},1} \tag{6.97a}$$

$$A_2 = p_{\text{rot},1} + p_{\text{rot},2} \tag{6.97b}$$

$$A_3 = p_{\text{rot},1} + p_{\text{rot},2} + p_{\text{vib},1} \tag{6.97c}$$

$$A_4 = p_{\text{rot},1} + p_{\text{rot},2} + p_{\text{vib},1} + p_{\text{vib},2} \tag{6.97d}$$

Here, the values of p are calculated in the same way as discussed earlier (Eqs. 6.95 and 6.96).

SUMMARY AND DISCUSSION

It may not be immediately evident that the algorithms of Gimelshein et al. (2002) and Zhang and Schwartzentruber (2013) do indeed lead to identical simulated relaxation processes. However, this has been analytically proven and numerically validated in the article by Zhang and Schwartzentruber (2013), which provides further details about these algorithms. Since the algorithm of Gimelshein et al. (2002) uses only one random number for each collision, it is more computationally efficient. However, selection procedures contribute negligible computational expense to typical DSMC simulations, and the choice between the above two selection procedures is a matter of preference between implementation styles. Specifically, the selection procedure of Zhang and Schwartzentruber has the advantage of using a standard acceptance–rejection procedure, with a specified probability, for each step of the algorithm. This general framework has recently been applied to vibrational relaxation in a polyatomic gas (CO_2) and also applied to gas-surface reaction probabilities (Poovathingal, Schwartzentruber, Murray, and Minton 2016). The selection procedure of Gimelshein et al. does not use a standard acceptance–rejection procedure, rather it evaluates a series of inequalities, and does so using only a single random number.

For any particle selection procedure that prohibits double relaxation, the constraint

$$p_{\text{rot},1} + p_{\text{rot},2} + p_{\text{vib},1} + p_{\text{vib},2} < 1 \tag{6.98}$$

must be satisfied (Gimelshein et al. 2002). While Eq. 6.98 is satisfied for the majority of nonequilibrium flow problems, as evident from Eq. 6.94, if $p_{\text{rot},i}$ were to approach 0.5 there would be a vanishing number of particles available to be tested for vibrational relaxation and $P_{\text{vib},i}$ may become larger than unity. Thus for generality, one could test for vibrational relaxation first, followed by rotational relaxation, since $p_{\text{vib},i}$ is typically much smaller than $p_{\text{rot},i}$. This ensures that vibrational relaxation remains accurate even in extreme cases with very fast rotational relaxation. Changing the order of rotational and vibrational relaxation only requires interchanging the subscripts (rot, vib) in all of the preceding equations.

Extension of the particle selection prohibiting double relaxation procedure to polyatomic species with additional internal energy modes is straightforward. For example, consider carbon dioxide (CO_2), a linear triatomic molecule, with two rotational degrees of freedom and multiple vibrational energy modes. The vibrational modes include a symmetric mode ($\theta_{\text{vib}} = 1890$ K), an asymmetric stretching mode ($\theta_{\text{vib}} = 3360$ K), and two degenerate bending modes ($\theta_{\text{vib}} = 954$ K). The SHO model (Eq. 3.132) could be used for each of these vibrational modes, so that depending on the ratio of gas temperature to characteristic temperature (θ_{vib}), each mode would be available to store the appropriate amount of energy. The selection procedure of Zhang and Schwartzentruber has been used to model CO_2 and further details can be found in Poovathingal et al. (2015) and Poovathingal et al. (2016). Essentially, a distinct Z_{vib} value can be specified for each vibrational energy mode and corresponding probabilities of energy transfer into each mode can be calculated using Eq. 6.96. The logical expressions in Eq. 6.94 (shown schematically in Fig. 6.14) simply include more probabilities, however, the procedure of testing for energy transfer into each mode sequentially remains identical. If four vibrational energy modes are considered, there would now be 10 probabilities to consider in Fig. 6.14, and the constraint

$$\begin{aligned} &p_{\text{rot},1} + (p_{\text{vib},1})_{\text{mode1}} + (p_{\text{vib},1})_{\text{mode2}} + (p_{\text{vib},1})_{\text{mode3}} + (p_{\text{vib},1})_{\text{mode4}} + \\ &p_{\text{rot},2} + (p_{\text{vib},2})_{\text{mode1}} + (p_{\text{vib},2})_{\text{mode2}} + (p_{\text{vib},2})_{\text{mode3}} + (p_{\text{vib},2})_{\text{mode4}} < 1 \end{aligned} \tag{6.99}$$

becomes more difficult to maintain. However, since the probability of vibrational energy transfer is typically $\ll 1$, the added constraint may not be significant. If a specific mode is accepted for energy exchange, the BL distribution is used to determine the post-collision energy of that mode where the energies and degrees of freedom used in the BL equations are consistent with the particle and internal energy mode selected. As an example, during a particular collision, the energies used in the BL equations may include the energy in the second vibrational mode of particle 1 ($\epsilon_i = (\epsilon_{\text{vib},1})_{\text{mode2}}$) and the translational energy of the particle pair (ϵ_{tr}). The corresponding degrees of freedom used in the BL equations would include the degrees of freedom of the selected vibrational mode ($\zeta_i = (\zeta_{\text{vib},1})_{\text{mode2}}$) evaluated at the current cell temperature (Eq. 3.132 using the appropriate value of θ_{vib}) and the translational degrees of freedom of the collision pair (ζ_{tr}). Recently, an alternative technique for polyatomic molecules, called the "multimode relaxation" procedure has been developed (Pfeiffer, Nizenkov, Mirza, and Fasoulas, 2016) where internal energy is redistributed among multiple internal modes during a single collision. Both multimode relaxation and particle selection prohibiting double relaxation procedures are verified to produce the same overall vibrational relaxation rates for CO_2 and further details and discussion regarding internal energy relaxation of polyatomic species can be found in the article (Pfeiffer, Nizenkov, Mirza, and Fasoulas, 2016).

In summary, once a pair of simulated particles is chosen to undergo a collision, a selection procedure must be employed that determines which internal energy mode(s) of which particle(s) may be updated during the collision. The selection procedure therefore determines both the portion of collision energy to be redistributed and the available degrees of freedom over which the energy is to be redistributed. This must be fully consistent with the degrees of freedom used in the BL energy redistribution model equations and consistent with the connection factor (Eq. 6.93) that links p to Z. As a result, in the limit of near-equilibrium flow, the DSMC simulation will satisfy equipartition of energy and will be consistent with continuum relaxation equations.

The particle selection procedure prohibiting double relaxation (using implementations from either Zhang and Schwartzentruber or Gimelshein et al.), is recommended, since this approach ensures that for the same set of pair-specific collision numbers ($Z_{\mathrm{rot},i|j}$, $Z_{\mathrm{vib},i|j}$), DSMC simulations in the near-equilibrium limit would agree precisely with CFD simulations using the continuum expressions in Eq. 6.58a and b.

Example 6.5 Isothermal Relaxation: Simulation compared to theory

To test the particle selection prohibiting double relaxation procedure, we conduct an isothermal relaxation simulation for a mixture of two species, where the translational temperature of the system is maintained at a constant value of $T_{\mathrm{tr}} = 10,000$ K. To maintain the translational temperature of the system at a constant value, each timestep, we regenerate the velocities of all particles contained in the simulation domain, following a Maxwell–Boltzmann distribution at $T_{\mathrm{tr}} = 10{,}000$ K. The rotational and vibrational energies of the particles are not changed during this process. Since the translational temperature is constant during an isothermal relaxation simulation, the mean collision time is also a constant, and the resulting Jeans equation has an analytical solution.

The relaxation equations for a gas mixture were presented previously in Eq. 6.58. For isothermal relaxations, $E^*_{\mathrm{rot},j}(t) = E_{\mathrm{rot},j}(\infty)$, and $E^*_{\mathrm{vib},j}(t) = E_{\mathrm{vib},j}(\infty)$. If rotational and vibrational relaxation times are assumed to be constant, or depend only on translational temperature, then $\tau_{\mathrm{rot}} = \tau_{\mathrm{coll}} Z_{\mathrm{rot}}$ and $\tau_{\mathrm{vib}} = \tau_{\mathrm{coll}} Z_{\mathrm{vib}}$ are constant, and Eqs. 6.58a and 6.58b have the following analytical solution:

$$\frac{E_{\mathrm{rot},j}(\infty) - E_{\mathrm{rot},j}(t)}{E_{\mathrm{rot},j}(\infty) - E_{\mathrm{rot},j}(0)} = \exp\left(-\sum_k \frac{t}{\tau_{\mathrm{coll},j|k} Z_{\mathrm{rot},j|k}}\right) \tag{6.100a}$$

$$\frac{E_{\mathrm{vib},j}(\infty) - E_{\mathrm{vib},j}(t)}{E_{\mathrm{vib},j}(\infty) - E_{\mathrm{vib},j}(0)} = \exp\left(-\sum_k \frac{t}{\tau_{\mathrm{coll},j|k} Z_{\mathrm{vib},j|k}}\right). \tag{6.100b}$$

Table 6.3 Simulation Parameters Specific to Each Collision Pair

$j \mid k$	Z_{rot}	Z_{vib}	ω	$d_{ref}(\times 10^{-10} m)$	T_{ref}(K)
1|1	5	40	0.74	4.17	273
1|2	8	60	0.755	4.12	273
2|1	10	80	0.755	4.12	273
2|2	15	60	0.77	4.07	273

Using $E_j = \frac{\zeta_j}{2} kT$, Eqs. 6.100a and 6.100b can be written in terms of temperature, as

$$\frac{T_{\text{rot},j}(\infty) - T_{\text{rot},j}(t)}{T_{\text{rot},j}(\infty) - T_{\text{rot},j}(0)} = \exp\left(-\sum_k \frac{t}{\tau_{\text{coll},j|k} Z_{\text{rot},j|k}}\right) \tag{6.101a}$$

$$\frac{\zeta_{\text{vib},j}(\infty) T_{\text{vib},j}(\infty) - \zeta_{\text{vib},j}(t) T_{\text{vib},j}(t)}{\zeta_{\text{vib},j}(\infty) T_{\text{vib},j}(\infty) - \zeta_{\text{vib},j}(0) T_{\text{vib},j}(0)} = \exp\left(-\sum_k \frac{t}{\tau_{\text{coll},j|k} Z_{\text{vib},j|k}}\right) \tag{6.101b}$$

where $\zeta_{\text{vib},j}(0)$, $\zeta_{\text{vib},j}(\infty)$ and $\zeta_{\text{vib},j}(t)$ are the effective vibrational degrees of freedom at time 0, ∞, and t, respectively.

The specified rotational and vibrational collision numbers for the two species are listed in Table. 6.3, together with the variable hard-sphere (VHS) parameters used in the DSMC simulations. The two species have mole fractions of 0.3 and 0.7, respectively. The VHS model parameters used for the two species (ω, d_{ref} in Table 6.3) correspond to those of N_2 and O_2; however, the rotational and vibrational collision numbers Z_{rot} and Z_{vib} do not correspond to the values for those gas species, and are set here for demonstration purpose only. The simulation results are shown in Fig. 6.15 for the relaxation history of the rotational and vibrational temperatures of each species in the mixture. It is evident that the particle selection prohibiting double relaxation procedure is able to accurately simulate the specified relaxation rate for the mixture. It is further noted that rotational relaxation is significantly faster than vibrational relaxation, as expected.

Example 6.6 Isothermal Relaxation: Pair Selection versus Particle Selection Prohibiting Double Relaxation

As a further demonstration, a similar isothermal relaxation simulation is conducted using the pair selection procedure (selection procedure (A)). Specifically the form of Eq. 6.93 is used. This simulation considers only rotational relaxation and the rotational collision numbers are modified from Table. 6.3 to be $Z_{\text{rot}} = 10$ for collisions 1|2 and 2|1, and to be $Z_{\text{rot}} = 20$ for collision 2|2. The results using the particle selection prohibiting double relaxation procedure are shown in Fig. 6.16(a) and the results from the pair selection

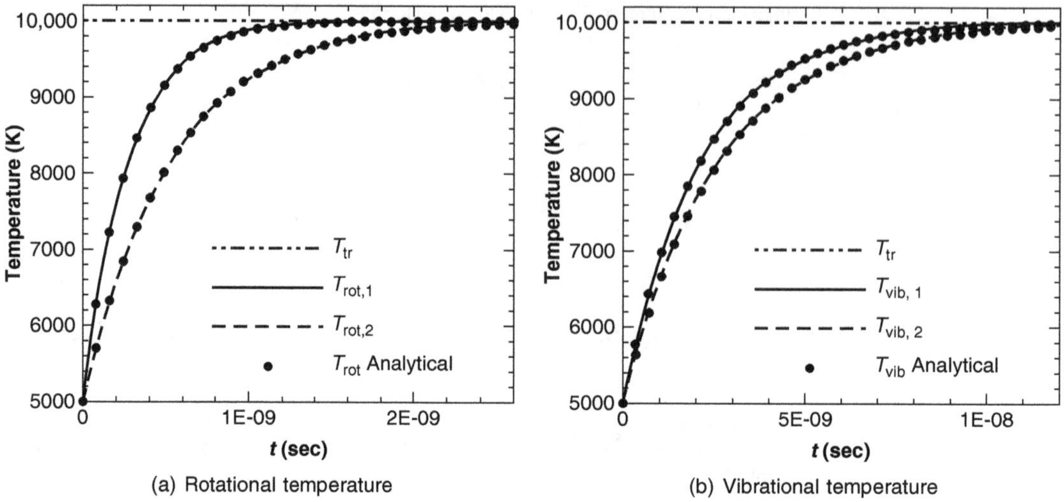

Figure 6.15 Rotational and vibrational relaxation temperature histories in an isothermal reservoir.

procedure are shown in Fig. 6.16(b). Clearly, the results using the pair selection procedure do not agree with the analytical solution for the mixture, whereas the particle selection prohibiting double relaxation procedure exactly reproduces the analytical solution.

6.4.4 Generalized Post-collision Energy Redistribution

In this section, an internal energy exchange model is derived generally in terms of the energies (ε_{tr}, ε_i) and degrees of freedom (ζ_{tr}, ζ_i) that *participate*

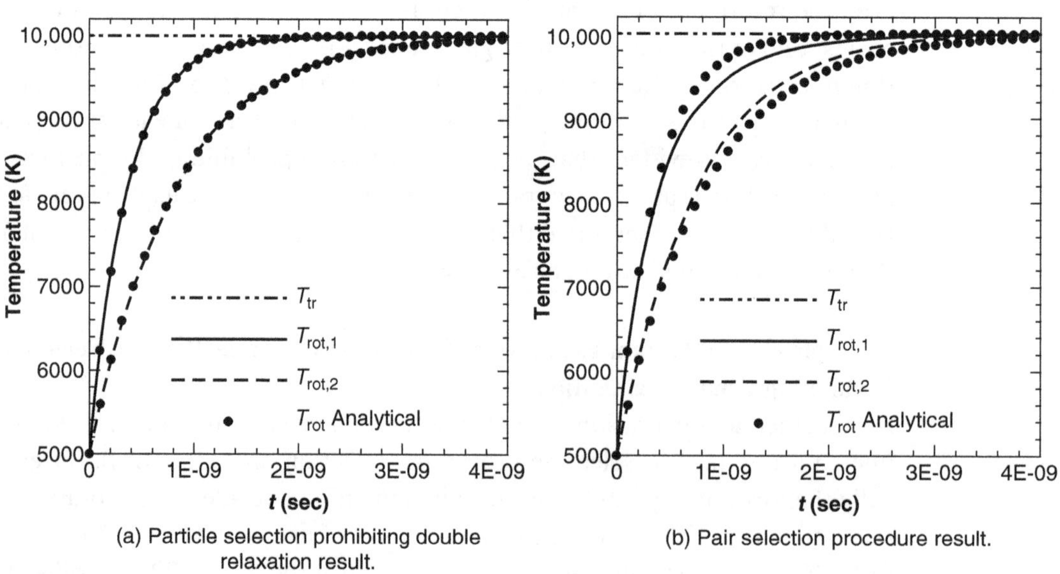

Figure 6.16 Comparison of two different selection procedures for rotational relaxation in a two species mixture in an isothermal reservoir simulation.

in selected collisions within the DSMC simulation. This model is an extension of the BL method and was developed by Zhang, Valentini, and Schwartzentruber (2014). The model is general to both translational-rotational energy exchange ($i = \text{rot}$) and translational-vibrational energy exchange ($i = \text{vib}$) for any species pair. The model equations derived in this section are valid for all types of inelastic collision selection procedures as long as the participating energy modes and degrees of freedom are assigned consistently. The model is general to any probability expression (p_{rot} or p_{vib}: p_i) and ensures detailed balance and the equipartition of energy at equilibrium. Finally, a general procedure to link DSMC inelastic collision models to continuum relaxation models is presented.

GENERALIZED POST-COLLISION EQUILIBRIUM DISTRIBUTION

As discussed earlier in Section 6.4.2, over successive collisions, and in the limit of equilibrium, molecular energies and distribution functions should relax towards an equilibrium state that satisfies the equipartition of energy. This can be achieved with the phenomenological BL model that samples post-collision energies from the same distributions as the pre-collision energies of molecules expected in collisions corresponding to an equilibrium state. In this manner, an equilibrium gas will remain in equilibrium as desired, and the rate of relaxation is controlled through a probability of inelastic collision, p_i.

At equilibrium, we can write the joint distribution of translational-internal energy expected within elastic collisions as $f(\epsilon_{\text{tr}};\, T_{\text{coll}})f(\epsilon_i;\, T_{\text{coll}})$, where these functions are given in Eq. 6.64. To account for translational energy bias within collisions, the translational degrees of freedom should be set to $\zeta_{\text{tr}} = 5 - 2\omega$, corresponding to VHS or VSS models. In previous sections, p_i was assumed to be constant, in which case there is no bias for the internal energies within collisions. However, in general, the probability of an inelastic collision may depend on the internal energy states of the molecules. In the most general case, the probability may depend on collision energies: $p_i = p_i(\epsilon_{\text{tr}}, \epsilon_i)$. In this case, the distributions of molecules expected within inelastic collisions at equilibrium is now $f(\epsilon_{\text{tr}};\, T_{\text{coll}})f(\epsilon_i;\, T_{\text{coll}})p_i(\epsilon_{\text{tr}}, \epsilon_i)$, and therefore, this is the distribution that should be sampled to determine post-collision energy states $(\epsilon'_{\text{tr}}, \epsilon'_i)$.

As highlighted in Section 6.4.3, the energies (ϵ_i) and degrees of freedom (ζ_i) used to evaluate $f(\epsilon_{\text{tr}};\, T_{\text{coll}})f(\epsilon_i;\, T_{\text{coll}})p_i(\epsilon_{\text{tr}}, \epsilon_i)$ should be those that participate in a redistribution process. Therefore, they should be set consistently with the particle selection procedure employed. As an example, for collisions involving translational-rotational energy transfer (performed according to p_{rot}), the pair selection procedure involves the combined rotational energies and degrees of freedom of both molecules, whereas the particle selection prohibiting double relaxation procedure involves only the rotational energy and degrees of freedom of one of the molecules in the pair (refer to Fig. 6.14).

For collisions involving translational-vibrational energy transfer (performed according to p_{vib}), the same logic applies for the vibrational energies and degrees of freedom.

At this point, all of the components required for the general model have been detailed. The remainder of this section presents the equations and acceptance–rejection sampling technique to be included in a DSMC implementation.

Post-collision properties $(\varepsilon'_{\text{tr}}, \varepsilon'_i)$ are sampled from an equilibrium distribution $f(\epsilon_{\text{tr}}; T_{\text{coll}})f(\epsilon_i; T_{\text{coll}})p_i(\epsilon_{\text{tr}}, \epsilon_i)$. The energies $(\epsilon_{\text{tr}},\epsilon_i)$ and degrees of freedom $(\zeta_{\text{tr}}, \zeta_i)$ are those participating in the energy redistribution (i.e., collision) process, and $T_{\text{coll}} = 2\varepsilon_{\text{coll}}/k$ can be thought of as the "collision temperature." Each collision process is subject to the constraint $\varepsilon_{\text{coll}} = \varepsilon_{\text{tr}} + \varepsilon_i$, and therefore it is equivalent to sample ε'_i from $f(\varepsilon_{\text{coll}} - \varepsilon_i; T_{\text{coll}})f(\varepsilon_i; T_{\text{coll}})p_i(\varepsilon_{\text{coll}} - \varepsilon_i, \varepsilon_i)$ with a constant $\varepsilon_{\text{coll}}$, and therefore $\varepsilon_{\text{coll}} = \varepsilon'_{\text{tr}} + \varepsilon'_i$.

The standard acceptance–rejection algorithm (refer to Appendix A) requires normalization by a constant that is greater than or equal to the maximum value of the distribution function being sampled. Thus, for a given $\varepsilon_{\text{coll}}$, we start by writing the following inequality,

$$\begin{aligned} &[f(\varepsilon_{\text{coll}} - \varepsilon_i; T_{\text{coll}})f(\varepsilon_i; T_{\text{coll}})p_i(\varepsilon_{\text{coll}} - \varepsilon_i, \varepsilon_i)]\,|_{\text{max}} \\ &\quad \le [f(\varepsilon_{\text{coll}} - \varepsilon_i; T_{\text{coll}})f(\varepsilon_i; T_{\text{coll}})]\,|_{\text{max}}\; p_i(\varepsilon_{\text{coll}} - \varepsilon_i, \varepsilon_i)|_{\text{max}} \end{aligned} \tag{6.102}$$

with $0 \le \varepsilon_i \le \varepsilon_{\text{coll}}$. Following this inequality, we define

$$M \equiv [f(\varepsilon_{\text{tr}}; T_{\text{coll}})f(\varepsilon_i; T_{\text{coll}})]\,|_{\text{max}}\; p_i(\varepsilon_{\text{tr}}, \varepsilon_i)|_{\text{max}} \tag{6.103}$$

We then define the generalized equilibrium distribution function as $I(\varepsilon_i; \varepsilon_{\text{coll}}, T_{\text{coll}})$, where

$$\begin{aligned} &I(\varepsilon_i; \varepsilon_{\text{coll}}, T_{\text{coll}}) \\ &\quad = \frac{f(\varepsilon_{\text{coll}} - \varepsilon_i; T_{\text{coll}})f(\varepsilon_i; T_{\text{coll}})p_i(\varepsilon_{\text{coll}} - \varepsilon_i, \varepsilon_i)}{M} \\ &\quad = \frac{f(\varepsilon_{\text{coll}} - \varepsilon_i; T_{\text{coll}})f(\varepsilon_i; T_{\text{coll}})p_i(\varepsilon_{\text{coll}} - \varepsilon_i, \varepsilon_i)}{[f(\varepsilon_{\text{coll}} - \varepsilon_i; T_{\text{coll}})f(\varepsilon_i; T_{\text{coll}})]|_{\text{max}}\; p_i(\varepsilon_{\text{coll}} - \varepsilon_i, \varepsilon_i)|_{\text{max}}} \\ &\quad = \frac{f(\varepsilon_{\text{coll}} - \varepsilon_i; T_{\text{coll}})f(\varepsilon_i; T_{\text{coll}})}{[f(\varepsilon_{\text{coll}} - \varepsilon_i; T_{\text{coll}})f(\varepsilon_i; T_{\text{coll}})]|_{\text{max}}} \times \frac{p_i(\varepsilon_{\text{coll}} - \varepsilon_i, \varepsilon_i)}{p_i(\varepsilon_{\text{coll}} - \varepsilon_i, \varepsilon_i)|_{\text{max}}} \\ &\quad = I_1(\varepsilon_i; \varepsilon_{\text{coll}}, T_{\text{coll}}) \times I_2(\varepsilon_i; \varepsilon_{\text{coll}}) \end{aligned} \tag{6.104}$$

Therefore the procedure to sample ε'_i from $f(\varepsilon_{\text{coll}} - \varepsilon_i; T_{\text{coll}})$ $f(\varepsilon_i; T_{\text{coll}})p_i(\varepsilon_{\text{coll}} - \varepsilon_i, \varepsilon_i)$ with a constant $\varepsilon_{\text{coll}}$, using an acceptance–rejection technique is as follows:

(1) Generate a random number R_1 uniformly distributed between (0, 1).
(2) Calculate the value of $I(\varepsilon_i; \varepsilon_{\text{coll}}, T_{\text{coll}})$ where $\varepsilon_i = R_1\varepsilon_{\text{coll}}$.

(3) Generate a different random number, R_2, uniformly distributed between (0, 1).
(4) If $R_2 \leq I(\varepsilon_i; \varepsilon_{\text{coll}}, T_{\text{coll}})$, then the sample is accepted and $\varepsilon_i' = R_1 \varepsilon_{\text{coll}}$. Else, if $R_2 > I(\varepsilon_i; \varepsilon_{\text{coll}}, T_{\text{coll}})$, then no sample is accepted and repeat the process (1)–(4) with new R_1 and R_2 values.

Finally, the resulting value of ϵ_{tr}' is used to determine the post-collision velocities using a desired scattering law (Appendix C). For the particle selection prohibiting double relaxation procedure, the value of ϵ_i' can be immediately associated with the molecule selected to undergo internal energy exchange (refer to Fig. 6.14). However, for the pair selection procedure (where ϵ_i' represents the combined internal energy of both molecules) an additional acceptance–rejection procedure may be required to divide ϵ_i' among the two molecules (similar to Eq. 6.79).

Example 6.7 Rotational Relaxation in a Simple Diatomic Gas (Using Pair Selection)
In this example, we will simplify the preceding generalized equilibrium distribution function for the case of a simple diatomic gas where pair selection is employed and compare the result to that derived in Section 6.4.2. Also, in this example, we assume that p_{rot} is a constant.

To start, using Eq. 6.64 for rotation, we have the following for a specific value of ϵ_{coll}:

$$\begin{aligned}
&f(\varepsilon_{\text{coll}} - \varepsilon_{\text{rot}}; T_{\text{coll}}) f(\varepsilon_{\text{rot}}; T_{\text{coll}}) p_{\text{rot}} \\
&= p_{\text{rot}} \frac{1}{\Gamma(\frac{\zeta_{\text{tr}}}{2}) k T_{\text{coll}}} \left(\frac{\varepsilon_{\text{coll}} - \varepsilon_{\text{rot}}}{k T_{\text{coll}}} \right)^{\frac{\zeta_{\text{tr}}}{2} - 1} \\
&\quad e^{-\frac{\varepsilon_{\text{coll}} - \varepsilon_{\text{rot}}}{k T_{\text{coll}}}} \frac{1}{\Gamma(\frac{\zeta_{\text{rot}}}{2}) k T_{\text{coll}}} \left(\frac{\varepsilon_{\text{rot}}}{k T_{\text{coll}}} \right)^{\frac{\zeta_{\text{rot}}}{2} - 1} e^{-\frac{\varepsilon_{\text{rot}}}{k T_{\text{coll}}}} \qquad (6.105) \\
&= p_{\text{rot}} \frac{e^{-\frac{\varepsilon_{\text{coll}}}{k T_{\text{coll}}}}}{\Gamma(\frac{\zeta_{\text{tr}}}{2}) \Gamma(\frac{\zeta_{\text{rot}}}{2}) (k T_{\text{coll}})^{\frac{\zeta_{\text{rot}}}{2}} (k T_{\text{coll}})^{\frac{\zeta_{\text{tr}}}{2}}} (\varepsilon_{\text{coll}} - \varepsilon_{\text{rot}})^{\frac{\zeta_{\text{tr}}}{2} - 1} \varepsilon_{\text{rot}}^{\frac{\zeta_{\text{rot}}}{2} - 1} \\
&= A(\varepsilon_{\text{coll}}, T_{\text{coll}}, \zeta_{\text{tr}}, \zeta_{\text{rot}}) p_{\text{rot}} \, (\varepsilon_{\text{coll}} - \varepsilon_{\text{rot}})^{\frac{\zeta_{\text{tr}}}{2} - 1} \varepsilon_{\text{rot}}^{\frac{\zeta_{\text{rot}}}{2} - 1}
\end{aligned}$$

The maximum value of $f(\varepsilon_{\text{coll}} - \varepsilon_{\text{rot}}; T_{\text{coll}}) f(\varepsilon_{\text{rot}}; T_{\text{coll}}) p_{\text{rot}}$ for a fixed $\varepsilon_{\text{coll}}$ should appear at a specific $\varepsilon_{\text{rot},0}$ where

$$0 = \frac{\partial}{\partial \varepsilon_{\text{rot}}} [f(\varepsilon_{\text{coll}} - \varepsilon_{\text{rot}}; T_{\text{coll}}) f(\varepsilon_{\text{rot}}; T_{\text{coll}}) p_{\text{rot}}] |_{\varepsilon_{\text{rot}} = \varepsilon_{\text{rot},0}} \qquad (6.106)$$

which is

$$\frac{\varepsilon_{\text{rot},0}}{\varepsilon_{\text{coll}}} = \frac{\zeta_{\text{rot}} - 2}{\zeta_{\text{tr}} + \zeta_{\text{rot}} - 4} \qquad (6.107)$$

Using Eqs. 6.105 and 6.107, we can write $I(\varepsilon_{\text{rot}}; \varepsilon_{\text{coll}}, T_{\text{coll}})$ (appearing in Eq. 6.104), as

$$\begin{aligned} I(\varepsilon_{\text{rot}}; \varepsilon_{\text{coll}}, T_{\text{coll}}) &= I_1(\varepsilon_{\text{rot}}; \varepsilon_{\text{coll}}, T_{\text{coll}}) \times I_2(\varepsilon_{\text{rot}}; \varepsilon_{\text{coll}}) \\ &= I_1(\varepsilon_{\text{rot}}; \varepsilon_{\text{coll}}, T_{\text{coll}}) = \frac{f(\varepsilon_{\text{coll}} - \varepsilon_{\text{rot}}; T_{\text{coll}}) f(\varepsilon_{\text{rot}}; T_{\text{coll}})}{[f(\varepsilon_{\text{coll}} - \varepsilon_{\text{rot}}; T_{\text{coll}}) f(\varepsilon_{\text{roll}}; T_{\text{coll}})]|_{\max}} \\ &= \left(\frac{\zeta_{\text{tr}} + \zeta_{\text{rot}} - 4}{\zeta_{\text{rot}} - 2}\right)^{\frac{\zeta_{\text{rot}}}{2} - 1} \left(\frac{\zeta_{\text{tr}} + \zeta_{\text{rot}} - 4}{\zeta_{\text{tr}} - 2}\right)^{\frac{\zeta_{\text{tr}}}{2} - 1} \left(\frac{\varepsilon_{\text{rot}}}{\varepsilon_{\text{coll}}}\right)^{\frac{\zeta_{\text{rot}}}{2} - 1} \left(1 - \frac{\varepsilon_{\text{rot}}}{\varepsilon_{\text{coll}}}\right)^{\frac{\zeta_{\text{tr}}}{2} - 1}. \end{aligned} \tag{6.108}$$

First, since $p_{\text{rot}}|_{\max} = p_{\text{rot}}$, then $I_2(\varepsilon_{\text{rot}}; \varepsilon_{\text{coll}}) = 1$. Furthermore, we can clearly see from the above expression, that I_1 does not explicitly depend on temperature, rather it depends only on ε_{rot}, $\varepsilon_{\text{coll}}$, ζ_{tr} and ζ_{rot}.

The distribution function in Eq. 6.108 is general to any selection procedure, as long as ϵ_{rot} and ζ_{rot} are set appropriately. If pair selection is used, then ϵ_{rot} corresponds to the combined rotational energy of the collision pair, and ζ_{rot} corresponds to the combined rotational degrees of freedom. With these values (i.e., ζ_{rot} is replaced with $2\zeta_{\text{rot}}$), Eq. 6.108 becomes equivalent to Eq. 6.76 derived in Section 6.4.2.

For the special case of a diatomic gas, where $\zeta_{\text{rot}} = 2$, Eq. 6.108 can be further simplified to

$$I(\varepsilon_{\text{rot}}; \varepsilon_{\text{coll}}, T_{\text{coll}}) = \left(1 - \frac{\varepsilon_{\text{rot}}}{\varepsilon_{\text{coll}}}\right)^{\frac{\zeta_{\text{tr}}}{2} - 1} \tag{6.109}$$

IMPLICATIONS FOR VARIOUS INTERNAL ENERGY MODELS

It is interesting to investigate how the generalized distribution function in Eq. 6.104, and specifically the term $I_2(\varepsilon_i; \varepsilon_{\text{coll}})$, simplifies for different types of internal energy transfer models. This analysis was first performed by Zhang et al. (2014). In doing so, many of the issues and corrections associated with detailed balance reported for prior DSMC models are clearly explained. In addition, this general analysis provides a framework for the development of future inelastic collision models.

Here, the exact form of $I_2(\varepsilon_i; \varepsilon_{\text{coll}})$ is analyzed for several types of relaxation models employing different functional forms for the inelastic collision probability, p_i.

$p_i = \text{constant}$: In this case, since $p_i|_{\max} = p_i$, then $I_2 = 1$. Thus, the inequality appearing in the acceptance-rejection method becomes $R_2 \leq I(\varepsilon_i; \varepsilon_{\text{coll}}, T_{\text{coll}}) = I_1(\varepsilon_i; \varepsilon_{\text{coll}}, T_{\text{coll}})$ (see Eq. 6.108 in Example 6.7), which is the same as that discussed by Borgnakke and Larsen (1975) and has been used for a number of prior relaxation models.

$p_i = p_i(\varepsilon_{\text{tr}})$: In this case, the probability of an inelastic collision is dependent on the relative translational energy of the collision pair, and $I_2 = p_i(\varepsilon_{\text{tr}})/\{p_i(\varepsilon_{\text{tr}})|_{\max}\}$. Without including I_2 in the inequality for acceptance–rejection sampling, there was difficulty in achieving equipartition of energy between ε_{tr} and ε_i in certain models from the literature (Boyd 1990b; Abe 1994; Choquet 1994; Wysong and Wadsworth 1998).

For translational-rotational energy exchange a cell-averaged probability, $\langle p_{rot} \rangle$, was used to correct this (Boyd 1990b), which is therefore constant for all collisions within a cell, resulting in $I_2 = 1$. This approach is similar to using a cell-averaged temperature to determine $p_{rot} = p_{rot}(T_{tr})$ and has the effect of relaxing all parts of the distribution function (within a given cell) at the same rate.

Later, Abe (1994) proposed to multiply I_1 in the inequality ($R_2 \leq I_1(\varepsilon_i; \varepsilon_{coll}, T_{coll})$) by a correction factor. This factor was demonstrated for hard-sphere molecules with the purpose of accounting for the bias introduced by the elastic collision rate on the selection of inelastic collision pairs. This correction factor is actually the same as I_2 (appearing in Eq. 6.104), that should indeed be included in the acceptance–rejection sampling inequality for this type of model.

$p_i = p_i(\varepsilon_{coll})$: In this case, the probability of inelastic collision is dependent on the total collision energy (ε_{coll}) (Boyd 1990a; Choquet 1994). When using only I_1 in the inequality ($R_2 \leq I_1(\varepsilon_i; \varepsilon_{coll}, T_{coll})$), this type of model does, in fact, satisfy detailed balance. In prior publications, this was explained based on the idea that ε_{coll} is a collision invariant (Bourgat, Desvillettes, Le Tallec, and Perthame 1994; Choquet 1994). An equivalent explanation is evident from Eq. 6.104, where since $p_i(\varepsilon_{coll})|_{max} = p_i(\varepsilon_{coll})$ for a fixed ε_{coll}, the result is that $I_2 = 1$. This is a benefit of using a probability model based on collision invariants, however, for general physical accuracy it may not be desirable to be restricted to only these collision quantities.

$p_i = p_i(\varepsilon_{tr}, \varepsilon_i)$: This is the most general expression for the probability of inelastic collision, that depends independently on both the translational and internal energy involved in the collision. This type of model has recently been proposed for translational-rotational energy transfer in nitrogen (Zhang et al., 2014). For any such model, $I_2(\varepsilon_i; \varepsilon_{coll}) = p_i(\varepsilon_{coll} - \varepsilon_i, \varepsilon_i)/\{p_i(\varepsilon_{coll} - \varepsilon_i, \varepsilon_i)|_{max}\}$. To sample the post collision energies ($\varepsilon'_{tr}, \varepsilon'_i$), the correct formulation for the acceptance–rejection sampling inequality, is $R_2 \leq I(\varepsilon_i; \varepsilon_{coll}, T_{coll}) = I_1(\varepsilon_i; \varepsilon_{coll}, T_{coll}) I_2(\varepsilon_i; \varepsilon_{coll})$ as shown in Eq. 6.104.

GENERAL CONNECTION FACTOR BETWEEN $Z_i(T_{tr}, T_i)$ AND $p_i(\epsilon_{tr}, \epsilon_i)$

It can be shown that for a general collision-quantity–based energy exchange model, for example $p_i(\epsilon_{tr})$ or $p_i(\epsilon_{tr}, \epsilon_i)$, that there is no analytical result for the connection factor C (given previously in Eq. 6.93 for constant p_i). This is a result of p_i being inside the integral expression in Eq. 6.89, with no analytical reduction possible. As shown by Zhang et al. (2014), the connection factor can be determined by numerically comparing DSMC relaxation rates with those using Jeans/Landau–Teller continuum equations. Since the basic dependence derived in Eq. 6.93 is expected for any model, one approach is to separate the connection factor as $C = C_a C_n$:

$$p_i = \frac{C_a C_n}{Z_i} = \left(\frac{\zeta_{tr} + \zeta_i}{\zeta_{tr}}\right) C_n \frac{1}{Z_i}. \tag{6.110}$$

Thus, an analytical factor $C_a = (\zeta_{tr} + \zeta_i)/\zeta_{tr}$ is maintained as in Eq. 6.93, and a numerical factor C_n is obtained numerically. In this manner, the dominant contribution to the connection factor may be captured by C_a, and the numerical factor C_n may have a weaker contribution. As an example, in a recent relaxation model for nitrogen (Zhang et al., 2014), a constant value of $C_n = 1.92$ was determined to be accurate. It is possible that such a value may apply to different selection procedures, a range of conditions, and even a range of species. However, this is presently not known and requires further study.

Finally, it is important to understand that both the BL molecular model and the Jeans/Landau–Teller continuum models, were independently developed phenomenological models. Therefore, it should not be expected that the models are analytically equivalent in the equilibrium limit. Of course, future nonequilibrium DSMC and continuum models could certainly be developed to have this consistency built into the models.

6.5 Summary

This chapter began by identifying the range of flow conditions and engineering applications for which the DSMC method is most appropriate. In general, this range includes transitional dilute gas flows between continuum and free-molecular regimes. The DSMC method achieves accuracy and efficiency by exploiting three fundamental characteristics of dilute gases. Specifically, molecules move in free-flight without interaction for the local mean collision time, impact parameters and initial orientations of colliding molecules are random, and there are an enormous number of molecules per cubic mean free path and a statistical representation of these molecules is accurate. Beyond these basic aspects of the DSMC method, this chapter described the various collision models used to determine the local collision rate and the outcomes of such collisions.

The collision rate is directly related to the gas transport properties (viscosity, thermal conductivity, and diffusivity). A number of collision models of varying complexity were discussed. The variable hard-sphere model, which captures the dependence of the total cross section on relative collision speed, was shown to be an elegant, efficient, and highly accurate model for a wide range of conditions.

Modeling internal energy transfer within DSMC includes a number of subtleties. Experimental data for molecular processes of interest are scarce and many DSMC models are parameterized using information obtained from near-equilibrium flow conditions. As detailed in this chapter, ensuring that molecular models used in DSMC are consistent with continuum models and experimental data requires understanding and careful implementation of the statistical models within DSMC.

Finally, there are many flows which do not include significant gradients in temperature and for which the vibrational energy of the gas can be safely neglected. For these flows, all required DSMC algorithms have been detailed in this chapter. For many such applications, the algorithms are straightforward and their accuracy has been well established. As computer resources continue to increase, the DSMC method will enable accurate solutions to many nonequiilibrium problems spanning from free-molecular to continuum regimes.

7 Models for Nonequilibrium Thermochemistry

7.1 Introduction

Chapter 6 outlined the Direct Simulation Monte Carlo (DSMC) method, as well as collision cross section models and their link to viscosity, thermal conductivity, and diffusivity. In addition, the general procedure for modeling internal energy transfer and ensuring consistency with continuum modeling was described. In this chapter, more advanced models for rotational and vibrational energy excitation are presented. These models are particularly important for high-temperature gas flows, where the vibrational energies of gas molecules become excited and chemical reactions begin to occur. This chapter describes the most widely used DSMC models for nonequilibrium thermochemistry and also focuses on consistency with continuum models for reacting flows. The DSMC models are formulated using theory from many chapters of the book: kinetic theory, quantum mechanics, statistical thermodynamics, and finite-rate processes. The chapter concludes by presenting DSMC simulation results for high-temperature chemically reacting air.

7.2 Rotational Energy Exchange Models

Various models for rotational relaxation have been proposed in the DSMC literature. Almost all widely used models are phenomenological and based on the approach of Borgnakke and Larsen (1975), referred to as the BL model. As described in Chapter 6, if a pair of simulated molecules is chosen for a collision (using the variable hard-sphere [VHS] or variable soft-sphere [VSS] model, for example), the pair is further tested for an inelastic collision, involving energy exchange between the translational and rotational energy modes (ε_{tr} and ε_{rot}).

The probability used within the DSMC method to perform an inelastic collision was given by Eq. 6.95, and more generally by Eq. 6.110, in conjunction with the selection procedure prohibiting double relaxation depicted schematically in Fig. 6.14. A physical model is required for what this probability should actually be. In general, there are three types of translational-rotational energy exchange models. The first approach uses a constant value for the rotational collision number ($Z_{\rm rot}$) everywhere in the flow. The second approach models $Z_{\rm rot}$ based on the gas temperature, typically the local cell temperature. In this manner, the value of $Z_{\rm rot}$ (and thus $p_{\rm rot}$) can be different in each DSMC cell, but is constant for all collisions within a cell. Finally, the third approach is to evaluate $p_{\rm rot}$ based on the collision quantities of each pair, for example, $p_{\rm rot} = p_{\rm rot}(\varepsilon_{\rm tr}, \varepsilon_{\rm rot})$. In this section, we present and discuss the most widely used translational-rotational energy exchange models.

7.2.1 Constant Collision Number

Although $Z_{\rm rot}$ has been shown to have a dependence on gas temperature, for some flow conditions, the approximation of constant $Z_{\rm rot}$ may be adequate. A typical value is $2 \leq Z_{\rm rot} \leq 6$, with a commonly used value of $Z_{\rm rot} = 4$. In this case, the probability used in the selection procedure (see Fig. 6.14) is simply obtained from Eq. 6.95. If the pair is selected for translational-rotational energy exchange then the post-collision rotational energy is sampled using the BL method using Eq. 6.104. Since the probability is constant within each cell, $I_2 = 1$ and the expression reduces to Eq. 6.108. For the common case of $\zeta_{\rm rot} = 2$, the post-collision rotational energy can be sampled by the simple expression in Eq. 6.109.

7.2.2 The Parker Model

The rotational inelastic collision number, $Z_{\rm rot}$, has been shown to depend on the gas temperature in previous theoretical (Parker 1959), computational (Nyeland and Billing 1988; Billing and Wang 1992; Lordi and Mates 1970), and experimental studies (Carnevale, Carey, and Larson 1967; Healy and Storvick 1969; Kistemaker, Tom, and De Vries 1970; Annis and Malinauskas 1971; Ganzi and Sandler 1971). The model of Parker (1959) gives a temperature-dependent expression for the rotational collision number:

$$Z_{\rm rot}^{\rm Parker}(T) = \frac{Z_{\rm rot}^{\infty}}{1 + a(T^*/T) + b(T^*/T)^{\frac{1}{2}}} \tag{7.1}$$

Here, $a = \pi(1 + \pi/4)$, $b = \pi^{3/2}/2$, T is the gas temperature, T^* is the characteristic temperature of the intermolecular potential, and $Z_{\rm rot}^{\infty}$ is the limiting value. The Parker model was parameterized by interpreting the value of $Z_{\rm rot}$ from experimental data using the mean-collision-time definition from kinetic theory, $\tau_{\rm coll}^{\rm KT} \equiv \pi\mu(T)/4p$. To ensure that the rotational relaxation time ($\tau_{\rm rot}$) simulated in DSMC is the same as that inferred from the experiments, we desire

$$\tau_{\rm rot} = Z_{\rm rot}^{\rm Parker}\tau_{\rm coll}^{\rm KT} = Z_{\rm rot}\tau_{\rm coll}^{\rm VHS} \tag{7.2}$$

Here, Z_{rot} is the appropriate collision number to use within a DSMC simulation, if for example, the VHS collision model is used. The resulting expression for Z_{rot} corresponding to the Parker model and the VHS model is

$$Z_{\mathrm{rot}} = \frac{\tau_{\mathrm{coll}}^{\mathrm{KT}}}{\tau_{\mathrm{coll}}^{\mathrm{VHS}}} Z_{\mathrm{rot}}^{\mathrm{Parker}} = \frac{15\pi}{2(6-2\nu)(4-2\nu)} Z_{\mathrm{rot}}^{\mathrm{Parker}} \tag{7.3}$$

Note that this is a near-unity, constant, correction factor that arises simply because the kinetic theory definition of mean collision time was used to infer $Z_{\mathrm{rot}}^{\mathrm{Parker}}$ from the experimental data. For collisions between diatomic nitrogen atoms, the parameters are determined as $Z_{\mathrm{rot}}^{\infty} = 23.5$ and $T^* = 91.5\,\mathrm{K}$ (Lordi and Mates 1970; Boyd 1990a).

The Parker model can be implemented within DSMC using a cell-based translational temperature ($T = T_{\mathrm{tr}}$, defined in Eq. 5.17 and in Appendix D) computed either at each timestep or using a running average of the cell temperature. Similar to the constant Z_{rot} case, the probability used in the selection procedure (see Fig. 6.14) is obtained from Eq. 6.95. If the pair is selected for transitional-rotational energy exchange then the post-collision rotational energy is sampled by means of the BL method using Eq. 6.104. Since the probability is still a constant within each cell, $I_2 = 1$ and the expression reduces to Eq. 6.108, and when $\zeta_{\mathrm{rot}} = 2$, to the simple expression in Eq. 6.109.

7.2.3 Variable Probability Exchange Model of Boyd

A collision-quantity–based model, where the probability of translational-rotational energy exchange is calculated directly from the collision energy of the colliding pair, has been proposed by Boyd (1990a). Whereas a temperature-based probability applies the same probability to all collision pairs within a cell, a collision-energy–based model applies a different energy exchange probability to each collision pair. Such a model has the potential to be more accurate for nonequilibrium flows since the model enables different parts of the velocity and rotational energy distribution functions to relax at different rates.

The probability expression is a function of the collision energy (ϵ_{coll}), which is the sum of the relative translational energy of the pair and the rotational energy (with degrees of freedom ζ_{rot}) *participating* in the collision as determined by the selection procedure (Section 6.4.3). The probability to be used in DSMC (analogous to Eq. 6.95), consistent with VHS model, is now directly obtained for each collision pair by the expression

$$\begin{aligned} \frac{p_{\mathrm{rot}}}{C}\frac{Z_{\mathrm{rot}}^{\infty}}{Z_{\mathrm{tr}}} = 1 &+ \left(\frac{\pi^2}{4}+\pi\right)\frac{\Gamma(\zeta_{\mathrm{rot}}+2-\nu)}{\Gamma(\zeta_{\mathrm{rot}}+1-\nu)}\left(\frac{kT^*}{\epsilon_{\mathrm{coll}}}\right) \\ &+ \left(\frac{\pi^{3/2}}{2}\right)\frac{\Gamma(\zeta_{\mathrm{rot}}+2-\nu)}{\Gamma(\zeta_{\mathrm{rot}}+(3/2)-\nu)}\left(\frac{kT^*}{\epsilon_{\mathrm{coll}}}\right)^{1/2} \end{aligned} \tag{7.4}$$

Here, C is the connection factor required for consistency with the Jeans equation, $C = (\zeta_{tr} + \zeta_{rot})/\zeta_{tr}$, and Z_{tr} is the translational collision number, which is taken to be unity. Upon integration of $p_{rot}(\epsilon_{coll})/C$ over the equilibrium distribution function for ϵ_{coll} (refer to Eq. 6.80), one recovers the rotational collision number, $Z_{rot}^{Parker}(T)$, in Eq. 7.1.

If a collision pair is selected for translational-rotational energy exchange (using the probability in Eq. 7.4) then the post-collision rotational energy is sampled using the BL method using Eq. 6.104. As discussed in Section 6.4.4, since the collision energy is a collision invariant, $I_2 = 1$, and the expression reduces to Eq. 6.108, and when $\zeta_{rot} = 2$, to the simple expression in Eq. 6.109.

7.2.4 Nonequilibrium Direction Dependent Model

In the Parker model, Z_{rot} is a function of the equilibrium translational temperature only. However, it has often been questioned if the rotational collision number, Z_{rot}, may depend not only on the equilibrium translational temperature, but also on the *direction* toward the equilibrium state (Lordi and Mates 1970; Boyd 1990b; Valentini, Zhang, and Schwartzentruber 2012). For example, compressing flows (such as a shock wave) involve regions where the rotational energy in the gas is much less than the translational energy, and thus rotational energy is being excited, whereas expanding flows involve regions where rotational modes can become partially frozen, containing more energy than the translation modes, and thus rotational energy is being deexcited. For the same equilibrium temperature, Z_{rot} may not be the same for both cases. These aspects of rotational energy modeling in direct simulation Monte Carlo are thoroughly assessed in a review article by Wysong and Wadsworth (1998). More recently, Valentini et al. (2012) quantified the direction dependence of Z_{rot} for nitrogen using molecular dynamics simulations of shock waves and expansions.

To account for the dependence of the rotational relaxation rate on the magnitude of nonequilibrium, and the direction to the equilibrium state (i.e., expansion vs. compression), Zhang et al. (2014) developed the nonequilibrium direction dependent (NDD) model for use in DSMC and computational fluid dynamics (CFD). The DSMC model uses the following probability of an inelastic collision:

$$p_{rot}(\varepsilon_{tr}, \varepsilon_{rot}) = \min\left\{C_n C_a\, \tilde{p}_{rot}(\varepsilon_{tr}, \varepsilon_{rot}), \frac{1}{2}\right\}, \tag{7.5}$$

where $\tilde{p}_{rot}(\varepsilon_{tr}, \varepsilon_{rot})$ is given by

$$\tilde{p}_{rot}(\varepsilon_{tr}, \varepsilon_{rot}) = \frac{\Gamma(\frac{\zeta_{tr}}{2})\Gamma(\frac{\zeta_{rot}}{2})}{\Gamma(\frac{\zeta_{tr}}{2}+n)\Gamma(\frac{\zeta_{rot}}{2}-n)Z_{rot}^{\infty}}\left[1+\left(\frac{\zeta_{tr}}{2}+n-1\right)\frac{kT^*}{\varepsilon_{tr}}\right]\left(\frac{\varepsilon_{tr}}{\varepsilon_{rot}}\right)^n \tag{7.6}$$

This probability is consistent with the VHS model and is used directly within the DSMC method (analogous to Eq. 6.95). As discussed earlier in Section 6.4.3, the energies (ε_{tr}, ε_{rot}) and associated degrees of freedom

(ζ_{tr}, ζ_{rot}) should be those that *participate* in selected collisions within the DSMC simulation. In this manner, the expression in Eq. 7.6 is valid for any inelastic collision selection procedure.

In Eq. 7.5, C_a is the analytical correction factor and C_n is the numerical correction factor (refer to Eq. 6.110) required for consistency with the Jeans equation. Specific for nitrogen, it was determined that $n = \frac{1}{2}$, $T^* = 180$ K and $Z_{rot}^{\infty} = 7.7$. The numerically determined connection factor was determined to be $C_n = 1.92$, and is specific to the particle selection prohibiting double relaxation procedure. However, it is possible that the value of C_n is rather insensitive to the selection procedure due to the separation of C_a.

The expression in Eq. 7.5 is directly used as the probability of an inelastic collision between a selected collision pair, and is also directly used for the BL post-collision energy redistribution using Eq. 6.104. As discussed in Section 6.4.4, $I_2 \neq 1$ and both terms in Eq. 6.104 must be used in the acceptance rejection algorithm. In the case of the NDD model, it is evident from Eq. 7.5 that the maximum probability is simply 1/2.

In the continuum limit of Maxwell–Boltzmann energy distributions (at T_{tr} and T_{rot}), the model reduces to the following rotational collision number $Z_{rot}(T_{tr}, T_{rot})$:

$$Z_{rot}(T_{tr}, T_{rot}) = \frac{Z_{rot}^{\infty}}{1 + \frac{T^*}{T_{tr}}} \left(\frac{T_{rot}}{T_{tr}} \right)^n \tag{7.7}$$

which is appropriate for use in multitemperature CFD solvers, where the Jeans equation appears as a source term in the transport equation for internal energy. In the limit of near equilibrium ($T_{tr} \approx T_{rot} = T$), this reduces to $Z_{rot}(T) = Z_{rot}^{\infty}/(1 + \frac{T^*}{T})$.

7.2.5 Model Results

Experimental data for translational-rotational relaxation in the literature is limited. As shown in Fig. 7.1, there is significant variability in the rotational collision number for nitrogen inferred from experiments at various temperatures. The functional form of the Parker model was derived analytically by invoking several physical assumptions regarding molecular collisions (Lordi and Mates 1970; Wysong and Wadsworth 1998) such as assuming two-dimensional collisions between initially nonrotating molecules. The resulting values of $Z_{rot}(T)$ are generally higher than predictions reported using more advanced computational methods (Nyeland and Billing 1988; Billing and Wang 1992; Valentini et al. 2012). Available experimental data and a number of computational predictions are overlaid in Fig. 7.1. In addition, the result of the NDD model, in the limit of near equilibrium ($T_{tr} \approx T_{rot} = T$) is shown, which, as described previously, was parameterized based on the molecular dynamics calculations of Valentini et al. (2012). For the NDD model, the dependence of Z_{rot} on the equilibrium temperature (T) is relatively weak, whereas the dependence of Z_{rot} on the magnitude of

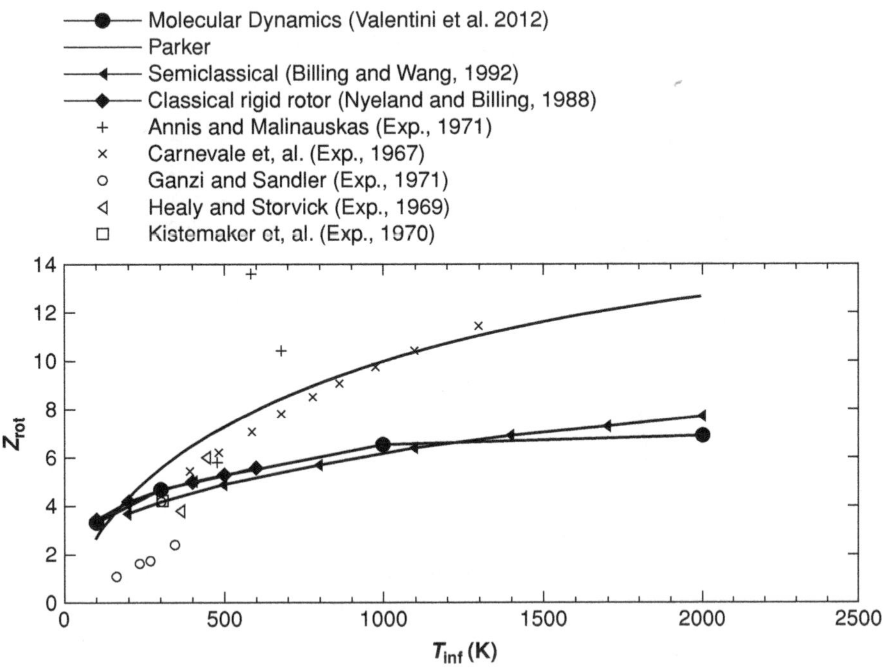

Figure 7.1 Experimental and computational data for the rotational collision number.

nonequilibrium and direction toward the equilibrium state is significant. Therefore the NDD model will have different effective values of Z_{rot} depending on the local values of both T_{tr} and T_{rot}, which Fig. 7.1 is not able to depict.

It is important to emphasize that any of the above described models can be reparameterized based on new experimental or computational data. However, the model formulations are different and represent the three basic options of using a constant cell-based temperature probability, a probability based on the total collision energy, or a probability based on the specific translational and rotational energies of molecules in a collision.

Figure 7.2 shows DSMC simulation results for a zero-dimensional isothermal heat-bath involving translational-rotational energy transfer. All simulations used the VHS model corresponding to nitrogen ($d_{ref} = 4.14 \times 10^{-10}$ m, $T_{ref} = 273$ K, $\omega = 0.74$). The translational energy of the gas was maintained at $T_{tr} = 2000$ K by resampling molecule velocities from a Maxwell–Boltzmann distribution after each timestep. The rotational energy of molecules was initially sampled from a Boltzmann distribution corresponding to either $T_{rot} = 200$ K or $T_{rot} = 3800$ K, and the rotational energy then increased or decreased due to translation–rotation energy transfer during the simulation. The pressure of the nitrogen gas was approximately $p = 4.7 \times 10^{-3}$ atm. The particle selection procedure of Zhang and Schwartzentruber (refer to Section 6.4.3) was used, and no vibrational degrees of freedom were considered.

Rotational excitation results are shown in Fig. 7.2(a). Results using a constant collision number ($Z_{rot} = 4$ and $Z_{rot} = 10$) are verified to agree exactly with the solution to the Jeans equation (refer to Eq. 6.101a where

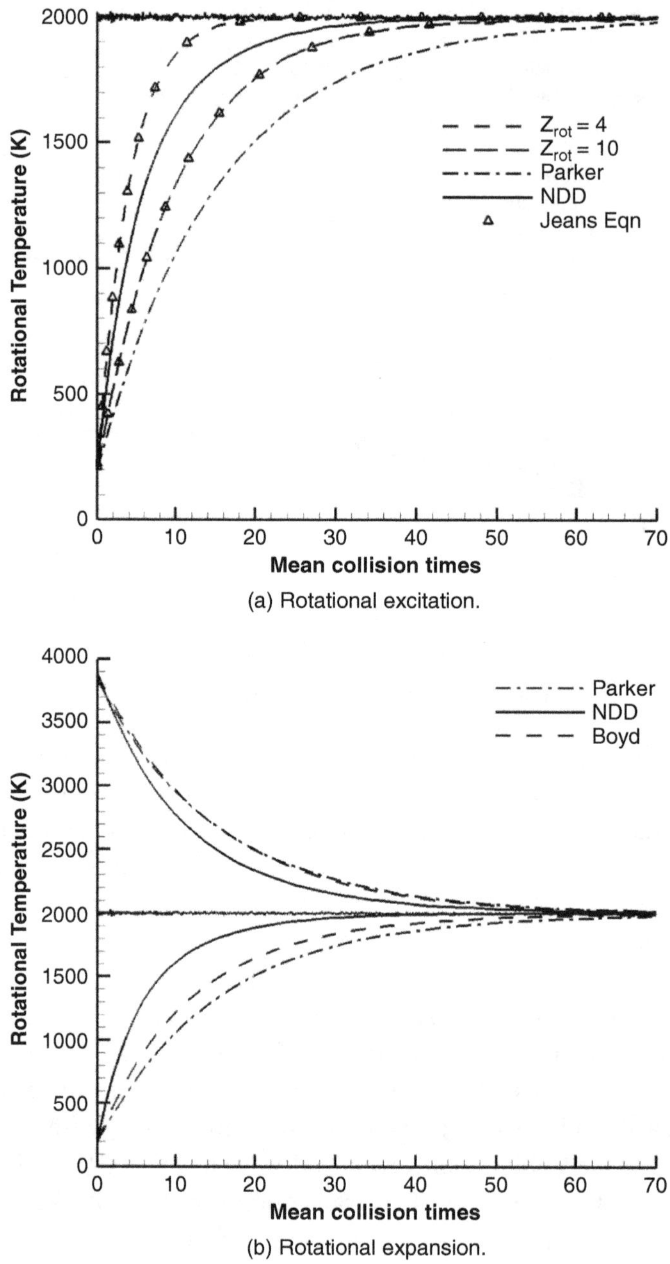

Figure 7.2 Rotational relaxation for different models.

$\tau_{\text{coll}} = \tau_{\text{coll}}^{\text{VHS}}$). As expected, based on the Parker model value for Z_{rot} at an equilibrium temperature of 2000 K (seen in Fig. 7.1), the Parker model predicts slightly slower rotational energy excitation compared to the $Z_{\text{rot}} = 10$ case. The NDD model predicts a rotational relaxation rate between the $Z_{\text{rot}} = 4$ and $Z_{\text{rot}} = 10$ cases. Also, since the NDD formulation models a dependence of the relaxation rate on the magnitude of nonequilibrium, the relaxation rate is seen to slow as T_{rot} approaches T_{tr}.

The results for rotational energy deexcitation are shown in the top half of Fig. 7.2(b). Since the Parker model has only a dependence on T_{tr}, which is constant, the deexcitation rate is identical to the excitation rate (bottom of Fig. 7.2(b)). The variable probability exchange model of Boyd, which is consistent with the Parker model in the limit of Maxwell–Boltzmann energy distributions, exhibits faster excitation than deexcitation. This is due to the fact that, for a given translational temperature, collision pairs with high rotational energy will have larger collision energies ϵ_{coll}. Since the probability of translational-rotational energy exchange is inversely proportional to ϵ_{coll} (Eq. 7.4), the gas will rotationally de-excite slower than it will excite. Therefore, the model of Boyd does capture a dependence on the relaxation rate due to the direction to the equilibrium state. The NDD model was formulated to capture this dependence and was parameterized with recent molecular dynamics data for nitrogen. For this case, the NDD model predicts an even larger difference between excitation and deexcitation rates.

7.3 Vibrational Energy Exchange Models

For many molecular species, internal energy is stored in vibrational modes only at high gas temperatures. As described in Chapters 2 and 3, in contrast to quantized rotational energy levels, quantized vibrational energy levels are widely spaced at low energies. As shown earlier in Fig. 3.9, the vibrational degrees of freedom begin to become active only above approximately 1000 K for air species. Indeed, at room temperature, the vast majority of molecules are in the ground vibrational energy level. Therefore, for many low-temperature flow conditions of interest, the vibrational energy of molecules can be safely neglected without loss of accuracy.

For gases at high temperatures, internal energy storage in the vibrational modes must be accounted for to simulate the correct specific heat and ratio of specific heats. Furthermore, as discussed in Chapter 4, the characteristic time associated with vibrational energy transfer is much larger compared to the characteristic time for rotational energy transfer ($\tau_{rot} < \tau_{vib}$). Thus it is more common for the vibrational relaxation timescale to become comparable to the flow timescale, resulting in nonequilibrium between the vibrational, rotational, and translational energies of the gas. This section describes the most widely used DSMC models for translational-vibrational energy transfer.

7.3.1 Constant Collision Number

The use of a constant vibrational collision number is identical to that for rotation described above. In this case, the probability used in the selection procedure (see Fig. 6.14) is simply obtained from Eq. 6.96. If the pair is selected for translational vibrational energy exchange then the post-collision vibrational energy is sampled by means of the BL method using Eq. 6.104. Since the probability is constant within each cell, $I_2 = 1$ and the expression simplifies similar to the case of rotational energy transfer. However, as described in the next section, the assumption of a constant vibrational collision number is inaccurate for most conditions.

7.3.2 The Millikan–White Model

It is well established that the vibrational collision number has a strong dependence on the gas temperature, often varying by orders of magnitude for temperature ranges of interest. Furthermore, unlike rotational degrees of freedom that become fully excited at very low temperature, the effective vibrational degrees of freedom, available for internal energy storage, are only partially excited over a large range of temperatures (refer to Fig. 3.9, for example). For this reason, the vibrational degrees of freedom are often modeled through a temperature dependence as well. Finally, due to the large energy spacing between quantized vibrational levels (compared to the spacing between rotational and translational levels), it is preferable to treat vibrational energy as quantized within a DSMC simulation. Each of these aspects along with appropriate DSMC models and algorithms are now described.

The time constant for vibrational energy relaxation (τ_{vib}) has a strong temperature dependence. Millikan and White (1963) developed the following correlation based on experimental data:

$$(p\tau_{\text{vib}})^{\text{MW}} = p^{\text{atm}} e^{A(T_{\text{tr}}^{-1/3} - B) - 18.42} \tag{7.8}$$

where $p^{\text{atm}} = 101,325$ Pa/atm is a conversion factor to convert the correlation to SI units (Pa sec), T_{tr} is the translational temperature, and A, B are constants related to molecular properties. The coefficients were determined by Millikan and White (1963) through fitting the experimental data as follows:

$$A = C\left(m_r \hat{N}\right)^{1/2} \theta_{\text{vib}}^{4/3} \tag{7.9}$$

$$B = 0.015\left(m_r \hat{N}\right)^{1/4} \tag{7.10}$$

where m_r is the reduced mass of the collision pair, θ_{vib} is the characteristic vibrational temperature, and C is a constant. The original paper by Millikan and White (1963) lists θ_{vib} and C values for a number of species, where C does not vary appreciably and has a value of approximately $C = 1.16 \times 10^{-3}$. The main dependence of the vibrational relaxation time constant, τ_{vib}, is on the temperature and the value of θ_{vib}, whereas the reduced mass of the collision pair, m_r, has only a small effect. Table 7.1 lists values of the model parameters for air species most commonly used in the DSMC and CFD communities. Note that experimental measurements were obtained only for $N_2 - N_2$ and $O_2 - O_2$ collisions, while the correlation is applied more generally for all

Table 7.1 Parameter Values for the Millikan–White Vibrational Relaxation Model for Air Species

Species pair	θ_{vib}	A	B
N_2–N_2^*, O_2, NO, N	3395	Eq. 7.9	Eq. 7.10
O_2–O_2^*, N_2, NO	2239	Eq. 7.9	Eq. 7.10
N_2–O	3395	72.4	0.015
O_2–N	2239	72.4	0.015
O_2–O	2239	47.7	0.059
NO–N_2, O_2, NO, N, O	2740	49.5	0.042

Species pairs for which experimental data are available are denoted by an asterisk.

collision pairs. For certain collisions involving O, N, and NO species, for which no experiments were performed, Park (1990) proposed alternate parameter values as listed in Table 7.1.

Analogous to Eq. 7.2, to ensure that the vibrational relaxation rate simulated in DSMC (using the VHS model) is the same as that inferred from the experiments, we desire

$$p\tau_{\text{vib}} = p\tau_{\text{coll}}^{\text{VHS}} Z_{\text{vib}}^{\text{MW}} = (p\tau_{\text{vib}})^{\text{MW}} \tag{7.11}$$

and therefore,

$$Z_{\text{vib}}^{\text{MW}} = \frac{(p\tau_{\text{vib}})^{\text{MW}}}{p\tau_{\text{coll}}^{\text{VHS}}} = \pi d_{\text{ref}}^2 \sqrt{\frac{8}{\pi m_r k T_{\text{tr}}}} \left(\frac{T_{\text{ref}}}{T_{\text{tr}}}\right)^{\nu} (p\tau_{\text{vib}})^{\text{MW}} \tag{7.12}$$

If this expression is extrapolated above 10,000 K, the vibrational relaxation time constant becomes unphysically low; lower than the expected mean collision time. As a result, high-temperature corrections have been added to the expression, in the form

$$Z_{\text{vib}} = Z_{\text{vib}}^{\text{MW}} + Z_{\text{vib}}^{\text{HT}} \tag{7.13}$$

Corrections were proposed by Park (1993) and Haas and Boyd (1993), both of which produce a very similar result:

$$Z_{\text{vib}}^{\text{HT-Park}} = \pi d_{\text{ref}}^2 \frac{T_{\text{tr}}^2}{\sigma_{\text{vib}}^{\text{Park}}} \left(\frac{T_{\text{ref}}}{T_{\text{tr}}}\right)^{\nu} \tag{7.14}$$

$$\sigma_{\text{vib}}^{\text{Park}} = 3 \times 10^{-21} \times (50{,}000)^2 \ (\text{m}^2\text{K}^2) \tag{7.15}$$

and

$$Z_{\text{vib}}^{\text{HT-Haas,Boyd}} = \pi d_{\text{ref}}^2 \frac{1}{\sigma_{\text{vib}}^{\text{Haas,Boyd}}} \left(\frac{T_{\text{ref}}}{T_{\text{tr}}}\right)^{\nu} \tag{7.16}$$

$$\sigma_{\text{vib}}^{\text{Haas,Boyd}} = 5.81 \times 10^{-21} \ (\text{m}^2) \tag{7.17}$$

The vibrational relaxation time constant (τ_{vib}) and the collision number (Z_{vib}) are plotted as a function of gas temperature for each expression in Fig. 7.3.

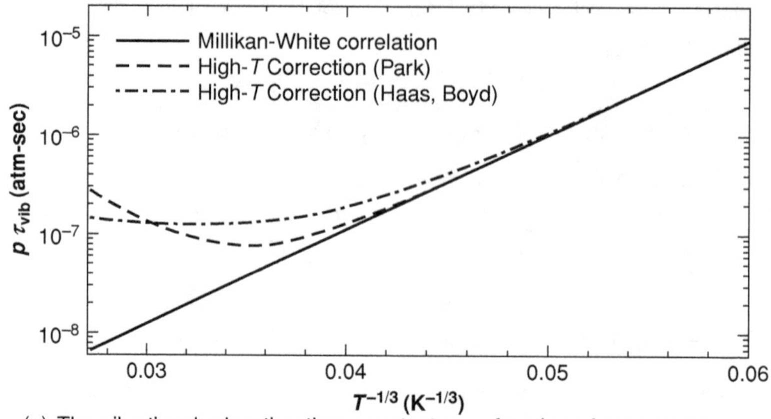

(a) The vibrational relaxation time constant as a function of temperature.

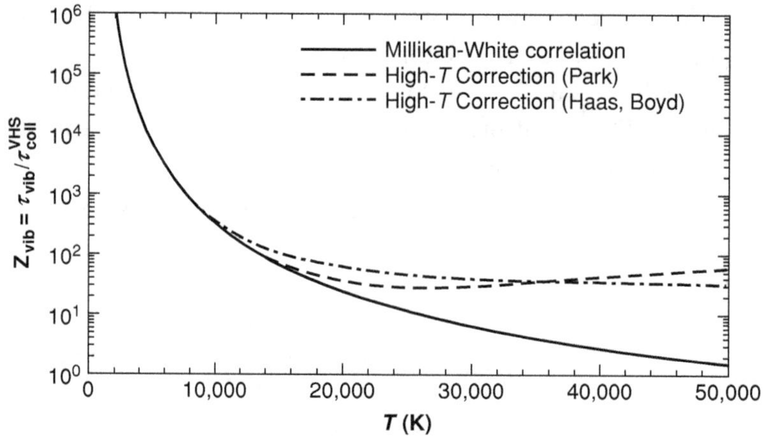

(b) The vibrational collision number as a function of temperature.

Figure 7.3 Temperature dependence of the vibrational relaxation time constant τ_{vib} and collision number Z_{vib}.

With the vibrational collision number specified in Eq. 7.13, this model can be implemented within DSMC using a cell-based translational temperature ($T = T_{\text{tr}}$, defined in Eq. 5.17 and Appendix D) either computed at each timestep or computed using a running average of the cell temperature. Similar to the constant Z_{vib} case, the probability used in the selection procedure (see Fig. 6.14) is obtained from Eq. 6.96, which also accounts for the temperature dependence of the available vibrational degrees of freedom, discussed earlier in Chapter 3 and shown in Fig. 3.9.

If the pair is selected for translational-vibrational energy exchange then, if a *continuous* vibrational energy model is used, the post-collision vibrational energy is sampled using the continuos BL distribution in Eq. 6.104. Since the probability is still a constant within each cell, $I_2 = 1$ and the expression simplifies in the same manner as described for rotation.

Currently, there is no widely accepted translational-vibrational energy exchange probability model that is based on collision quantities. The Millikan and White (1963) experimental data was obtained for near-equilibrium conditions and parameterized only as a function of the gas temperature. For this reason, also discussed by Bird (2013) on page 65, it is currently recommended to model the vibrational collision number using the macroscopic (cell-based) translational temperature as detailed in this section.

Recent research has focused on using computational chemistry to study internal energy excitation and its coupling to dissociation in high-temperature air. Accurate potential energy surfaces (PESs) for air species have been developed based on first-principles quantum chemistry calculations (Paukku et al., 2013; Bender et al., 2015; Lin et al., 2016; Varga et al., 2016). These PESs determine the interaction forces between nitrogen and oxygen atoms in any configuration. Collisions involving the species pairs listed in Table 7.1 can be performed at conditions corresponding to a range of gas temperatures and probabilities of vibrational energy exchange can be predicted. Recent simulations for nitrogen have revealed that Eq. 7.13 (with parameters from Table 7.1) is accurate for N_2-N_2 collisions, however, for N_2-N collisions the Millikan–White model over predicts $p\tau_{\text{vib}}$ by almost an order of magnitude (Kim and Boyd 2013; Panesi, Jaffe, Schwenke, and Magin 2013; Valentini et al. 2015; Valentini et al. 2016). Therefore the parameters in Table 7.1, and possibly the Millikan–White model itself, may be replaced by new models (temperature based or collision-quantity based) in coming years, as such first-principles data become available.

7.3.3 Quantized Treatment for Vibration

The analysis in Chapters 2 and 3 has shown that vibrational energy is quantized with relatively large energy spacing between each discrete level. Therefore, it may be more accurate to model vibrational energy as quantized within DSMC calculations. In this case, DSMC particles should have quantized energy levels and post-collision vibrational energies must be sampled from discrete (quantized) energy distributions. In this section we describe the discrete vibrational energy model of Bergemann and Boyd (1994).

For the case of a simple harmonic oscillator, the equilibrium distribution function of discrete vibrational every levels (ϵ_{vib}), relative to the ground state, can be written as

$$f_{\text{vib}}(\epsilon_{\text{vib}}, i; T_{\text{coll}}) = \delta\left(\frac{\epsilon_{\text{vib}}}{kT_{\text{coll}}} - i\frac{\theta_{\text{vib}}}{T_{\text{coll}}}\right)\frac{1}{kT_{\text{coll}}}\left[1 - e^{-\theta_{\text{vib}}/T_{\text{coll}}}\right]e^{-\epsilon_{\text{vib}}/kT_{\text{coll}}}$$
$$i = 0, 1, \ldots, \infty \tag{7.18}$$

where δ is the Dirac delta function, θ_{vib} is the characteristic temperature of vibration corresponding to the molecule undergoing vibrational energy exchange (see values in Table 7.1, for example), and T_{coll} represents an equilibrium temperature.

Within an equilibrium gas, the joint distribution function for a selected collision pair (involving translational-vibrational energy exchange), that has relative translational energy ϵ_{tr} and one molecule in quantized vibrational energy level i, is generally expressed as $f(\epsilon_{tr}; T_{coll})f_{vib}(\epsilon_{vib}, i; T_{coll})p_{vib}$. As described in Section 6.4.4, this is the distribution that post-collision energies should be sampled from. Here, $f(\epsilon_{tr}; T_{coll})$ corresponds to the continuous equilibrium distribution (Eq. 6.64) for the translational energy and $f_{vib}(\epsilon_{vib}, i; T_{coll})$ is the discrete equilibrium distribution given in Eq. 7.18. Finally, p_{vib} is the probability expression used to select a collision for translational-vibrational energy exchange (Eq. 6.96), which typically involves a vibrational collision number Z_{vib} (Eq. 7.13).

For the particle selection prohibiting double relaxation technique, the energy being redistributed (the collision energy) is the sum of the relative translational energy and the vibrational energy of one of the molecules, $\epsilon_{coll} = \epsilon_{tr} + \epsilon_{vib} = \epsilon_{tr} + ik\theta_{vib}$. Since this collision energy is conserved, it follows that $\epsilon_{coll} = \epsilon_{tr} + ik\theta_{vib} = \epsilon'_{tr} + i'k\theta_{vib}$, where primed values indicate post-collision values. Analogous to Eqs. 6.70–6.72, for a specific ϵ_{coll}, we can write a joint distribution function,

$$
\begin{aligned}
f_{coll,i}(\epsilon_{coll}, i'; T_{coll}) &= \int_0^{\epsilon_{coll}} f(\epsilon'_{tr}; T_{coll}) f_{vib}(\epsilon'_{vib}, i'; T_{coll}) d\epsilon'_{vib} \\
&= \int_0^{\epsilon_{coll}} f(\epsilon_{coll} - \epsilon'_{vib}; T_{coll}) f_{vib}(\epsilon'_{vib}, i'; T_{coll}) d\epsilon'_{vib}
\end{aligned}
\tag{7.19}
$$

Ultimately, the expression required in the acceptance–rejection algorithm, analogous to Eq. 6.104, requires a maximum value for the distribution function. The maximum of $f_{coll,i}(\epsilon_{coll}, i'; T_{coll})$ is realized when $i = 0$. Furthermore, for the case where p_{vib} is based on the cell-averaged temperature (for example, the Millikan–White model described earlier), or where p_{vib} is based on a collision invariant quantity, then $I_2 = p_{vib}/[p_{vib}]_{max} = 1$. The resulting expression, used directly in the acceptance–rejection algorithm within the DSMC simulation is

$$
I = I_1 = \frac{f_{coll,i}(\epsilon_{coll}, i')}{[f_{coll,i}]_{max}} = \left(1 - \frac{i'k\theta_{vib}}{\epsilon_{coll}}\right)^{\zeta_{tr}/2-1} \tag{7.20}
$$

Note that it is the collision energy ϵ_{coll} (not the equilibrium temperature T_{coll}) that appears in the final expression. Similar to the procedure to sample the generalized continuous expression (given previously in Eq. 6.104), the procedure to sample vibrational energy levels (i) from the discrete expression in Eq. 7.20 involves the following steps:

(1) Calculate the integer: $i_{max} = \lfloor \epsilon_{coll}/(k\theta_{vib}) \rfloor$, where $\lfloor\ \rfloor$ indicates truncation.
(2) Randomly select a level $0 < i' < i_{max}$ using the expression: $i' = \lfloor (i_{max} + 1) R_1 \rfloor$, where R_1 is a random number uniformly distributed between 0 and 1.
(3) Calculate the value of I (Eq. 7.20) using the randomly selected value i'.

(4) Generate a different random number, R_2, uniformly distributed between 0 and 1.
(5) If $R_2 \leq I$, then the sample is accepted, $\epsilon'_{\mathrm{vib}} = i'k\theta_{\mathrm{vib}}$ is the post-collision energy of the molecule undergoing vibrational energy exchange, and the new translational energy of the collision pair is $\epsilon'_{\mathrm{tr}} = \epsilon_{\mathrm{coll}} - \epsilon'_{\mathrm{vib}}$. Else, if $R_2 > I$, then no sample is accepted and repeat the process (1)–(5) with new R_1 and R_2 values.

Finally, the resulting value of ϵ'_{tr} is used to determine the post-collision velocities using a desired scattering law (refer to Appendix C).

7.3.4 Model Results

DSMC simulation results for a zero-dimensional isothermal heat bath involving translational-rotational and translational-vibrational energy transfer are shown in Fig. 7.4. All simulations use the VHS model corresponding to nitrogen ($d_{\mathrm{ref}} = 4.14 \times 10^{-10}$ m, $T_{\mathrm{ref}} = 273$ K, $\omega = 0.74$) and the pressure of the nitrogen gas was approximately $p = 4.7 \times 10^{-3}$ atm. The translational energy of the gas was maintained at $T_{\mathrm{tr}} = 8000$ K and $T_t = 20{,}000$ K by resampling molecule velocities from a Maxwell–Boltzmann distribution after each timestep. The rotational and vibrational energies of molecules were initially sampled from Boltzmann energy distributions corresponding to $T_{\mathrm{rot}} = T_{\mathrm{vib}} = 1000$ K. Rotational and vibrational energies then increase due to translation–rotation and translation–vibration energy transfer during the simulation. The Parker model (Section 7.2.2) was used for the probability of translational-rotational energy transfer and the Millikan–White model combined with a quantized treatment for vibrational energy was used for translational-vibrational energy exchange. The particle selection procedure prohibiting double relaxation, following the implementation of Zhang and Schwartzentruber (refer to Section 6.4.3), was used for all results.

At a translational temperature of 8000 K, vibrational excitation is much slower than rotational excitation. At this temperature, the Millikan–White expression for the vibrational collision number (Eq. 7.13) is approximately $Z_{\mathrm{vib}} = 720$. As seen in Fig 7.4, the DSMC solution using the Millikan–White model is verified to agree with the solution to the Landau–Teller equation (Eq. 6.101b, where $\tau_{\mathrm{coll}} = \tau_{\mathrm{coll}}^{\mathrm{VHS}}$). Note that the number of vibrational degrees of freedom (ζ_{vib}) is constant since it is a function of the translational temperature, which is constant in such isothermal relaxations. At a temperature of 20,000 K, vibrational excitation is much faster and approaches the rotational energy relaxation rate predicted by the Parker model. Indeed, the Millikan–White model predicts a vibrational collision number of approximately $Z_{\mathrm{vib}} = 31$ and the DSMC simulation result is verified to reproduce the solution of the Landau–Teller equation for this value.

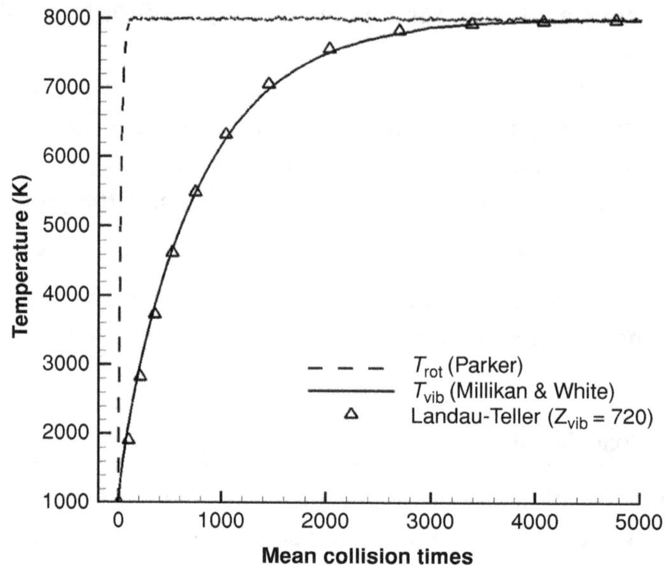

(a) Rotational and vibrational excitation to 8000 K.

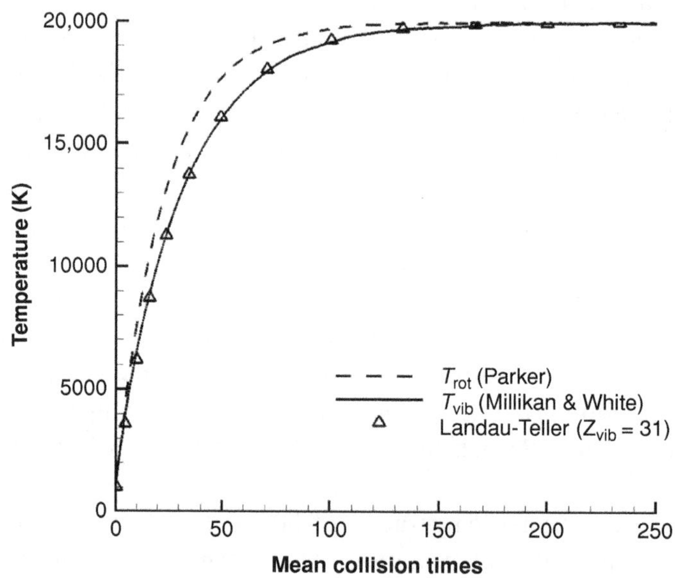

(b) Rotational and vibrational excitation to 20,000 K.

Figure 7.4 Rotational and vibrational excitation using various models.

In summary, for air species at low temperatures (typically below 1000 K) most molecules are in the ground vibrational state and the vibrational energy in the gas is negligible. Above these temperatures, vibrational energy is excited/deexcited at a finite rate that is much slower than rotational energy transfer. Unlike the rate of rotational energy transfer, the rate of vibrational energy transfer has a strong temperature dependence. At very high temperatures, vibrational relaxation is fast and begins to approach the rotational relaxation rate. Since the dissociation process of molecular species is coupled to the internal energy state, for example, molecules with high vibrational energy tend to dissociate rapidly, when studying reacting flows it is important to accurately model internal energy relaxation rates. DSMC modeling of dissociation and general chemical reactions is the focus of the next two sections.

7.4 Dissociation Chemical Reactions

Collisions within a DSMC simulation that involve sufficient energy to break a chemical bond can result in dissociation. For such collisions, a probability that the pair undergoes a dissociation reaction must be specified. Ideally the dissociation probability should be physically realistic for individual collisions, and therefore accurate for any degree of nonequilibirum, yet should also be consistent with standard reaction rate laws, that use for example, Arrhenius rate coefficients. If a pair is selected for a dissociation reaction, the properties of all product atoms and molecules must be determined. This section presents the most widely used DSMC dissociation models and the manner in which they are parametrized.

7.4.1 Total Collision Energy Model

The most common reaction probability model used in DSMC is the Total Collision Energy (TCE) model of Bird (1994). In this model, the reaction probability is a function of the total collision energy (ϵ_{coll}) and the activation energy (ϵ_a) for the reaction being considered:

$$P_{\text{react}} = \begin{cases} 0 & \text{if } \epsilon_{\text{coll}} \leq \epsilon_a \\ C_1(\epsilon_{\text{coll}} - \epsilon_a)^{C_2}(1 - \epsilon_a/\epsilon_{\text{coll}})^{C_3} & \text{if } \epsilon_{\text{coll}} > \epsilon_a \end{cases} \tag{7.21}$$

where C_1, C_2, and C_3 are constants related to the species collision parameters and the dissociation reaction rate constant k_f. In the case of dissociation, the activation energy is simply the dissociation energy. Such reaction rates are typically specified for each possible species pair (A colliding with B), where A and B can be the same species or different species.

In the limit of thermal equilibrium ($T_{\text{tr}} = T_{\text{rot}} = T_{\text{vib}} = T$) we can link this reaction probability directly to the reaction rate $k_f(T)$. In this manner, we can determine the model parameters (C_1, C_2, C_3) to ensure consistency between the simulated reaction rate in DSMC and the reaction rate expression $k_f(T)$. Specifically, we can write the time rate of change of species A due to reactive collisions with species B that is simulated in DSMC as

$$\frac{dn_{\text{A}}}{dt} = -n_{\text{A}}\nu_{\text{A,B}} \int_0^\infty P_{\text{react}}(\epsilon_{\text{coll}})f(\epsilon_{\text{coll}})d(\epsilon_{\text{coll}}) \tag{7.22}$$

In this equation n_{A} is the number density of species A and $\nu_{\mathrm{A,B}}$ is the collision rate between molecules of species A and B, per molecule of species A (units of $1/s$). For the VHS collision model we have

$$\nu_{\mathrm{A,B}} = \frac{n_{\mathrm{B}}\,(d_{\mathrm{ref}})^2_{\mathrm{A,B}}}{(1+\delta_{\mathrm{A,B}})}\sqrt{\frac{8\pi k\,(T_{\mathrm{ref}})_{\mathrm{A,B}}}{m_r}}\left(\frac{T}{(T_{\mathrm{ref}})_{\mathrm{A,B}}}\right)^{1-\omega_{\mathrm{A,B}}} \tag{7.23}$$

Here $\delta_{\mathrm{A,B}}$ is the symmetry factor that accounts for the fact that collisions between like-species are counted twice ($\delta_{\mathrm{A,B}} = 1$ for like-species where A = B, and is zero otherwise). For the remainder of the section, the subscripts (A, B) have been dropped from the VHS model parameters, since all equations are specific to the collision pair under consideration. The integral expression in Eq. 7.22 represents the probability of dissociation (given collision energy ϵ_{coll}) averaged over the distribution function of ϵ_{coll} found in collisions, given by

$$f(\epsilon_{\mathrm{coll}}) = \frac{1}{\Gamma(\zeta_T/2)}\frac{1}{kT}\left(\frac{\epsilon_{\mathrm{coll}}}{kT}\right)^{\zeta_T/2-1} e^{-\epsilon_{\mathrm{coll}}/kT} \tag{7.24}$$

This is the general equilibrium distribution function given previously in Eq. 6.64. Specifically, the total collision energy is the sum of the relative translational energy of the collision pair and the energies of all internal modes (i) for each of the two colliding particles:

$$\epsilon_{\mathrm{coll}} = \epsilon_{\mathrm{tr}} + \sum_i (\epsilon_{i,1} + \epsilon_{i,2}) \tag{7.25}$$

The total number of degrees of freedom (referred to as ζ_T) corresponding to this total collision energy is therefore the sum of the translational degrees of freedom and all internal degrees of freedom for both particles:

$$\zeta_T = \zeta_{\mathrm{tr}} + \sum_i (\zeta_{i,1} + \zeta_{i,2}) \tag{7.26}$$

Recall that the translational degrees of freedom (ζ_{tr}) depends on the elastic collision model used, for example $\zeta_{\mathrm{tr}} = 5 - 2\omega$ for the VHS model.

The exponent C_3 in the TCE model is directly related to the total degrees of freedom,

$$C_3 = \zeta_T/2 - 1 \tag{7.27}$$

To directly compare the simulated dissociation rate to a standard reaction rate equation, and therefore determine C_1 and C_2, we first combine Eqs. 7.21 and 7.22:

$$\frac{dn_{\mathrm{A}}}{dt} = -n_{\mathrm{A}}\nu_{\mathrm{A,B}}\int_{\epsilon_a}^{\infty} C_1(\epsilon_{\mathrm{coll}}-\epsilon_a)^{C_2}(1-\epsilon_a/\epsilon_{\mathrm{coll}})^{\zeta_T/2-1} f(\epsilon_{\mathrm{coll}})d(\epsilon_{\mathrm{coll}}) \tag{7.28}$$

Next, by rewriting this equation in terms of $(\epsilon_{\text{coll}} - \epsilon_a)/kT$ we have

$$\frac{dn_{\text{A}}}{dt} = -n_{\text{A}}\nu_{\text{A,B}}C_1\frac{(kT)^{C_2}}{\Gamma(\zeta_T/2)}e^{-\epsilon_a/kT}\int_0^{\infty}\left(\frac{\epsilon_{\text{coll}}-\epsilon_a}{kT}\right)^{C_2+\zeta_T/2-1} \times e^{-\left(\frac{\epsilon_{\text{coll}}-\epsilon_a}{kT}\right)}d\left(\frac{\epsilon_{\text{coll}}-\epsilon_a}{kT}\right) \tag{7.29}$$

The integral is now the standard Γ function, $\Gamma(Y) \equiv \int_0^{\infty} x^{Y-1}e^{-x}dx$, where, in this case, $x = (\epsilon_{\text{coll}} - \epsilon_a)/kT$ and $Y = C_2 + \zeta_T/2$. Therefore the expression reduces to

$$\frac{dn_{\text{A}}}{dt} = -n_{\text{A}}\nu_{\text{A,B}}C_1\frac{\Gamma(C_2+\zeta_T/2)}{\Gamma(\zeta_T/2)}(kT)^{C_2}e^{-\epsilon_a/kT} \tag{7.30}$$

Finally, by substitution of the collision rate (Eq. 7.23) for the VHS collision model, we have

$$\frac{dn_{\text{A}}}{dt} = -\frac{n_{\text{A}}n_{\text{B}}d_{\text{ref}}^2}{(1+\delta_{\text{A,B}})}\sqrt{\frac{8\pi kT_{\text{ref}}}{m_r}}C_1\frac{\Gamma(C_2+\zeta_T/2)}{\Gamma(\zeta_T/2)}(k)^{C_2}\frac{T^{C_2+1-\omega}}{T_{\text{ref}}^{1-\omega}}e^{-\epsilon_a/kT} \tag{7.31}$$

We can now compare this to the general rate expression,

$$\frac{dn_{\text{A}}}{dt} = -k_f(T)n_{\text{A}}n_{\text{B}} \tag{7.32}$$

where the rate constant is given in modified Arrhenius form:

$$k_f(T) = AT^{\eta}e^{-\epsilon_a/kT} \tag{7.33}$$

This rate expression was presented earlier in Chapter 4 in a slightly different form (Eq. 4.56) with units of cm^3/mol/sec. Since Eq. 7.32 uses species number densities, the units of k_f in Eq. 7.33 are m^3/molecule/sec.

By equating the result from the TCE model evaluated in the limit of thermal equilibrium (Eq. 7.31) with the Arrhenius rate model (Eqs. 7.32 and 7.33), the remaining TCE model parameters C_1 and C_2 are determined to be

$$C_2 = \eta - 1 + \omega \tag{7.34}$$

and

$$C_1 = A\left\{\frac{d_{\text{ref}}^2}{(1+\delta_{A,B})}\frac{\Gamma(C_2+\zeta_T/2)}{\Gamma(\zeta_T/2)}\sqrt{\frac{8\pi kT_{\text{ref}}}{m_r}}\frac{(k)^{C_2}}{T_{\text{ref}}^{1-\omega}}\right\}^{-1} \tag{7.35}$$

Therefore, if Arrhenius rate model parameters (A, η, and ϵ_a) are known, one can determine the corresponding TCE model parameters (C_1, C_2, and C_3) such that under thermal equilibrium conditions, the simulated dissociation rate is fully consistent with the continuum expression.

An alternate form of Eq. 7.35 can be derived by separating the translational degrees of freedom from the internal degrees of freedom. Specifically, by grouping the internal degrees of freedom as

$$\overline{\zeta} \equiv \frac{\sum_i (\zeta_{i,1} + \zeta_{i,2})}{2} \tag{7.36}$$

we can express the total degrees of freedom, corresponding to the VHS model, as

$$\frac{\zeta_T}{2} = \overline{\zeta} + \frac{5}{2} - \omega \tag{7.37}$$

In this case, the expression (equivalent to Eq. 7.35) for C_1 becomes

$$C_1 = A \frac{\sqrt{\pi}\,(1+\delta_{A,B})}{2\pi d_{\text{ref}}^2} \frac{\Gamma(\overline{\zeta} + 5/2 - \omega)}{\Gamma(\overline{\zeta} + \eta + 3/2)} \sqrt{\frac{m_r}{2kT_{\text{ref}}}} \frac{T_{\text{ref}}^{1-\omega}}{k^{\eta-1+\omega}} \tag{7.38}$$

This is the form derived by Bird (1994, p. 127), corresponding to the VHS collision model.

In this manner, in addition to VHS model parameters, standard Arrhenius rate parameters (A, η, and ϵ_a) can form the input for DSMC simulations of reacting flows. Within the DSMC algorithm, for each pair chosen to collide (using the VHS model, for example), the total collision energy, total degrees of freedom, and TCE model parameters (C_1, C_2, and C_3) can be determined. The pair is then selected to undergo a dissociation reaction using a standard acceptance–rejection technique with the probability P_{react} given in Eq. 7.21. Within DSMC, the value of P_{react} is specific to the properties of an individual collision, yet for a gas where velocity and internal energy distributions are close to Maxwell–Boltzmann and the gas is in thermal equilibrium ($T_{\text{tr}} = T_{\text{rot}} = T_{\text{vib}} = T$), the simulated dissociation rate will be fully consistent with an equilibrium rate coefficient, $k_f(T)$, used for example in CFD calculations.

As presented earlier in Chapter 4, one may interpret a steric factor (P_s) from such a rate expression. Analogous to the earlier theoretical derivation, the steric factor can be thought of as the probability of dissociation, given that $\epsilon_{\text{coll}} > \epsilon_a$. Therefore, for the TCE model, the steric factor is simply equal to P_{react}. Unlike the *constant* steric factor derived in Chapter 4 (Eq. 4.96), the steric factor in the TCE model is applied on a per-collision basis and is dependent on the collision energy. After substituting the expressions for C_1, C_2, and C_3 into Eq. 7.21, the TCE model probability (i.e., steric factor), for $\epsilon_{\text{coll}} > \epsilon_a$, is

$$P_s^{\text{TCE}} \equiv P_{\text{react}} = C_1 \frac{(\epsilon_{\text{coll}} - \epsilon_a)^{\eta + \overline{\zeta} + 1/2}}{\epsilon_{\text{coll}}^{\overline{\zeta} + 3/2 - \omega}} \tag{7.39}$$

On inspection, it is evident that the TCE probability expression should be used only for the following range of parameters:

$$-(\overline{\zeta} + 1/2) < \eta < -(1-\omega) \tag{7.40}$$

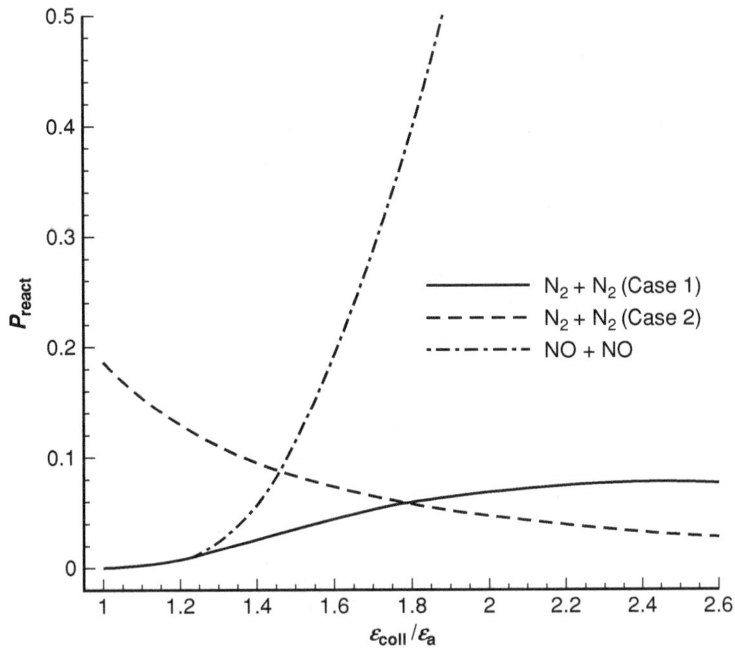

Figure 7.5 Dissociation probability as a function of collision energy for various model parameter values.

Parameters for many reacting gases, do in-fact lie within this range. For example, typical parameters for nitrogen dissociation ($N_2 + N_2 \rightarrow N + N + N_2$) are $A = 1.162 \times 10^{-8}$ (in units of m^3/molecule/sec), $\eta = -1.5$, and $\epsilon_a/k = 113{,}000$ K. Since both rotation and vibration are typically excited before significant dissociation occurs, a value of $\overline{\zeta} = 4$ is appropriate. Also, standard VHS parameter values for nitrogen are $\omega = 0.75$, $d_{\text{ref}} = 4 \times 10^{-10}$ m, and $T_{\text{ref}} = 273$ K. Referring to this set of parameters as Case 1, the values of P_{react} are plotted for values of $\epsilon_{\text{coll}}/\epsilon_a \geq 1$ in Fig. 7.5. In this case, we see that $P_{\text{react}} = 0$ for $\epsilon_{\text{coll}} \leq \epsilon_a$. The probability increases with increasing ϵ_{coll}, but remains relatively constant for $\epsilon_{\text{coll}}/\epsilon_a < 3$. In fact, the probability reaches a maximum at some value of $\epsilon_{\text{coll}}/\epsilon_a$ and then decreases afterward. Case 1 parameter values satisfy Eq. 7.40 and therefore this result is an example of the desired trend for P_{react} in function of ϵ_{coll}.

Previously, in the theoretical derivation presented in Chapter 4, a hard-sphere gas ($\omega = 1/2$) and an even number of degrees of freedom ($\zeta_T = 2\zeta$) was assumed. With these assumptions, Eq. 7.37 becomes $\overline{\zeta} = \zeta - 2$, and furthermore as used previously in Eq. 4.93, $\eta = 3/2 - \zeta = -1.5$. Substituting these specific values for ω, $\overline{\zeta}$, and η into Eq. 7.39 leads to

$$P_s^{\text{TCE}} = \frac{(1 + \delta_{\text{A,B}})}{\pi d_{\text{ref}}^2} \sqrt{\frac{\pi m_r}{8k}} \frac{\Gamma(\zeta)}{(\epsilon_{\text{coll}}/k)^{\zeta-1}} A \tag{7.41}$$

This result is the same as that derived in Eq. 4.96 except, instead of a constant dependence on $\theta_d^{\zeta-1} = (\epsilon_a/k)^{\zeta-1}$ for the steric factor, there is now a dependence on $(\epsilon_{coll}/k)^{\overline{\zeta}-1}$. The result is plotted in Fig. 7.5 (referred to as Case 2). Here, $\eta = -(\overline{\zeta} + 1/2)$, which leads to a finite value for P_{react} when $\epsilon_{coll} = \epsilon_a$. Note that this value ($P_{react} = 0.19$) is the same value calculated in Eq. 4.98 in Chapter 4, where the theoretical expression for P_s was a constant (a function only of ϵ_a). In this case, we see that P_{react} immediately decreases with increasing ϵ_{coll}, which is likely not physically correct on a per-collision basis. However, as mentioned previously, this particular set of parameters has $\eta = -(\overline{\zeta} + 1/2)$, and so (via Eq. 7.40) these parameters would not typically be used within the TCE model. In fact, if $\eta < -(\overline{\zeta} + 1/2)$ then as $\epsilon_{coll} \rightarrow \epsilon_a$ the expression results in $P_{react} \rightarrow \infty$, which is clearly unphysical.

Finally, if $\eta > -(1 - \omega)$, then $P_{react} = 0$ for $\epsilon_{coll} = \epsilon_a$ and increases for $\epsilon_{coll} > \epsilon_a$. However, the probability no longer reaches a maximum; rather, it may rapidly increase and tend to infinity with increasing ϵ_{coll}. An example involving such a range of parameters is the dissociation of nitric oxide (NO) where, in many CFD rate models, $\eta = 0$. A common set of parameters for the reaction NO + NO → N + O + NO includes $A = 8.302 \times 10^{-15}$ (in units of m^3/molecule/sec), $\eta = 0$, $\epsilon_a/k = 75{,}500$ K, and $\overline{\zeta} = 4$. Using the same set of VHS parameters as for Cases 1 and 2 discussed previously, the result is plotted in Fig. 7.5. Clearly, the probability is seen to rapidly increase for large $\epsilon_{coll}/\epsilon_a$ and at some point $P_{react} > 1$. However, it is important to note that for $\epsilon_{coll}/\epsilon_a < 2$ the probabilities are physically reasonable. In fact, previous analysis in Chapter 4 assumed a constant steric factor of $P_s = 1/3$ for this reaction and found good agreement between the theoretical rate expression and experimental data (refer to Fig. 4.12). Furthermore, the number of collisions where $\epsilon_{coll}/\epsilon_a > 2$ drops quickly as ϵ_{coll} is further increased. Therefore, even if probabilities become greater than unity, they will be applied infrequently and may have little effect on the dissociation process. In this manner, some parameter combinations that have $\eta > -(1 + \omega)$ (such as this NO dissociation example), may still result in probabilities that are both physically accurate on a per-collision basis and are also consistent with the equilibrium rate expression, with rate coefficient $k_f(T)$.

In summary, as long as model parameters within the proper range are utilized, the TCE model is a simple model that accurately captures reactive collision physics under nonequilibrium conditions, while also integrating exactly to standard Arrhenius reaction rate coefficient models in the limit of equilibrium velocity and internal energy distribution functions. The TCE model is phenomenological and different sets of parameters (A, η, ϵ_a, and choice of $\overline{\zeta}$) may be able to match the same equilibrium dissociation rate. Ideally, one desires a set of parameters that lead to a physically realistic trend for P_{react} versus ϵ_{coll}, while also ensuring consistency with the desired equilibrium reaction rate. As described earlier, the TCE model is implemented by using Eqs. 7.21, 7.27, 7.34, and 7.38, while satisfying Eq. 7.40. Full parameter sets for reacting air species are presented in Section 7.5 along with example simulation results.

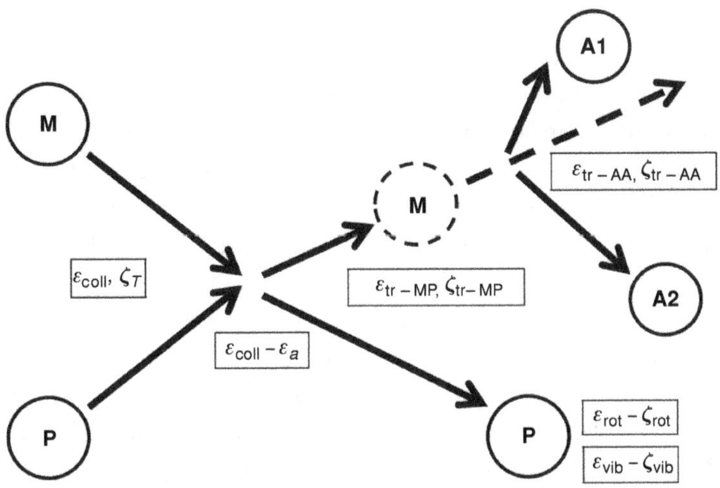

Figure 7.6 Schematic showing the procedure to determine post-collision energies following a dissociation reaction.

7.4.2 Redistribution of Energy Following a Dissociation Reaction

Within a typical DSMC simulation, a majority of collision pairs are not selected for a chemical reaction, since P_{react} is zero for many pairs ($\epsilon_{coll} < \epsilon_a$) and otherwise is typically small (see Fig. 7.5, for example). The strategy most often employed in DSMC is, after a pair is selected for a collision (using the VHS model for example), the pair is first tested for a chemical reaction. If the pair is not accepted for a chemical reaction, then it is considered for translational-rotational-vibrational energy transfer exactly as outlined in Section 6.4 and depicted in Fig. 6.14. However, for pairs that are accepted for a reaction, the energies of all product atoms and molecules must be assigned. In general, the bond energy removed due to a dissociation reaction (ϵ_a) is subtracted from the total collision energy (ϵ_{coll}), and the remaining energy is then redistributed to the various energy modes of the product species using the Borgnakke–Larsen (BL) technique, which samples energies from equilibrium distributions.

Recall that for internal energy transfer within nonreactive collisions, the particle selection procedure prohibiting double relaxation (portrayed in Fig. 6.14) was recommended due to it's consistency with the Jeans and Landau–Teller equations in the near-equilibrium limit (refer to Figs. 6.15 and 6.16). In contrast, for reactive collisions where new species are formed, the properties of all product species are updated during each reactive collision. Therefore, the redistribution of energy in a reactive collision is processed in a different manner. In this section, we outline how the translational and internal energies of product atoms and molecules are computed. Specifically, we will consider a dissociation reaction that results in either one molecule (i.e., $N_2 + N_2 \rightarrow N + N + N_2$) or no molecules (i.e., $N + N_2 \rightarrow N + N + N$). A schematic of the procedure is shown in Fig. 7.6.

The procedure to redistribute energy among the available energy modes of product species may now include the relative translational energies and degrees of freedom associated with both molecules and atoms. This is handled by first considering the relative translational degrees of freedom (ζ_{tr}) between the dissociated molecule (M) and the other particle (P), denoted as $\zeta_{\text{tr}-\text{MP}}$. The expression corresponding to the VHS model is $\zeta_{\text{tr}-\text{MP}} = 5 - 2\omega$. For the above nitrogen reactions, if the other particle is a molecule then $\omega = \omega_{\text{N}_2,\text{N}_2}$, whereas, if the other particle is an atom then $\omega = \omega_{\text{N}_2,\text{N}}$. Next, we consider the additional translational degrees of freedom within the dissociated molecule itself (soon to become two atoms, $A1$ and $A2$), denoted as $\zeta_{\text{tr}-\text{AA}}$. For the example of nitrogen dissociation, we have $\zeta_{\text{tr}-\text{AA}} = 5 - 2\omega_{\text{N,N}}$. In this manner, the total translational degrees of freedom to consider is $\zeta_{\text{tr}-\text{MP}} + \zeta_{\text{tr}-\text{AA}}$. Following this notation, the procedure to determine the properties of the reaction products is:

(1) Determine the total collision energy (ϵ_{coll}) using Eq. 7.25.

(2) Determine the translational degrees of freedom corresponding to particles M and P ($\zeta_{\text{tr}-\text{MP}}$) and corresponding to the atoms, $A1$ and $A2$, resulting from the dissociated molecule ($\zeta_{\text{tr}-\text{AA}}$).

(3) Subtract the dissociation energy (ϵ_a), such that the remaining collision energy, to be redistributed among product species, is $\epsilon_{\text{coll}} = \epsilon_{\text{coll}} - \epsilon_a$.

(4) If all product species are atoms, then proceed to (5).
Otherwise, if the product species include one molecule, first redistribute a portion of ϵ_{coll} into the rotational and vibrational modes of the molecule:

(4a) Redistribute a portion of ϵ_{coll} between ϵ_{vib} (with ζ_{vib} degrees of freedom) and all other remaining energy modes of the products ($\zeta_P = \zeta_{\text{rot}} + \zeta_{\text{tr}-\text{MP}} + \zeta_{\text{tr}-\text{AA}}$). Here, the remaining modes include the rotational mode of the molecule and all translational modes as described previously.

(4b) Determine the new value of ϵ_{vib} using the quantized BL method (Eq. 7.20 and steps (1)–(5) beneath). Since Eq. 7.20 considered only translational-vibrational energy exchange, ζ_{tr} was previously just the translational degrees of freedom of the collision pair. Since we now desire redistribution between the vibrational mode and all other energy modes of the products, Eq. 7.20 should be used where ζ_{tr} is replaced with ζ_P.

(4c) Now that the new value of ϵ_{vib} has been determined, update the remaining collision energy as $\epsilon_{\text{coll}} = \epsilon_{\text{coll}} - \epsilon_{\text{vib}}$.

(4d) Redistribute a portion of ϵ_{coll} between ϵ_{rot} (with ζ_{rot} degrees of freedom) and all other remaining energy modes of the products ($\zeta_P = \zeta_{\text{tr}-\text{MP}} + \zeta_{\text{tr}-\text{AA}}$). Here, the remaining modes include only the translational modes of the products.

(4e) Determine the new value of ϵ_{rot} using the continuous BL method (Eq. 6.104 and steps (1)–(4) beneath). Since Eq. 6.104 considered only translational-rotational energy exchange, ζ_{tr} was previously just the translational degrees of freedom of the collision pair. Since we now desire redistribution between the rotational mode and all other energy modes of the products, Eq. 6.104 should be used where ζ_{tr} is replaced with ζ_P from (4d). Furthermore, note that since the probability used to select this reactive collision was a function of ϵ_{coll} (a collision-invariant quantity), that $I_2 = 1$ in Eq. 6.104.

(4f) Now that the new value of ϵ_{rot} has been determined, update the remaining collision energy as $\epsilon_{coll} = \epsilon_{coll} - \epsilon_{rot}$.

(5) Redistribute the remaining collision energy (ϵ_{coll}) among the translational modes of the product species.

(5a) First redistribute a portion of ϵ_{coll} between ϵ_{tr-MP} (the relative translational energy between the dissociated molecule and the other particle, associated with ζ_{tr-MP} degrees of freedom) and all the other remaining energy modes of the products ($\zeta_P = \zeta_{tr-AA}$). Here the only remaining mode corresponds to the relative translational energy between atoms of the dissociated molecule.

(5b) Determine the new value of ϵ_{tr-MP} by using the continuous BL method (Eq. 6.104 and steps (1)–(4) beneath). Since Eq. 6.104 is a general expression to redistribute energy between any two modes of a gas, with specified degrees of freedom, this equation should be used where ζ_i is replaced with ζ_{tr-MP} and ζ_{tr} is replaced with ζ_P from 5(a). Furthermore, similar to (4e), $I_2 = 1$ in Eq. 6.104.

(5c) Now that the new value of ϵ_{tr-MP} has been determined, use this value to update the center-of-mass velocities of *both* the dissociated molecule and the other particle. This is achieved by using the standard VHS elastic collision procedure (Appendix C) where the center-of-mass velocity is that of the original collision pair, and the new relative translational energy is ϵ_{tr-MP}. All properties of the other particle, whether it is an atom or molecule, are now updated.

(5d) Determine the remaining collision energy as $\epsilon_{coll} = \epsilon_{coll} - \epsilon_{tr-MP}$. Create new DSMC particles (two atoms). Determine their velocities using the standard VHS elastic collision procedure (Appendix C) where the center-of-mass velocity is that of the dissociated molecule, just determined in (5c), and the new relative translational energy is ϵ_{coll}. All properties of the atoms produced by the dissociated molecule are now updated.

(6) All properties of the atoms and molecules comprising the reaction products have now been updated using equilibrium energy distributions based on the collision energy and conserving total energy. The collision process for this pair has ended and the algorithm advances to the next collision pair.

Although this procedure may seem lengthy, it simply involves acceptance–rejection sampling from the continuous and quantized BL distributions, both of which are standard subroutines in a DSMC code that can be used as is with the relevant energy values and degrees of freedom. Also, as detailed later in Section 7.5.4, this procedure can be readily extended to handle any type of chemical reaction.

7.4.3 Vibrationally Favored Dissociation Model

One deficiency of the TCE model is that it does not explicitly account for the fact that molecules in high vibrational energy states are much more likely to dissociate compared to molecules in low vibrational energy states, given the same total collision energy. Models that attempt to capture this effect are often referred to as coupled vibration–dissociation (CVD) models (Wadsworth and Wysong 1997). One such model is the vibrationally favored dissociation (VFD) model proposed by Haas and Boyd (1993). Two versions of the model were developed for vibrational energy levels corresponding to a bounded anharmonic oscillator (AHO) and also an unbounded simple harmonic oscillator (SHO). In this subsection we present the SHO–VFD model, which has a functional form very similar to the TCE model. Specifically, the probability expression for the SHO–VFD model, written in terms of VHS parameters, is

$$P^{\mathrm{VFD}}_{\mathrm{react}} = C_1^{\mathrm{VFD}} \frac{(\epsilon_{\mathrm{coll}} - \epsilon_a)^{\eta+\bar{\zeta}+1/2}}{\epsilon_{\mathrm{coll}}^{\phi+\bar{\zeta}+3/2-\omega}} \epsilon_{\mathrm{vib}}^{\phi} \tag{7.42}$$

where

$$C_1^{\mathrm{VFD}} = A \frac{\sqrt{\pi}(1+\delta_{\mathrm{A,B}})}{2\pi d_{\mathrm{ref}}^2} \frac{\Gamma(\bar{\zeta}+5/2-\omega+\phi)}{\Gamma(\bar{\zeta}+\eta+3/2)} \sqrt{\frac{m_r}{2kT_{\mathrm{ref}}}} \frac{T_{\mathrm{ref}}^{1-\omega}}{k^{\eta-1+\omega}} \frac{\Gamma(\zeta_{\mathrm{vib}}/2)}{\Gamma(\zeta_{\mathrm{vib}}/2+\phi)} \tag{7.43}$$

In this probability expression, ϕ is a modeling parameter that controls the strength of vibrational favoring, ϵ_{vib} is the vibrational energy of the molecule being considered for dissociation, and ζ_{vib} represents the effective vibrational degrees of freedom of the molecule. Recall that for the SHO model, for example, ζ_{vib} can be calculated using Eq. 3.132. All other parameters are the same as those used for the TCE model.

The VFD probability expression was originally derived by Haas and Boyd (1993) for inverse power-law collision models and therefore included the exponent parameter α and reference cross section σ_{ref} (refer to Eqs. 42 and 43 in Haas and Boyd (1993)). As previously derived in Eqs. 6.17 and 6.21, for the VHS model, $\sigma_{\mathrm{ref}} = \pi d_{\mathrm{ref}}^2 g_{\mathrm{ref}}^{2\nu}$, where $g_{\mathrm{ref}}^{2\nu} = (2kT_{\mathrm{ref}}/m_{\mathrm{r}})^{\nu} / \Gamma(2-\nu)$. Using this reference cross section, as well as the relations between exponent parameters α, ν, and ω given in Eq. 6.24, and using the notation for energies and degrees of freedom from this chapter, Eqs. 42 and 43 in Haas and Boyd (1993) become identical to Eqs. 7.42 and 7.43.

A notable attribute of the SHO–VFD probability expression, evident by comparison with Eqs. 7.38 and 7.39, is that with no vibrational favoring the expression reduces exactly to the TCE probability expression:

$$P_{\text{react}}^{\text{VFD}}(\phi = 0) = P_{\text{react}}^{\text{TCE}} \tag{7.44}$$

The SHO–VFD model is parameterized for O_2 dissociation in argon ($\phi = 1$) and N_2 dissociation due to both N_2 and N collisions ($\phi = 3$). As described and verified in Haas and Boyd (1993), the VFD models capture a number of phenomena involved in the coupling between vibrational energy and dissociation. The model accurately reproduces the incubation time and nonequilibrium dissociation rates measured by Wray (1962) for oxygen dissociation in argon, as well as the incubation time measured by Hornung (1972) for nitrogen dissociation. In addition, the VFD model predicts a quasi steady state (QSS) during the dissociation process. This QSS is characterized by non-Boltzmann vibrational energy distributions where the high-energy vibrational levels are significantly depleted. Such depletion results from the fact that molecules in high vibrational energy states are strongly favored to dissociate and the repopulation of these high energy levels due to collisional processes is not as rapid as their depletion due to dissociation. This effect is discussed in more detail in Section 7.5.5, which presents DSMC solutions for high-temperature reacting flows. Finally, owing to the depleted population of high vibrational energy states, dissociation rates during QSS are found to be lower than equilibrium dissociation rates corresponding to Boltzmann internal energy distributions.

Recently all of these phenomena have been confirmed by state-resolved calculations (Kim and Boyd 2013; Panesi et al. 2013) and direct molecular simulations (Valentini et al. 2015, 2016) that employ potential energy surfaces constructed using quantum-mechanical, electronic structure calculations. In addition to experimental data, as more first-principles data becomes available, the VFD model parametrization could be improved.

7.5 General Chemical Reactions

In this section, we extend the theory and numerical methods for dissociation chemistry to general chemical reaction rate sets. This includes methods to simulate both forward and reverse reactions, including exchange reactions, as well as three-body recombination reactions. A more general implementation of DSMC chemistry models will therefore be described and example results for high-temperature air will be presented.

7.5.1 Reaction Rates and Equilibrium Constant

A common set of reactions widely used for high-temperature air, is the five-species model (N_2, N, O_2, O, NO) listed in Table 7.2. The rate coefficient parameters for all reactions are those proposed by Park (1993), except for

Table 7.2 Forward Reaction Rate Coefficients (m^3/molecule/s) for five-Species High-Temperature Air

Number	Reaction	Rate Coefficient (k_f)
1M	$N_2 + M \Leftrightarrow N + N + M$	$1.162 \times 10^{-8}\, T^{-1.6} \exp(-113{,}200/T)$
1A	$N_2 + A \Leftrightarrow N + N + A$	$4.980 \times 10^{-8}\, T^{-1.6} \exp(-113{,}200/T)$
2M	$O_2 + M \Leftrightarrow O + O + M$	$3.321 \times 10^{-9}\, T^{-1.5} \exp(-59{,}400/T)$
2A	$O_2 + A \Leftrightarrow O + O + A$	$1.660 \times 10^{-8}\, T^{-1.5} \exp(-59{,}400/T)$
3M	$NO + M \Leftrightarrow N + O + M$	$8.302 \times 10^{-15} \exp(-75{,}500/T)$
3A	$NO + A \Leftrightarrow N + O + A$	$1.826 \times 10^{-13} \exp(-75{,}500/T)$
4	$O + NO \Leftrightarrow N + O_2$	$1.389 \times 10^{-17} \exp(-19{,}700/T)$
5	$O + N_2 \Leftrightarrow N + NO$	$1.069 \times 10^{-12}\, T^{-1.0} \exp(-37{,}500/T)$

M refers to diatomic molecular species (N_2, O_2, NO). A refers to atomic species (N,O).

the parameters for reaction 5 which are those proposed by Bose and Candler (1996).

Reactions 1 through 3 describe the dissociation of N_2, O_2, and NO, due to collisions with molecules in the system (denoted by M) and collisions with atoms in the system (denoted by A). Dissociation rates can be inferred from shock-tube experiments where a strong shock wave is used to heat the gas to a known post-shock temperature. During this process, the composition of the gas is measured either directly through optical diagnostics or indirectly through pressure traces on the shock-tube end-wall, for example. Experimental results (Appleton, Steinberg, and Liquornik 1968; Hanson and Baganoff 1972; Kewley and Hornung 1974) and the Park model (Park 1993) for the rate coefficient corresponding to nitrogen dissociation are shown in Fig. 7.7(a) for $N_2 + N_2$ collisions, and in Fig. 7.7(b) for $N_2 + N$ collisions. Most of the experimental measurements were made at or below 10,000 K and are extrapolated to higher temperatures for use in thermochemical models. As seen in Fig. 7.7, rate coefficients inferred from such experiments can differ by an order of magnitude, especially at higher temperatures where the data have been extrapolated. The dissociation energy for NO is significantly lower than that for N_2, and the dissociation energy for O_2 is lower still. Therefore, as temperature is increased oxygen dissociation occurs well before nitrogen dissociation. Reactions 4 and 5 are exchange reactions, referred to as the Zeldovich reactions. These reactions play a key role in the finite-rate dissociation of N_2, O_2, and the production of NO. As seen in Table 7.2, once oxygen dissociation has initiated, reaction 5 provides a mechanism to produce NO molecules which can subsequently react to form N atoms through reaction 4, and both N and O atoms through reactions 3M and 3A. The Zeldovich reactions lead to some interesting nonequilibrium chemistry behavior in air that will be discussed later in this section. However, the overall chemical nonequilibrium state of the gas is governed by both forward and backward reactions

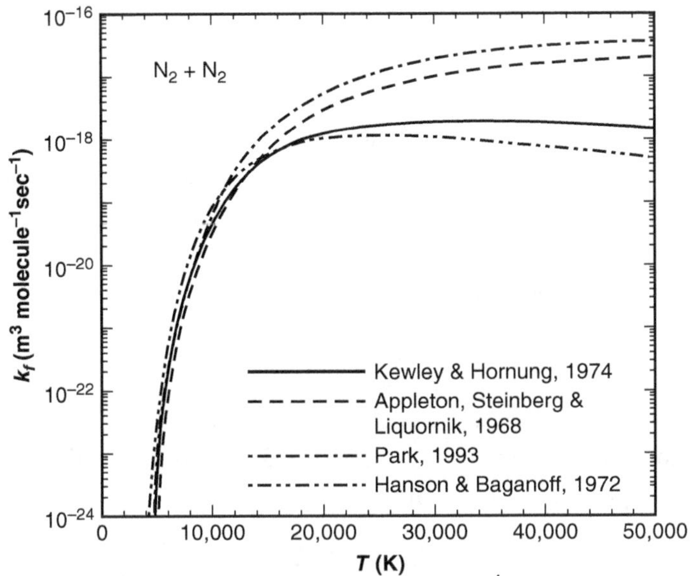

(a) Dissociation rate coefficient for $N_2 + N_2 \xrightarrow{k_f} N + N + N_2$.

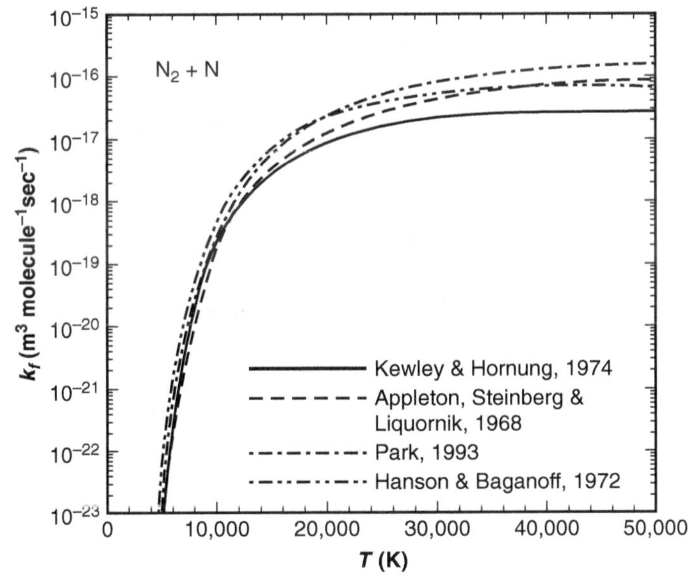

(b) Dissociation rate coefficient for $N_2 + N \xrightarrow{k_f} N + N + N$.

Figure 7.7 Comparison of existing experimental dissociation rate coefficients for N_2.

rates. This was introduced in Chapter 4. We now present the details of general chemical reaction rates as they pertain to DSMC.

Consider an endothermic reaction represented symbolically as

$$AB + C \rightarrow AC + B \tag{7.45}$$

The time rate of change of the number density, n_{AB}, of reactant AB is written

$$\frac{dn_{\mathrm{AB}}}{dt} = -k_f(T)n_{\mathrm{AB}}n_{\mathrm{C}}, \tag{7.46}$$

where the forward rate coefficient, k_f, is a function of temperature, T. Of course, the reaction may also proceed in the backward direction,

$$\mathrm{AC} + \mathrm{B} \rightarrow \mathrm{AB} + \mathrm{C} \tag{7.47}$$

for which the time rate of change of species AB number density is

$$\frac{dn_{\mathrm{AB}}}{dt} = k_b(T)n_{\mathrm{AC}}n_{\mathrm{B}} \tag{7.48}$$

where again the backward rate coefficient, k_b, is a function of temperature.

Using statistical mechanics (Chapter 3), it was shown that the number densities of species, at equilibrium, satisfy

$$\frac{n_{\mathrm{AC}}n_{\mathrm{B}}}{n_{\mathrm{AB}}n_{\mathrm{C}}} = \frac{Q_{\mathrm{AC}}Q_{\mathrm{B}}}{Q_{\mathrm{AB}}Q_{\mathrm{C}}} \exp\left(-\frac{\epsilon_f}{kT}\right), \tag{7.49}$$

where the partition functions, Q, are a function of temperature and volume, ϵ_f is the activation energy for the forward rate process, and k is the Boltzmann constant. A form of this equation was previously derived in Eq. 3.144 in Chapter 3.

For a case such as Eq. 7.45 where the order of the reaction is the same in the forward and backward directions, the volume dependence of the various species partition functions is canceled out on the right-hand side of Eq. 7.49 and it is convenient to introduce the equilibrium constant

$$K_e(T) = \frac{Q_{\mathrm{AC}}Q_{\mathrm{B}}}{Q_{\mathrm{AB}}Q_{\mathrm{C}}} \exp\left(-\frac{\epsilon_f}{kT}\right), \tag{7.50}$$

which is the same expression as derived in Eq. 4.32 of Chapter 4.

It is a simple matter to evaluate the equilibrium constant for a particular reaction knowing the quantized energy states of the participating atoms and molecules. In general, the overall partition function is a product of the partition functions of all of the activated energy modes (i),

$$Q = \prod_i Q_i \tag{7.51}$$

The partition function of each mode (i) is obtained by summing over all quantized energy states (i),

$$Q_i = \sum_j g_{ji} \exp\left(-\epsilon_{ji}/kT\right) \tag{7.52}$$

where g_{ji} is the degeneracy of the j energy level of mode i, and ϵ_{ji} is the energy of that level. The partition functions of each energy mode may then be evaluated using results from quantum mechanics and statistical mechanics (Chapter 3). Specifically, the partition functions for translational, electronic, rotational, and vibrational energies were given in Eqs. 3.86, 3.115, 3.121, and 3.127, respectively.

Statistical mechanics also shows that the ratio of the forward and backward rate coefficients of a reaction is given by its equilibrium constant,

$$K_e(T) = \frac{k_f(T)}{k_b(T)} \tag{7.53}$$

Continuum-based analysis of chemically reacting systems usually specifies the rate coefficients in the forward, endothermic, direction, using the modified Arrhenius form,

$$k_f(T) = A_f T^{\eta_f} \exp\left(-\frac{\epsilon_f}{kT}\right) \tag{7.54}$$

Thus, the rate coefficient of the backward reaction step is calculated as

$$k_b(T) = \frac{k_f(T)}{K_e(T)} \tag{7.55}$$

and, based on Eqs. 7.50 and 7.54 this coefficient clearly has a complex dependence on temperature. Note that the finite-rate analysis for dissociation/recombination, presented in Chapter 4, assumed an Arrhenius form for the equilibrium constant (Eq. 4.66) and therefore for the backward rate coefficient. This is analyzed in the next section.

As discussed previously, the probability of reaction used in DSMC is determined by integrating over the microscopic equilibrium distribution function for the total collision energy and equating it to the rate coefficient. Therefore, given any reaction with a rate coefficient in Arrhenius form (Eq. 7.54), the reaction probabilities used in DSMC can be determined using the TCE model outlined in the previous section. For example, all of the dissociation and forward exchange reactions listed in Table 7.2 enable the direct calculation of TCE model parameters in Eqs. 7.27, 7.34, and 7.38, which can then be used to determine the reaction probability, Eq. 7.21. The VFD model could also be used for any of these forward reactions. We now discuss techniques to simulate backward reactions within DSMC.

7.5.2 Backward Reaction Rates in DSMC

As discussed earlier, the standard continuum approach for computing the backward rate of a chemical reaction uses Eq. 7.55. The DSMC TCE chemistry model requires that any reaction rate coefficient, whether in the forward or backward direction, be specified in modified Arrhenius form. As an example, for a backward reaction step this would require

$$k_b = A_b T^{\eta_b} \exp\left(-\frac{\epsilon_b}{kT}\right) \tag{7.56}$$

where ϵ_b is the activation energy of the reaction. As discussed previously, the standard way of evaluating the backward reaction rate coefficient does not

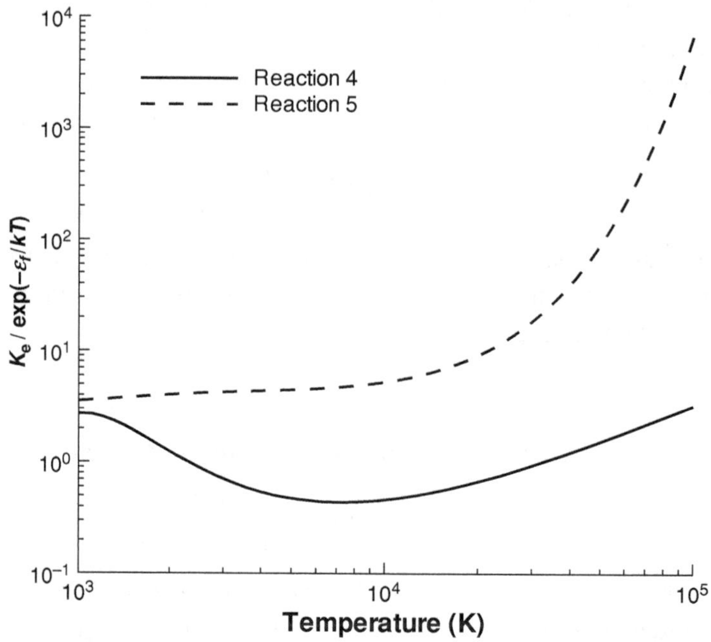

Figure 7.8 The left-hand side of Eq. 7.57 for reactions 4 and 5 as a function of temperature.

lead to a rate that has the simple temperature dependence of the modified Arrhenius form required by the TCE model. For k_b in Eq. 7.55 to have the modified Arrhenius form, for a forward reaction rate expressed in modified Arrhenius form, the equilibrium constant must also be expressible in modified Arrhenius form. Put another way, to satisfy the condition of the TCE model to have k_b expressed in modified Arrhenius form, it must be possible to write

$$\frac{K_e}{\exp(-\epsilon_f/kT)} = A_e T^{\eta_e} \tag{7.57}$$

Using the detailed evaluations of equilibrium constants provided by Park (1990), the quantity on the left-hand side of Eq. 7.57 is shown in Fig. 7.8 as a function of temperature for reactions 4 and 5 in Table 7.2. On a log–log plot, these curves will appear as straight lines if these functions can be expressed in the simple manner indicated on the right hand side of Eq. 7.57. Clearly, that is not the case, and this is the source of problems for application of the TCE chemistry model for simulating backward reactions.

One approach taken in DSMC to address this problem is to fit the backward rate coefficients over a "reasonable" temperature range to a modified Arrhenius form. The backward rate coefficients provided in Table 7.3 were obtained in this way for temperatures in the range of 5000 to 20,000 K (Boyd and Gokcen 1994). At this point, we are omitting recombination reactions for simplicity. This best fit approach appears satisfactory only for a limited temperature range. If there is a need to simulate a reaction over a wider temperature range, new best-fits must be formulated for all reaction

Table 7.3 Backward Reaction Rate Coefficients (m^3/molecule/sec) Fit in Modified Arrhenius Form for Possible Use in DSMC

Number	Reaction	Rate Coefficient
4B	$N + O_2 \rightarrow O + NO$	$4.601 \times 10^{-15}\ T^{-0.546}$
5B	$N + NO \rightarrow O + N_2$	$4.059 \times 10^{-12}\ T^{-1.359}$

mechanisms. For example, sets of curve fits may be required for simulating hypersonic entry into different planetary atmospheres, such as Mars, Venus, and so forth. Finally, there is considerable activity in the development of hybrid continuum-particle methods that use either a continuum or particle approach locally in a flow field based on local flow field information (Schwartzentruber and Boyd 2006; Schwartzentruber et al. 2007, Schwartzentruber et al. 2008a, 2008b, 2008c; Deschenes and Boyd 2011). Such methods are based on the premise that all flow phenomena are simulated consistently by both continuum and particle methods at the locations in a flow domain where the solution methodology changes. For chemically reacting flows, it is therefore essential that at these interfaces, the particle and continuum approaches simulate equivalent rates of all reactions. Due to the problems discussed previously, it is possible to achieve this requirement only to an approximate extent with this DSMC curve-fitting approach.

The problem with the TCE model is illustrated in Figs. 7.9 and 7.10, where the backward rate coefficients for the Zeldovich exchange reactions (reactions 4 and 5 in Tables 7.2 and 7.3) are shown as a function of temperature. The lines labeled "exact" are obtained using accurate equilibrium constants evaluated using partition functions (Eq. 7.50) combined with the forward rate coefficients listed in Table 7.2. The lines labeled "Arrhenius fit" are those listed in Table 7.3 that provide good agreement with the "exact" profiles for temperatures between 5000 and 20,000 K. These fits were performed for analysis of spacecraft entering the Earth's atmosphere from low Earth orbit at a velocity around 8 km/sec. However, at larger temperatures associated for example with Earth entry after return from the Moon, particularly for reaction 5, there are orders of magnitude difference between the "exact" and "Arrhenius fit" reaction rates.

To address these issues, a new DSMC model for simulating backward rates was developed by Boyd (2007) that represents a simple extension of the TCE model. In this approach, the backward rate coefficient is written as

$$k_b(T) = \frac{k_f(T)}{\exp\left(-\frac{\varepsilon_f}{kT}\right)} \frac{\exp\left(-\frac{\varepsilon_f}{kT}\right)}{K_e(T)} = \left[A_f T^{\eta_f}\right] \times \left[\frac{\exp\left(-\frac{\varepsilon_f}{kT}\right)}{K_e(T)}\right]$$
$$= \left[A_f T^{\eta_f}\right] \times \left[\frac{\Pi_p Q_p}{\Pi_r Q_r}\right] \tag{7.58}$$

where Q_p and Q_r are the total partition functions for the products and reactants, respectively. The first term on the right-hand side of this equation

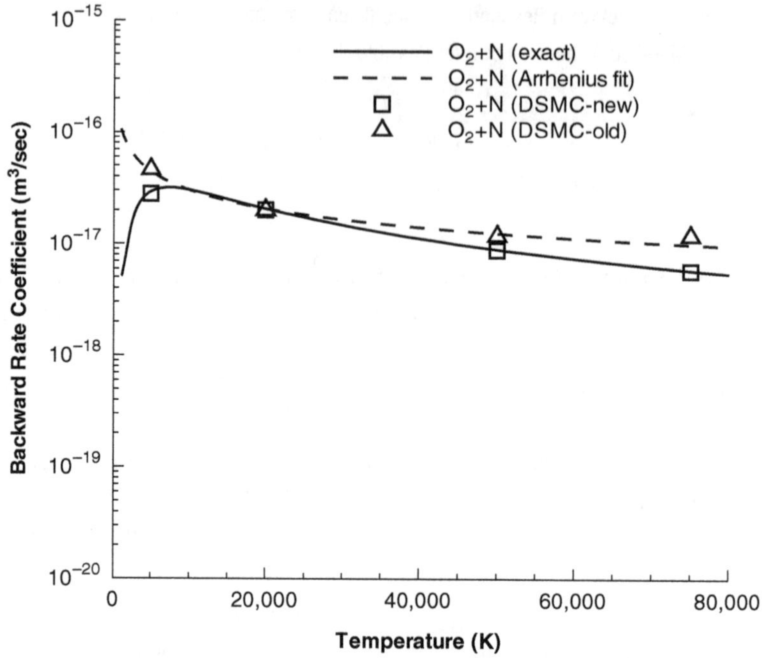

Figure 7.9 Backward reaction rates for the Zeldovich exchange reaction, involving $O_2 + N$, as a function of temperature.

represents an Arrhenius expression for a reaction with zero activation energy. The model is constructed in this way to be consistent with continuum modeling of exothermic backward reactions in which zero activation energy is also assumed. For this model, the rate coefficient is a function of temperature,

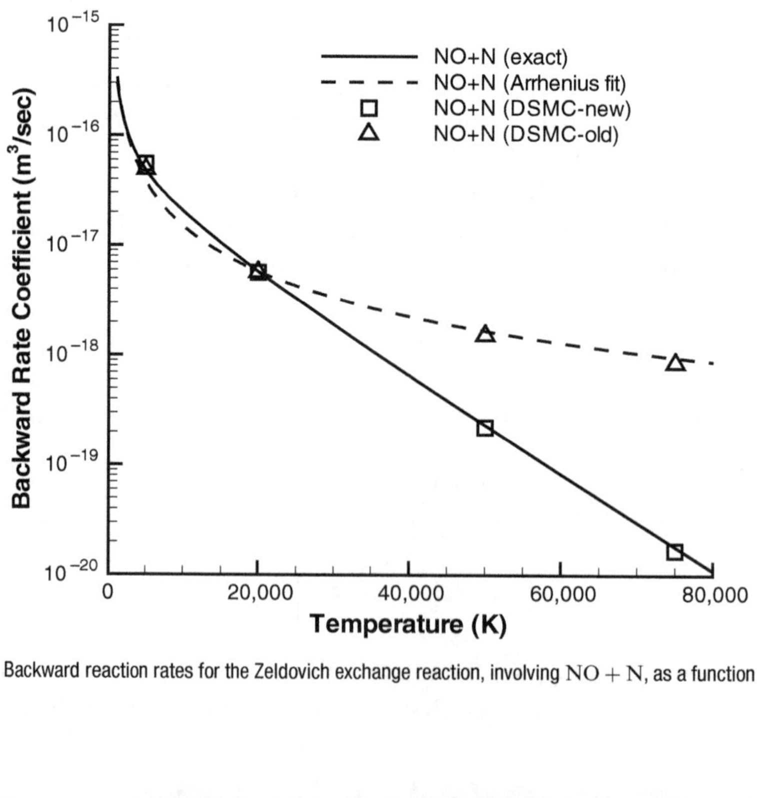

Figure 7.10 Backward reaction rates for the Zeldovich exchange reaction, involving $NO + N$, as a function of temperature.

and the cell-averaged temperature within the DSMC cell is used. The probability used for a specific reaction is therefore a constant within each cell (not dependent on the collision energy of each particular particle pair). In this case, by comparing Eqs. 7.22 and 7.32 for constant P_{back}, we have

$$P_{back} = \frac{k_b(T)}{\nu_{A,B}/n_B} \tag{7.59}$$

and therefore,

$$P_{back} = A_f T^{\eta_f} \times \left[\frac{\Pi_p Q_p}{\Pi_r Q_r}\right] \times \frac{1}{\nu_{A,B}/n_{AB}} \tag{7.60}$$

where $\nu_{A,B}$ was given in Eq. 7.23. In this manner, given the modified Arrhenius parameters for the forward reaction (A_f and η_f) and the cell-averaged temperature, one can compute the partition functions corresponding to the backward reaction of interest and directly compute P_{back} to be applied to the particle pair using the standard acceptance–rejection technique in DSMC. It is important to note that Eq. 7.60 assumes zero activation energy for the backward reaction and is only correct for binary collision reactions. The probability expression for three-body recombination reactions is very similar, however, and is presented in the next section.

Although it may seem like a step in the wrong direction to introduce temperature-dependent modeling into the DSMC technique, recall that this is also the recommended approach for evaluating the probability of vibrational energy exchange (Section 7.3).

Zero dimensional heat bath simulations serve as a useful test to verify correct implementation of forward and backward reactions rates within a DSMC code. In this case, we assess the ability of the backward step DSMC chemistry model to accurately simulate the desired chemistry rates. Rotational relaxation is modeled using the variable probability exchange model of Boyd (Section 7.2.3) and vibrational relaxation is modeled using the quantized Millikan–White model (Section 7.3.3).

Figures 7.9 and 7.10 include evaluations of the average backward rate coefficients obtained using Eq. 7.60 within heat bath, equilibrium DSMC calculations. In these heat bath simulations, all energy modes are in equilibrium at a single temperature. Pre-collision particle properties are sampled from the relevant equilibrium energy distribution functions (refer to Appendix A). Particle pairs are formed and used to evaluate the probability of reaction for each collision. Data are accumulated by averaging over billions of collisions to obtain an average reaction probability that is subsequently converted into an average rate coefficient using Eq. 7.59. Note that post-collision energy redistribution is not performed in these calculations since only the rate of reactions, based on equilibrium distributions corresponding to T, is of interest.

The symbols labeled "DSMC-old" are obtained using the TCE model along with the Arrhenius-fit backward rates listed in Table 7.3. These results

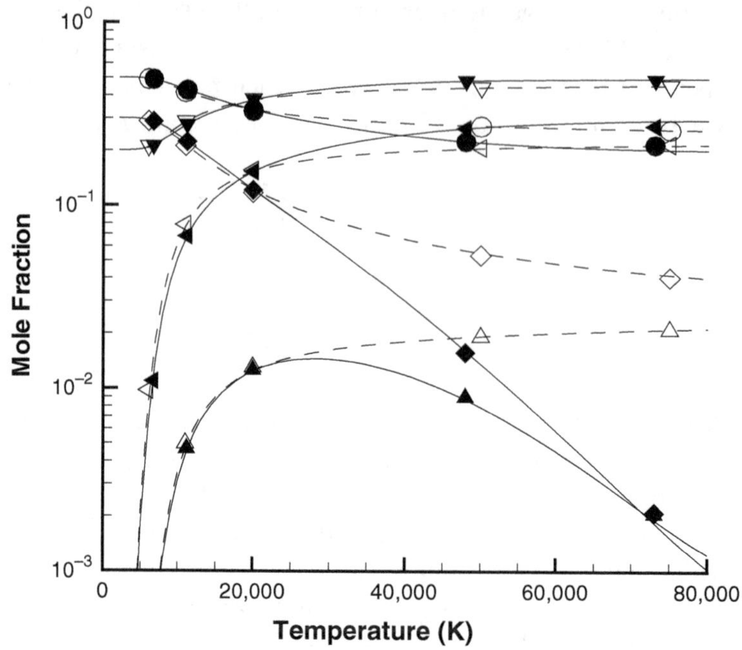

Figure 7.11 Equilibrium composition as a function of temperature when only the Zeldovich exchange reactions are included. Plotting scheme: solid line = forward rates from Table 7.2 and backward rates evaluated from accurate equilibrium constants; dashed line = rates from Table 7.3. ○ = N_2, Δ = O_2, ∇ = N, ◇ = *O*, ◁ = NO.

illustrate that the TCE model is able to accurately reproduce the specified backward rates, even though those rates themselves are inaccurate at high temperature. The symbols labeled "DSMC-new" are obtained using the probability model in Eq. 7.60. Clearly, this approach is able to accurately simulate the backward rates employed in the continuum approach, and these rates are in some cases orders of magnitude different from the fitted backward rates in Table 7.3 used with the TCE model approach.

In the next set of tests, a heat bath is again considered in which millions of DSMC particles are initialized to air at very high temperature. The particles are allowed to collide and react over millions of iterations until a steady-state chemical composition is established. During the reacting collisions, the Borgnakke–Larsen method is applied to sample post-collision properties. This was described for dissociation reactions in Section 7.4.2, and is extended to general chemical reactions in Section 7.5.4.

In the first series of tests, only the Zeldovich exchange reactions (reactions 4 and 5 of Tables 7.2 and 7.3) are included. The initial pressure is 1032 Pa and the initial chemical composition by moles is 40% N_2, 40% N, 10% O_2, and 10% O. Continuum results are obtained through simultaneous solution of the appropriate set of rate equations (Eq. 7.32 for both forward and backward rates), using the rates listed in Table 7.2. In Fig. 7.11, four sets of simulation results are compared. The solid lines are the continuum results that employ the forward rates from Table 7.2 and the backward rates using the accurate equilibrium constants. The dashed lines are the continuum results using the forward rates in combination with the backward rates listed in Table 7.3. The open symbols represent the DSMC results obtained using the backward rates listed in Table 7.3 within the TCE model. The filled symbols represent the DSMC results obtained using Eq. 7.60 for simulating the backward rate step.

Comparing the dashed lines with the open symbols, it is found that use of the TCE model for the backward rates does accurately compute the equilibrium compositions obtained with the associated, and physically inaccurate, equilibrium constants. Similarly, comparison of the solid lines with the filled symbols illustrates that the backward step DSMC chemistry model employing Eq. 7.60 does accurately compute the equilibrium compositions obtained with the physically accurate equilibrium constants. Comparison of the dashed and solid lines (or open and filled symbols) shows that, particularly at high temperature, the use of inaccurate equilibrium constants can lead to orders of magnitude differences in the mole fractions of some of the chemical species under the conditions considered.

7.5.3 Three-Body Recombination Reactions

To simulate recombination reactions, three-body collisions must be considered. Although the theory and methods discussed in this text have so far been for binary collisions, the standard DSMC approach for three-body collisions is a simple extension of the procedures already discussed for binary collisions.

Once a pair of particles is selected for a collision, just as the pair must be considered for a binary-collision reaction, it must also be considered for a three-body collision reaction with some probability. Three-body reaction probabilities are typically small compared to two-body reaction probabilities, but clearly cannot be neglected for flows in which recombination is important.

One way to handle three-body collisions is to simply set aside a small fraction of the number of particles within each cell (typically 5% or less) and reserve these particles for three-body collisions only. Recall that when using the NTC algorithm, only a fraction of the particles in each cell are tested for collision ($N_{\text{pairs-tested}}$ in Eqs. 6.10 and 6.11), and thus, typically there are a number of particles left in the cell that could be set aside and reserved for this purpose. In this manner, on starting the collision routine within a cell, a number ($N_{3\text{B}}$) of particles are set aside into a short list of potential third-body (3B) particles. For example, it may be reasonable to set aside $N_{3\text{B}} \approx \text{floor}\left[\max(0.05 \times N_p, 1)\right]$ particles. The NTC method for binary collisions is then performed, as previously described (Eqs. 6.9 through 6.13), except that the number pairs to be tested is no longer constrained by Eq. 6.11, but instead by

$$1 \leq N_{\text{pairs-tested}} \leq \text{floor}(N_p - N_{3\text{B}})/2 \tag{7.61}$$

Using the NTC algorithm (now with Eq. 7.61) will result in a certain number of collision pairs accepted for a binary collision. As described in the previous section, each selected collision pair would then be considered for a chemical reaction with some probability P_{react}. The only difference is that now, each selected collision pair is first considered for a three-body recombination reaction with some probability $P_{\text{3B-react}}$. This probability is typically much lower than P_{react} and therefore each pair is typically tested for a recombination reaction first, to ensure that all pairs have the opportunity to experience a three-body recombination reaction. Furthermore, just as all binary pairs were selected randomly from within the cell, all three-body particle sets (binary pair plus 3B particle) must be randomly chosen from within the cell, since all particles must have the chance to undergo binary and three-body collisions. This is achieved by simply selecting a 3B particle randomly from the reserved list of 3B particles for each binary collision pair considered. Similar to the backward reaction rate probability for binary collisions (P_{back} in Eq. 7.60), $P_{\text{3B-react}}$ is set to achieve the desired backwards reaction rate (i.e., recombination rate) for the reaction being considered.

The probability expression is identical to the expression for a binary pair, except that it depends on the number density of 3B particles. Since all particles in the cell are potential 3B particles, $n_{\text{3B}} = n$, and the result is simply

$$P_{\text{3B-react}} = n \times A_f T^{\eta_f} \times \left[\frac{\Pi_p Q_p}{\Pi_r Q_r} \right] \times \frac{1}{\nu_{\text{A,B}}/n_{\text{AB}}} \tag{7.62}$$

where n is the total number density in the cell (i.e., the sum over all species i: $n = \sum_i n_i$), A_f and η_f are the rate parameters for the dissociation (forward) reaction, and the partition functions are those of the product and reactant species as defined by the recombination reaction being considered. It is typical to use the instantaneous number density, which is simply $n = N_p W_p / V_{\text{DSMC}}$, where W_p is the particle weight in the cell and V_{DSMC} is the volume of the cell. Equation 7.62 also implies that the units of the rate coefficient parameters are different than those for binary reactions. In fact the product of number density and the three-body reaction rate coefficient has the same units as the binary reaction rate coefficient.

Therefore, given the cell-averaged temperature, the probability that a particular three-body set is accepted for a recombination reaction can be calculated directly from the forward Arrhenius dissociation rate expression, such as listed in Table 7.2. If the three-body set is accepted, then the correct product species must be formed and the remaining collision energy must be redistributed among the various energy modes of the product species. Similar to binary reactions, after all energy modes have been updated, the particles are finished interacting. In this case, the properties of the 3B particle are also updated and it should be removed from the available 3B list so that it is not considered for any further collisions during the current timestep.

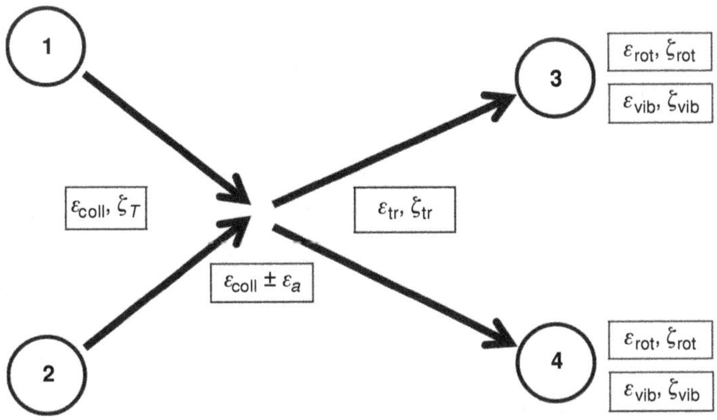

Figure 7.12 Schematic showing the procedure to determine post-collision energies following an exchange reaction.

The method to assign post-collision properties is similar to the method outlined for dissociation in Section 7.4.2, except that, since the reaction is now exothermic, the bond energy must be added to the energy available for redistribution to the products, and one of the simulation particles will be deleted from the simulation. The general procedure to assign the post-reaction states of product molecules, general to any type of chemical reaction is described next.

7.5.4 Post-Reaction Energy Redistribution and General Implementation

This section summarizes the general procedure for determining post-reaction energies of product species. Schematics that show the process for determining post-reaction energies are shown in Fig. 7.12 for exchange reactions, and in Fig. 7.13 for recombination reactions. A schematic for dissociation reactions was presented earlier in Fig. 7.6. Similar schematics are also presented in Appendix C, which describes how post-reaction velocity vectors are determined for all reaction product species.

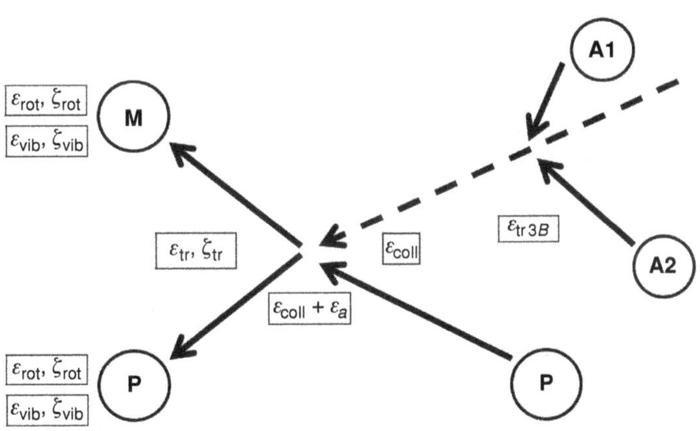

Figure 7.13 Schematic showing the procedure to determine post-collision energies following a recombination reaction.

A general implementation strategy for simulating the collision step within a DSMC cell during a single DSMC timestep, for a multispecies reacting gas flow involves the following steps:

(1) Use the NTC algorithm to form random pairs of particles within the cell and accept each pair for a binary collision with probability P_{DSMC} (Eq. 6.13). For each collision pair, consisting of species i and species j, allow for chemical reactions and internal energy transfer as described in steps 2–6.

(2) Given a pair of particles (species i and j), determine all of the possible reactions for which the two particles are the primary reactants (i.e., determine a *reaction list*).

(2a) For example, using Table 7.2 for reference, a selected collision pair consisting of N_2 and O_2 can be involved in either reaction 1M or reaction 2M (both dissociation reactions). Whereas a pair consisting of N and NO can be involved in NO dissociation (reaction 3A) or a backward exchange reaction (reaction 5 in the reverse direction). Finally, a collision pair consisting of N and O can be involved only in a recombination reaction to produce NO (either reaction 3M or 3A depending on the species type of the third body, in the reverse direction).

(2b) If the reaction list for the collision pair contains more than one reaction, then an order in which to test for the reactions must be specified. Although somewhat arbitrary, reactions are typically ordered with recombination tested first, followed by exchange reactions, followed by dissociation reactions.

(3) Test for each type of reaction on the list. If the end of the list has been reached without accepting the pair for a reaction, then consideration for a chemical reaction has failed. In this case, exit the current algorithm and consider the particle pair for internal energy exchange (an inelastic collision) as described in Section 6.4 of Chapter 6 (Fig. 6.14). Otherwise, if further reactions remain on the list, test for the next reaction.

(3a) If the reaction under consideration is a recombination reaction, choose a third-body (3B) particle randomly from the 3B particle list. Identify precisely which recombination reaction is being considered and identify the corresponding Arrhenius rate parameters for the forward dissociation reaction. Use the cell averaged temperature to determine $P_{3B-\mathrm{react}}$ (Eq. 7.62) and use the standard acceptance–rejection algorithm to accept or decline the recombination reaction. If the reaction is declined, return to step (3) and test for the next reaction on the list. If the reaction is accepted, proceed to step (4).

(3b) If the reaction under consideration is a backward exchange reaction, identify the corresponding Arrhenius rate parameters for the forward exchange reaction and use the cell averaged temperature to determine P_{back} (Eq. 7.60). If the reaction under consideration is a forward exchange reaction, identify the corresponding Arrhenius rate parameters for the forward exchange reaction and use the TCE model to determine P_{react} (Eq. 7.21 or Eq. 7.42 for the VFD model). For both forward and backward exchange reactions, use the standard acceptance–rejection algorithm to accept or decline the exchange reaction. If the reaction is declined, return to step (3) and test for the next reaction on the list. If the reaction is accepted, proceed to step (4).

(3c) If the reaction under consideration is a dissociation reaction, identify the corresponding Arrhenius rate parameters for the dissociation reaction and use the TCE model to determine P_{react} (Eq. 7.21 or Eq. 7.42 for the VFD model). Use the standard acceptance–rejection algorithm to accept or decline the dissociation reaction. If the reaction is declined, return to step (3) and test for the next reaction on the list. If the reaction is accepted, proceed to step (4).

(4) Evaluate the collision energy (ϵ_{coll}) available for redistribution among product species.

(4a) For recombination or backward exchange reactions: $\epsilon_{coll} = \epsilon_{coll}^{reactants} + \epsilon_f$. Here, ϵ_f is the activation energy of the forward reaction. For recombination reactions, $\epsilon_{coll}^{reactants}$ involves calculating the relative translational energy of the three-body system (refer to Appendix C).

(4b) For dissociation or forward exchange reactions: $\epsilon_{coll} = \epsilon_{coll}^{reactants} - \epsilon_f$. Here, ϵ_f is the activation energy of the forward reaction.

Note: It is typical that $\epsilon_{coll}^{reactants}$ is calculated using all energy modes of all reactant species and that ϵ_{coll} is redistributed among all energy modes of all product species. However, this is somewhat arbitrary and not strictly required. If there are reaction mechanisms for which only certain energy modes are expected to participate, then the collision energy (and energy redistribution to products) can be restricted to only these energy modes, as desired.

(5) Redistribute the available collision energy (ϵ_{coll}) among the energy modes of product species using the Borgnakke–Larsen (BL) model. This is similar to steps (4) and (5) described in Section 7.4.2 specific to dissociation reactions. As an example, for a reaction that results in two product molecules (denoted by subscripts 1 and 2), such as the recombination reactions in Table 7.2 (1M, 2M, and 3M in the reverse direction), the redistribution process would involve:

(5a) Use the quantized BL method (Section 7.3.3) to redistribute a portion of ϵ_{coll} between $\epsilon_{vib,1}$ (with $\zeta_{vib,1}$ degrees of freedom) and all other remaining energy modes of the products ($\zeta_P = \zeta_{rot,1} + \zeta_{vib,2}+$

$\zeta_{rot,2} + \zeta_{tr}$). Determine and assign $\epsilon_{vib,1}$ to molecule 1 and update the available collision energy as $\epsilon_{coll} = \epsilon_{coll} - \epsilon_{vib,1}$.

(5b) Use the continuous BL method (Section 6.4.4) to redistribute a portion of ϵ_{coll} between $\epsilon_{rot,1}$ (with $\zeta_{rot,1}$ degrees of freedom) and all other remaining energy modes of the products ($\zeta_P = \zeta_{vib,2} + \zeta_{rot,2} + \zeta_{tr}$). Determine and assign $\epsilon_{rot,1}$ to molecule 1 and update the available collision energy as $\epsilon_{coll} = \epsilon_{coll} - \epsilon_{rot,1}$.

(5c) Use the quantized BL method (Section 7.3.3) to redistribute a portion of ϵ_{coll} between $\epsilon_{vib,2}$ (with $\zeta_{vib,2}$ degrees of freedom) and all other remaining energy modes of the products ($\zeta_P = \zeta_{rot,2} + \zeta_{tr}$). Determine and assign $\epsilon_{vib,2}$ to molecule 2 and update the available collision energy as $\epsilon_{coll} = \epsilon_{coll} - \epsilon_{vib,2}$.

(5d) Use the continuous BL method (Section 6.4.4) to redistribute a portion of ϵ_{coll} between $\epsilon_{rot,2}$ (with $\zeta_{rot,2}$ degrees of freedom) and all other remaining energy modes of the products ($\zeta_P = \zeta_{tr}$). Determine and assign $\epsilon_{rot,2}$ to molecule 2 and update the available collision energy as $\epsilon_{coll} = \epsilon_{coll} - \epsilon_{rot,2}$.

(5e) The remaining collision energy is to be distributed to the translational modes of the product species and post-collision particle velocities are calculated using the appropriate scattering model (i.e., VHS or VSS, for example) as described in Appendix C.

Note: If the product species involve only one molecule, then $\zeta_{vib,2} = \zeta_{rot,2} = 0$, and steps (5c)–(5d) are omitted. If the product species contain no molecules, then $\zeta_{vib,1} = \zeta_{rot,1} = \zeta_{vib,2} = \zeta_{rot,2} = 0$, and steps (5a)–(5d) are omitted. Finally, for dissociation reactions, the remaining collision energy in step (5e) must be redistributed among the three product particles, as previously described in steps (5c) and (5d) of Section 7.4.2.

Note: As mentioned earlier, if it is desired to exclude certain energy modes from participation in the chemical reaction, then these energy modes should not contribute to $\epsilon_{coll}^{reactants}$, nor should the energies or degrees of freedom of these modes be included in steps (5a)–(5d) above. To give an example, some DSMC implementations may only allow the translational energy of a third body to contribute to a recombination reaction. In this case, the internal energy modes of the third body (if it is a molecule) would not contribute to $\epsilon_{coll}^{reactants}$ and would be omitted in steps (5a)–(5d) above. However, it is stressed for this example, that there is no physical rationale that supports this one way or the other.

(6) All properties of the atoms and molecules comprising the reaction products have now been updated using equilibrium energy distributions based on the collision energy and conserving total energy. The collision process for this pair has ended and the algorithm advances to the next collision pair, return to step (2) above.

Finally, it is important to remember that the Borgnakke–Larsen energy redistribution model is a phenomenological model. This model drives molecular distribution functions toward local equilibrium distribution functions as desired, however, it is generally not known what the post-reaction translational and internal energy distributions are. If these distributions were known for a given reaction mechanism, then these distributions could be sampled directly to obtain post-reaction properties instead of sampling the BL equilibrium distribution functions. This is an area of ongoing research.

7.5.5 DSMC Solutions for Reacting Flows

In this section, we present DSMC simulation results for high-temperature chemically reacting flows involving air species. Zero-dimensional simulations are performed under isothermal conditions and also under adiabatic conditions. In addition, solutions for two-dimensional hypersonic flow over a cylinder are presented.

The reaction mechanisms and rate coefficients are those listed in Table 7.2. Specifically, dissociation and forward exchange reactions are performed using the TCE model probability expression. Recombination and backward exchange reactions are performed using reaction probabilities calculated using the cell averaged temperature and partition functions.

Vibrational energy is treated as quantized and available vibrational degrees of freedom are calculated using the SHO model (see Eq. 3.132) based on the cell averaged temperature. Translational-vibrational energy exchange is performed using the Millikan–White model outlined in Section 7.3.2 with the model parameters listed in Table 7.1. Rotational energy is treated as continuous with two degrees of freedom for each diatomic species. Translational-rotational energy exchange is performed using the Parker model outlined in Section 7.2.2. For all inelastic collisions, the particle selection procedure prohibiting double relaxation, following the implementation of Zhang and Schwartzentruber, outlined in Section 6.4.3 is used.

Finally, the NTC collision rate algorithm is used, combined with the VHS cross section model. The required species parameters are listed in Table 7.4. The required VHS species-pair parameters are listed in Table 7.5. Note that for unlike species pairs, the values of ω, T_{ref}, and d_{ref} are simply the averages of the values for like-species pairs. The Molecular Gas Dynamic Simulator (MGDS) code (Gao et al. 2011; Nompelis and Schwartzentruber 2013) is used for all computations presented in this section.

We perform zero-dimensional heat-bath simulations, first under isothermal conditions and then under adiabatic conditions. Isothermal conditions maintain a constant translational temperature for the system by simply resetting (resampling) all particle velocities from Maxwell–Boltzmann velocity distribution functions corresponding to a specific translational temperature (T_{tr}). Adiabatic conditions conserve the total system energy and allow the

Table 7.4 DSMC Model Parameters for Each Species

Species	N_2	O_2	NO	N	O
Mw (kg/kmol)	28	32	30	14	16
d (Å)	4.17	4.07	4.20	3.00	3.00
ζ_{rot}	2	2	2	0	0
ζ_{vib}	Eq. 3.132	Eq. 3.132	Eq. 3.132	0	0
T_{ref}^{rot}–Parker model	91.5 K	90.0 K	91.5 K	N/A	N/A
Z_{ref}^{rot}–Parker model	18.1	14.4	18.1	N/A	N/A
θ_{rot}	2.88 K	2.07 K	2.44 K	N/A	N/A
θ_{vib}	3390 K	2270 K	2740 K	N/A	N/A

translational energies of system molecules to evolve without any artificial resampling. Both types of simulations begin with a system of molecules sampled from a high translational temperature (T_{tr}) but low vibrational temperature (T_{vib}) and low rotational temperature (T_{rot}). This situation is representative of conditions immediately behind a strong shock wave. As the DSMC method iterates, collisions between simulation particles begin to excite rotational energy, followed by vibrational energy, eventually leading to dissociation, exchange, and finally recombination reactions in the resulting multi-species mixture. We will observe the evolution of system internal energy and chemical composition over short timescales and also over long timescales where we may expect to see equilibrium conditions. In some cases, the DSMC simulation results will be directly compared to analytical results from Chapter 4.

The first simulation involves isothermal relaxation to a temperature of $T_{tr} = 6500$ K. Results are shown in Fig. 7.14. At $t = 0$ seconds, the system comprises diatomic nitrogen with $\rho = 0.022$ kg/m^3, $T_{rot} = T_{vib} = 200$ K, and the specified translational temperature T_{tr}. Figure 7.14(a) plots T_{rot}, T_{vib}, as well as the number densities of N_2 and N over a time of 6 msec. When plotted over this timescale, the gas is seen to reach thermal equilibrium ($T_{rot} = T_{vib} = T_{tr} = 6500$ K) almost instantaneously. Dissociation, however, occurs over a much longer timescale and an equilibrium, partially dissociated, state is achieved after approximately 4 msec. At this equilibrium state,

Table 7.5 DSMC VHS Model Parameters for Each Species Pair

Collision Pair	ω	$T_{ref}[K]$	d_{ref} [Å]
N_2–N_2	0.74	273	4.17
O_2–O_2	0.77	273	4.07
NO–NO	0.79	273	4.20
N–N	0.80	273	3.00
O–O	0.80	273	3.00

dissociation and recombination reactions continue to occur, however, the rates of these processes are balanced. This result (Fig. 7.14(a)) can be directly compared with the analytical result in Fig. 4.13 of Chapter 4, which involved similar isothermal heat bath conditions. Figure 7.14(b) plots the same gas properties over a much shorter time scale of 20 μsec. Here, we see that rotational and vibrational energy excitation do occur at finite rates, with vibrational excitation being far slower than rotational excitation. This result is expected based on the collision numbers (Z_{rot} and Z_{vib}) corresponding to $T_{\mathrm{tr}} = 6500$ K, and is very similar to the solution presented in Fig. 7.4(a) of Section 7.3.4 for $T_{\mathrm{tr}} = 8000$ K. It is also important to note that, at these relatively low temperatures, no significant nitrogen dissociation occurs until the the gas is in thermal equilibrium. As the system temperature is increased, we will see that this is no longer the case.

The next two simulations involve isothermal relaxation to translational temperatures of $T_{\mathrm{tr}} = 13{,}000$ K (Fig. 7.15) and $T_{\mathrm{tr}} = 20{,}000$ K (Fig. 7.16). At $t = 0$ seconds, the system comprises diatomic nitrogen with $\rho = 0.022$ kg/m^3, $T_{\mathrm{rot}} = T_{\mathrm{vib}} = 200$ K, and the specified translational temperature T_{tr}. At such extreme temperatures, dissociation is rapid. Complete nitrogen dissociation occurs in approximately 6 μsec for $T_{\mathrm{tr}} = 13{,}000$ K and in approximately 0.6 μsec for $T_{\mathrm{tr}} = 20{,}000$ K. Moreover, rotational and vibrational excitation now occur over similar timescales as dissociation.

Specifically, Fig. 7.15 shows that rotational excitation is significantly faster than vibrational excitation. Furthermore, the rotational temperature equilibrates with the translation temperature before significant dissociation occurs, however, dissociation begins to occur during vibrational energy excitation. In Fig. 7.16, rotational and vibrational energy excitation rates are now quite similar and dissociation begins well before these energy modes are equilibrated with the translational temperature.

Another important result, seen clearly in both Fig. 7.15 and Fig. 7.16, is that neither the rotational energy nor the vibrational energy come into equilibrium with the translational temperature. Rather, the temperatures reach steady-state values where $T_{\mathrm{vib}} < T_{\mathrm{rot}} < T_{\mathrm{tr}}$. This trend is due to the fact that as the molecules in the gas dissociate, rotational and vibrational energy is removed from the system. Therefore, as inelastic collisions act to increase both the rotational and vibrational energies of molecules toward equilibrium with the translational energy of the gas (which is artificially held fixed), rapid dissociation reactions act to remove internal energy. The conditions under which these two processes balance is referred to as "quasi steady state" (QSS). By comparing Figs. 7.15 and 7.16, it is evident that as the value of T_{tr} is increased, the QSS state ($T_{\mathrm{vib}} < T_{\mathrm{rot}} < T_{\mathrm{tr}}$) becomes more pronounced.

Recent research has investigated internal energy transfer and dissociation physics using molecular dynamics to simulate high-energy collisions between air species. Accurate potential energy surfaces (PESs) fit to large databases of electronic structure calculations (Paukku et al. 2013; Bender et al. 2015; Lin

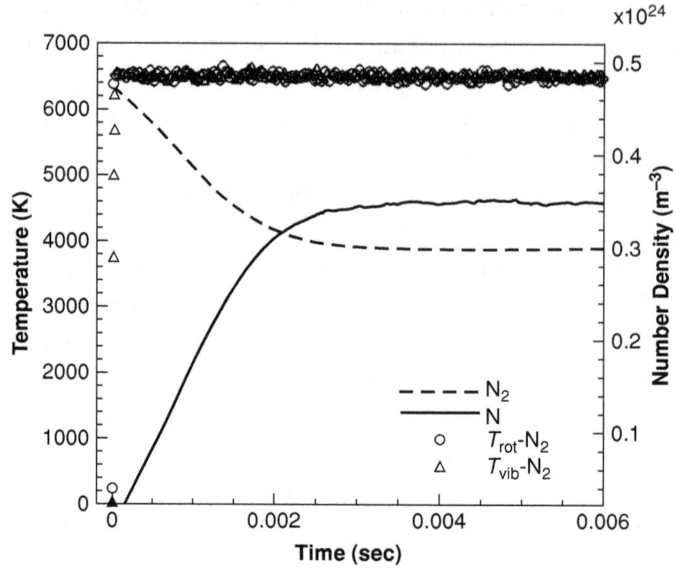

(a) System temperatures and composition over long timescales.

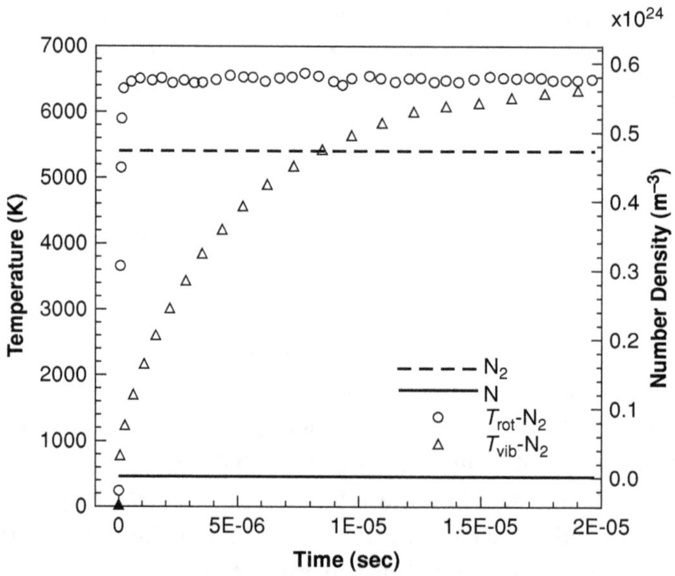

(b) System temperatures and composition over short timescales.

Figure 7.14 Isothermal relaxation of nitrogen to $T_{tr} = 6500$ K.

et al. 2016; Varga et al. 2016) are employed for these studies. In some cases, many individual collisions are analyzed to determine the degree of vibrational favoring for dissociation and also to study the internal energy distribution functions of reactants and products (Kim and Boyd 2013; Panesi et al. 2013). In other studies, molecular dynamics collisions are embedded within DSMC simulations, producing heat bath relaxation results similar to Figs. 7.15 and 7.16. This approach, originally proposed by Koura as classical trajectory calculation (CTC) DSMC (Koura 1997, 1998; Matsumoto and Koura

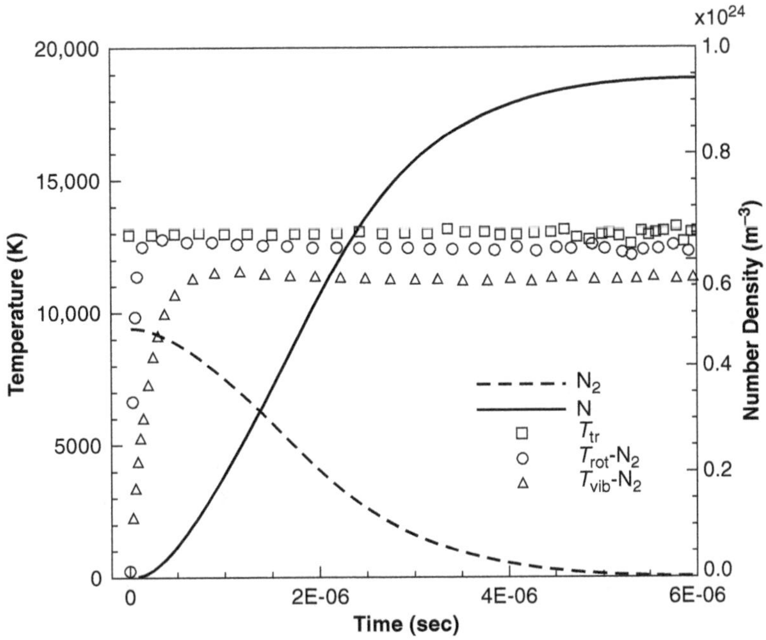

Figure 7.15 Isothermal relaxation of nitrogen to $T_{tr} = 13{,}000$ K showing system temperatures and composition.

1991), has now been extended to reacting systems using ab initio PESs and is referred to as direct molecular simulation (DMS) (Norman, Valentini and Schwartzentruber 2013; Valentini et al. 2015, 2016).

These studies are able to quantify vibrational favoring in dissociation processes. In general, these studies show that the vibrational energy lost due to dissociation is (on average) higher compared to the rotational energy lost. This is a contributing factor to the relative ordering of temperatures in QSS ($T_{vib} < T_{rot} < T_{tr}$). Furthermore, these studies reveal that the rotational and vibrational energy distributions are non-Boltzmann during such QSS dissociation. It is interesting that standard DSMC calculations naturally predict such phenomena, as evident from Figs. 7.15 and 7.16. However, it is important to recall that the TCE model, used for this example, does not specifically model vibrationally favored dissociation. Also, the BL model used for post-collision energy redistribution (in both inelastic collisions and chemical reactions) is phenomenological. Yet, as more results using ab initio PESs to simulate reactive collision dynamics become available, improved DSMC models (such as the VFD model) could be parameterized, leading to an increase in accuracy over the existing phenomenological models. This is an area of ongoing research.

We now switch to adiabatic relaxation calculations. In some sense, adiabatic conditions are more representative of real flow conditions, since as internal energy is excited and dissociation reactions occur, translational energy is removed from the gas. Two adiabatic relaxation simulations are performed, one starting with $T_{tr} = 13{,}000$ K, and the other starting with

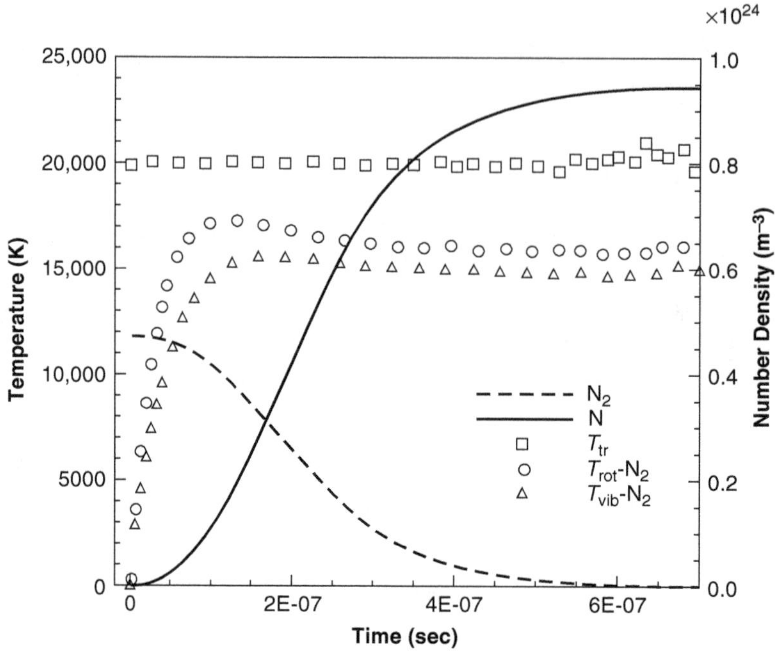

Figure 7.16 Isothermal relaxation of nitrogen to $T_{tr} = 20{,}000$ K showing system temperatures and composition.

$T_{tr} = 20{,}000$ K. For both simulations, at $t = 0$ seconds, the system is composed of diatomic nitrogen with $\rho = 0.022$ kg/m^3, $T_{rot} = T_{vib} = 200$ K, and the specified translational temperature T_{tr}.

For the case where $T_{tr} = 13{,}000$ K, Fig. 7.17(a) shows the system temperatures and species number densities as they evolve over a timescale of 15 msec. When viewed at this timescale, we see that the gas reaches thermal equilibrium ($T_{rot} = T_{vib} = T_{tr}$) almost instantaneously, resulting in a system temperature just below 6000 K. Dissociation then proceeds slowly as seen by the number densities of N_2 and N. Since energy is required to break the N_2 bonds, the system temperature is reduced from approximately 6000 K to 5000 K, as the gas approaches chemical equilibrium. This DSMC solution can be compared to the analytical result presented earlier in Fig. 4.14 of Chapter 4, which used similar adiabatic conditions. Specifically, the result in Fig. 4.14 was obtained for an adiabatic system initialized in thermal equilibrium, $T_{rot} = T_{vib} = T_{tr} = 7500$ K. This system energy is higher than that of the current DSMC simulation, which quickly reached a thermal equilibrium temperature just below 6000 K as discussed previously. Therefore, compared to Fig. 4.14, the final system temperature (after chemical equilibrium is achieved) in the current DSMC simulation (Fig. 7.17(a)) is lower, the degree of dissociation is lower, and the time to reach equilibrium is larger. When the solution is viewed over a timescale of 10 μsec, Fig. 7.17(b) shows how the gas quickly reaches thermal equilibrium. Specifically, rotational and translational energies equilibrate rapidly to a transrotational temperature of $T_{tr} = T_{rot} = 7500$ K. Vibrational energy takes much longer to excite and

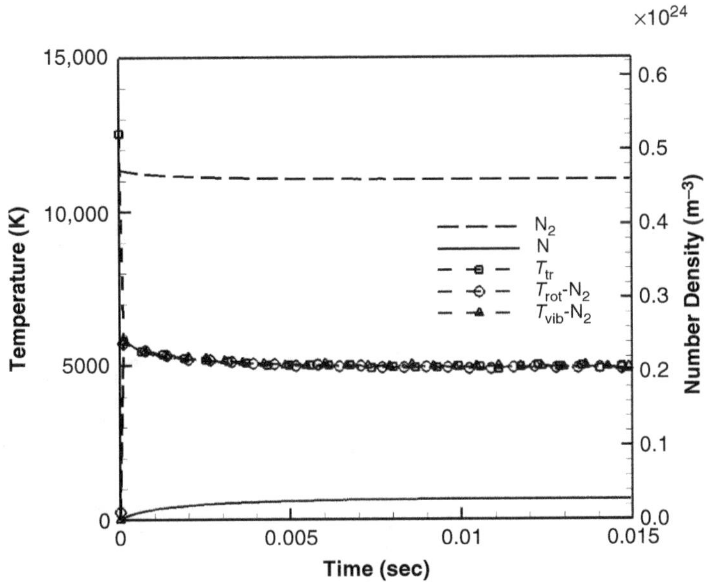

(a) System temperatures and composition over long timescales.

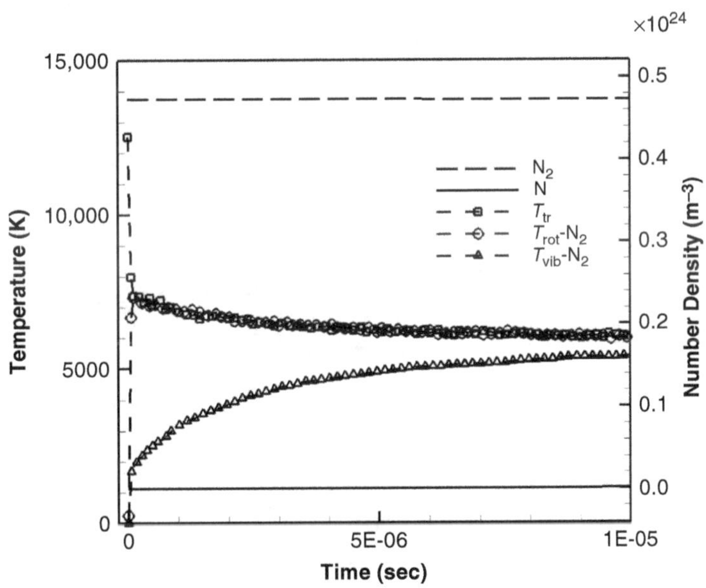

(b) System temperatures and composition over short timescales.

Figure 7.17 Adiabatic relaxation of nitrogen, initialized with $T_{tr} = 13{,}000$ K and $T_{rot} = T_{vib} = 200$ K.

removes energy from both the translational and rotational energies of the system molecules. As discussed above, all energy modes reach thermal equilibrium where $T_{rot} = T_{vib} = T_{tr} \approx 6000$ K well-before significant dissociation begins.

For the adiabatic system where initially $T_{tr} = 20{,}000$ K, Figs. 7.18(a) and 7.18(b) show the same trends as the case where $T_{tr} = 13{,}000$ K. All system temperatures are higher, the degree of dissociation is larger,

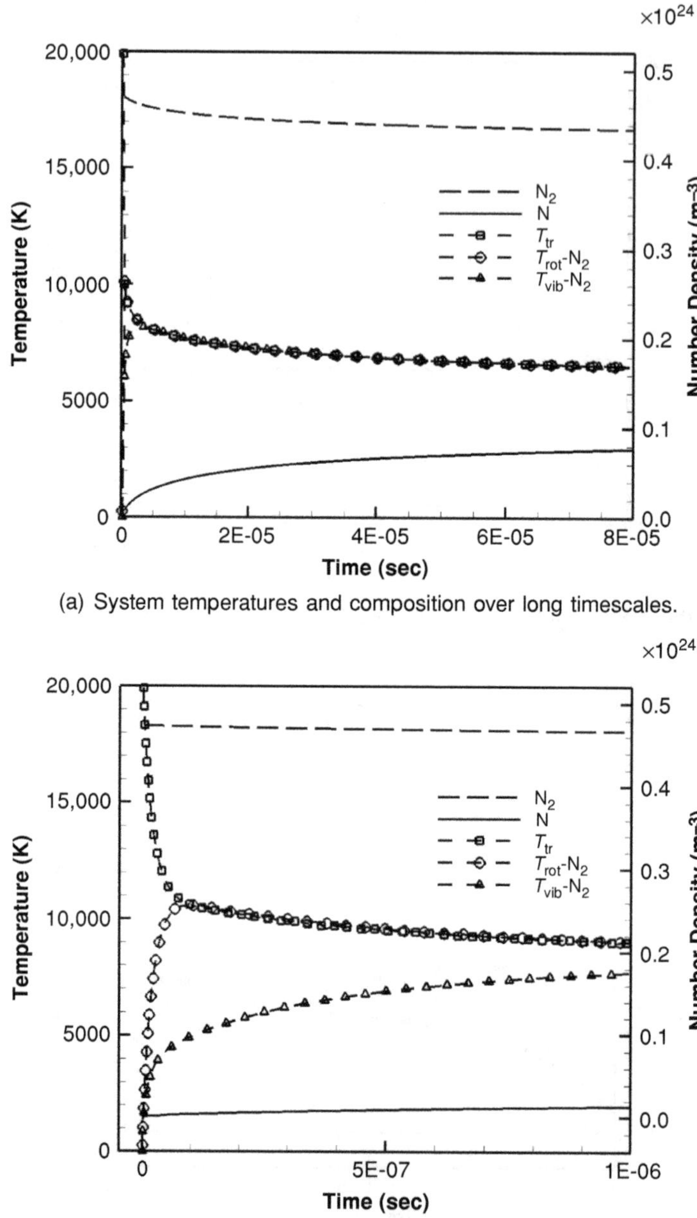

(a) System temperatures and composition over long timescales.

(b) System temperatures and composition over short timsecales.

Figure 7.18 Adiabatic relaxation of nitrogen, initialized with $T_{tr} = 20,000$ K and $T_{rot} = T_{vib} = 200$ K.

and the time required to reach equilibrium is shorter compared to Figs. 7.17(a) and 7.17(b). When the solution is viewed over a short timescale of 1 μsec, Fig. 7.18(b) shows that rotational excitation is still much faster than vibrational excitation; however, some degree of dissociation occurs during the internal energy excitation.

Next, we repeat the isothermal nitrogen relaxation simulations (Figs. 7.14 through 7.16), however, now for a five-species air mixture. Therefore, in these DSMC calculations, all reaction mechanisms in Table 7.2 are active. Recall

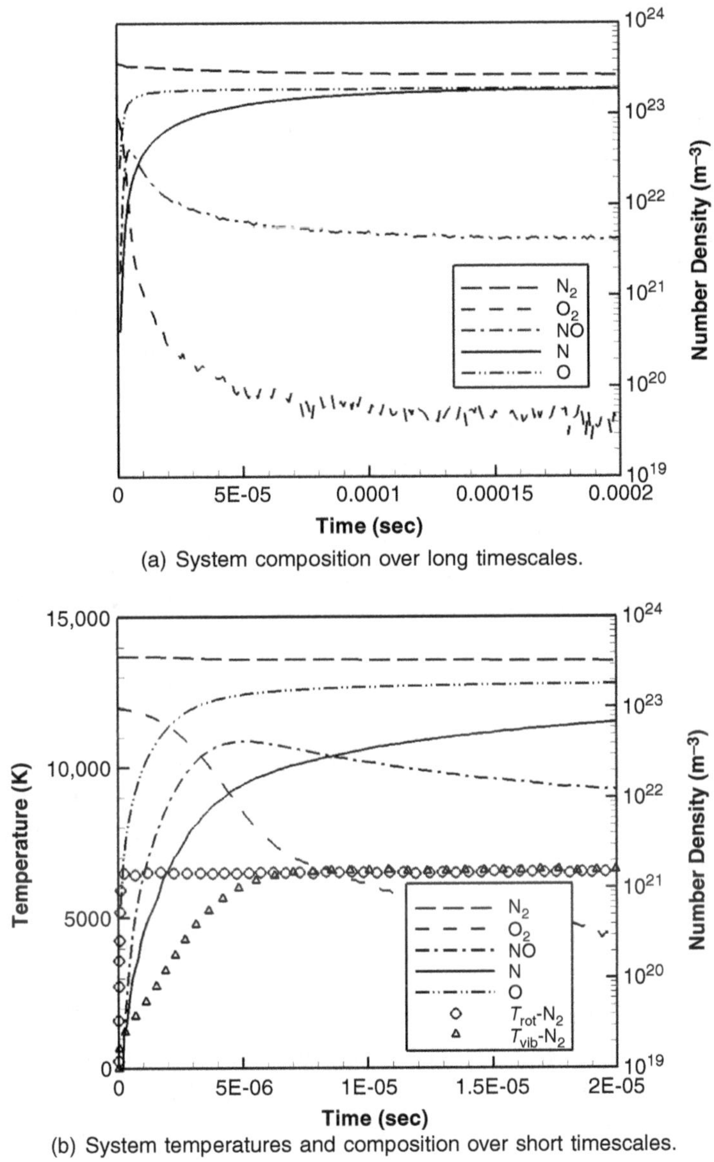

Figure 7.19 Isothermal relaxation of air to $T_{tr} = 6500$ K.

that for isothermal conditions, the translational temperature is maintained at a constant value throughout the relaxation. For each case, at $t = 0$ seconds, the systems are comprised of 76.7% diatomic nitrogen and 23.3% diatomic oxygen (by mass fraction) with $\rho = 0.022$ kg/m^3, $T_{rot} = T_{vib} = 200$ K, and the specified translational temperature T_{tr}. Similar to the isothermal nitrogen relaxations above, we investigate three system translational temperatures of 6500 K, 13,000 K, and 20,000 K.

The results for isothermal relaxation of air to $T_{tr} = 6500$ K are plotted over a timescale of 0.2 msec in Fig. 7.19. Over this timescale, the gas is seen to approach a state of chemical equilibrium consisting of fully dissociated oxygen, partially dissociated nitrogen, and a small amount of nitric oxide. This result can be directly compared to the analytical result presented in Fig. 4.15(b) of Chapter 4, which used the same isothermal relaxation conditions in five-species air. Both results (Figs. 7.19(a) and 4.15(b)) are the same, and furthermore, the predicted equilibrium compositions agree precisely with the analytical prediction in Fig. 4.5(a) of Chapter 4 (evaluated at $T = 6500$ K). When plotted over a short timescale of 20 μsec, Fig. 7.19(b) shows that nitrogen dissociation does not begin until the nitrogen molecules in the system have been sufficiently excited into equilibrium with the translational temperature. However, oxygen dissociation begins immediately and is almost fully dissociated by the time significant nitrogen dissociation begins. Another interesting result is the peak in NO number density at $t = 5\mu$sec. On close inspection, this peak in NO species can also be seen in Figs. 7.19(a) and 4.15(b), at very early times. The peak in NO number density is a nonequilibrium phenomenon resulting from the Zeldovich exchange reaction mechanisms (reactions 4 and 5 in Table 7.2). After O atoms are created through O_2 dissociation, these O atoms produce NO through the Zeldovich exchange reaction (reaction 5 in Table 7.2). This leads to rapid increase in the NO number density. Over longer timescales the NO number density is governed by equilibrium chemistry where dissociation reactions involving all molecular species (O_2, N_2, and NO) are fully participating, leading to an overall decrease in the amount of NO. This nonequilibrium behavior of nitric oxide is also observed in shock layer calculations of hypersonic flow over a cylinder in the text that follows.

The results for isothermal relaxation of air to $T_{tr} = 13{,}000$ K are plotted over both long and short timescales in Figs. 7.20(a) and 7.20(b), respectively. Additionally, the results for isothermal relaxation of air to $T_{tr} = 20{,}000$ K are plotted over both long and short timescales in Figs. 7.21(a) and 7.21(b), respectively. As expected from the equilibrium composition of air at such a high temperatures (refer to Fig. 4.5(a)), the DSMC results show near complete dissociation of all molecular species when the gas reaches chemical equilibrium. As the system translational temperature is increased, the time required for the air to dissociate is reduced. Also, as seen in Figs. 7.20(b) and 7.21(b), as the system temperature is increased, internal energy excitation overlaps with the dissociation processes. Even as translational-rotational and translational-vibrational energy transfer act to increase the system internal energy, rapid dissociation removes internal energy from the system molecules. The result is the QSS region that becomes more pronounced as temperature is increased. Although rotational and vibrational temperatures are only plotted for N_2, results for O_2 and NO species would have similar trends. However, the system is quickly depleted of O_2 and NO under these conditions. Owing to the finite number of simulated particles present in DSMC simulations, trends for trace species are affected by statistical scatter, which can be seen for both O_2 and NO species in the aforementioned results.

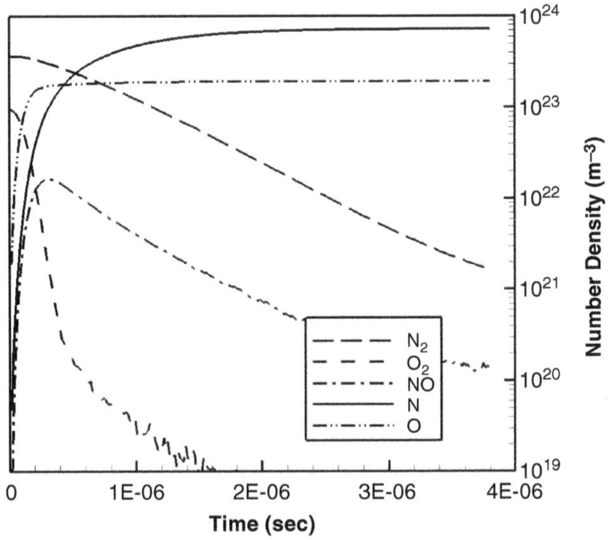

(a) System composition over long timescales.

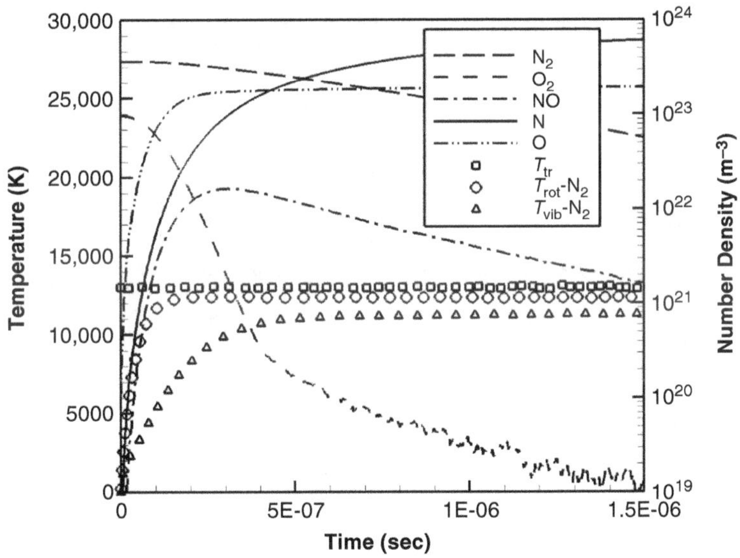

(b) System temperatures and composition over short timescales.

Figure 7.20 Isothermal relaxation of air to 13,000 K.

The above zero-dimensional relaxation calculations provide much physical insight into internal energy relaxation and nonequilibrium chemistry, as well as how these processes are coupled. Such simulations are also useful as they eliminate the complexity of moving particles through a flow field and evaluating gas-surface collisions with complex geometry. As such, zero-dimensional simulations are useful to validate the implementation of DSMC collision models and are also a useful first step to investigate new collision models. Ultimately, however, the degree of nonequilibrium in a flow is determined by a characteristic timescale or length-scale of interest. For example,

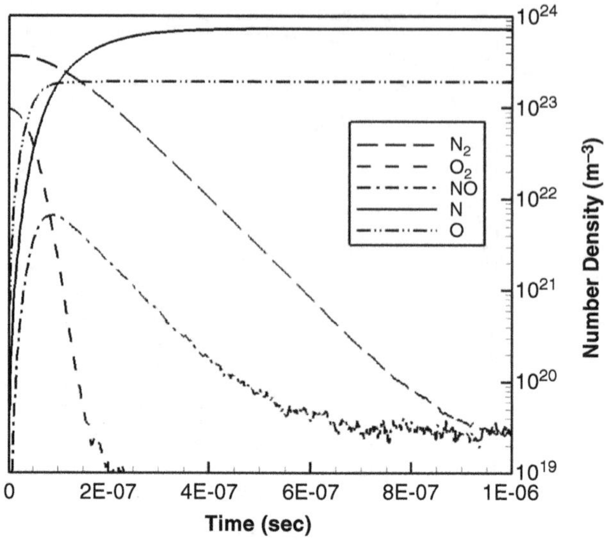

(a) System composition over long timescales.

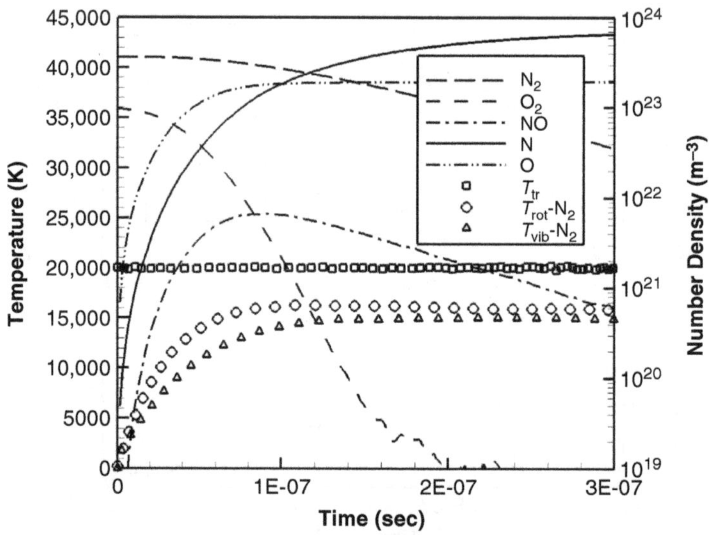

(b) System temperatures and composition over short timescales.

Figure 7.21 Isothermal relaxation of air to 20,000 K.

despite equilibrium compositions corresponding to full dissociation in many of the examples given earlier, in a hypersonic flow the timescales of internal energy excitation and chemical reactions become comparable to the characteristic flow time. As a result, the gas does not reach equilibrium before interacting with the vehicle surface. Furthermore, since the surface of a vehicle must be maintained at a temperature well below the shock layer temperature, the gas cools within the boundary layer as it approaches the surface. In this region, backward reactions including recombination may further influence the nonequilibrium state of the gas. These processes are investigated next by

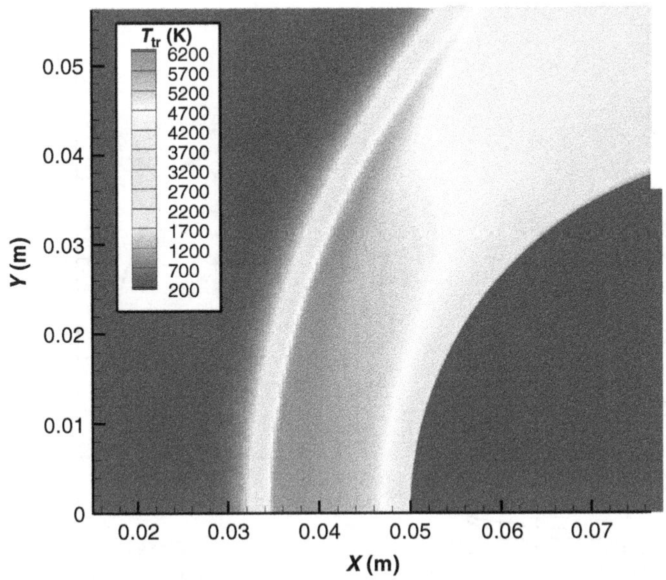

Figure 7.22 Contours of translational temperature for Mach 12 nitrogen flow over a 8 cm diameter cylinder at 70 km altitude. Chemical reactions are not considered.

performing DSMC simulations of hypersonic flow over an 8 cm diameter cylinder for a range of free-stream conditions. All DSMC parameters and models are the same as those used above for the zero-dimensional calculations.

The first simulation involves Mach 12 nitrogen flow at an altitude of 70 km over a two-dimensional cylinder with a diameter of 8 cm. The freestream boundary conditions are $\rho = 7.48 \times 10^{-5}$ kg/m^3, $T = 217.45$ K, $V = 3608$ m/sec. The boundary condition at the cylinder surface is diffuse reflection and full thermal accommodation to a wall temperature of $T_{\text{wall}} = 1000$ K. Supersonic outflow boundary conditions are applied to the upper and downstream boundaries by simply removing particles. Rotational and vibrational energies are considered, but no chemical reactions are allowed.

The flow field solution is shown in Fig. 7.22 with contours of translational temperature. This flow involves large density gradients. Specifically, the density increases across the shock wave and also increases dramatically in the boundary layer as the gas cools to the wall temperature. This is seen in Fig. 7.23, which plots the gas temperatures, bulk x-velocity (u), and density, along the stagnation streamline. Since DSMC collision cells must be sized to the local mean free path, which is inversely proportional to density, such a calculation requires grid-refinement. For the present calculations, an initial solution is obtained on a uniform grid (sized to the free-stream mean free path). Based on this initial solution a refined grid is created and a new simulation is performed using the new grid. This process is repeated (in this case three refinements are performed) until the grid is verified to be sized to the local mean free path of the final solution. For all cylinder simulations presented,

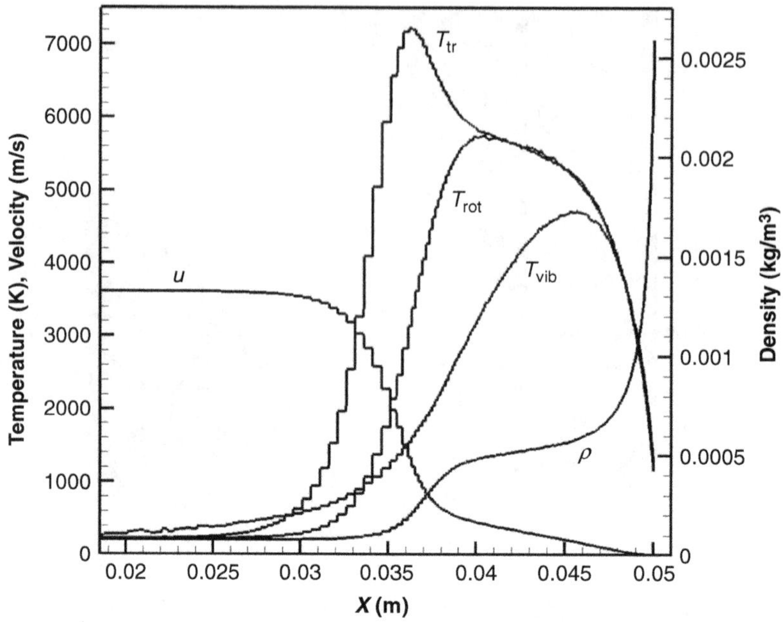

Figure 7.23 Stagnation line profiles for non-reacting nitrogen flow over a cylinder.

the final cell size is close to one-half of the local mean free path everywhere in the flow. Likewise, the local mean collision time becomes small near the stagnation region. A global DSMC timestep (constant timestep) is used for the simulations that is verified to be approximately one-fifth of the smallest mean collision time found in the flow. There are at least 20 particles in each computational cell (in many cells there are far more than 20). These are the recommended numerical parameters to ensure an accurate DSMC solution. Of course, it may be possible that these simulations are over-resolved and that calculations using fewer cells, fewer particles per cell, and larger timesteps may obtain the same results.

The solution along the stagnation streamline, plotted in Fig. 7.23, shows that rotational energy equilibrates with translational energy first, whereas vibrational energy does not equilibrate with translation and rotation until much closer to the cylinder surface. One notable aspect of the solution is the behavior of the vibrational temperature in the freestream region of the flow. This is an artifact of the quantized treatment of vibration and the finite number of simulation particles employed by DSMC. For nitrogen, $\theta_{vib} = 3390$ K, and if even one particle in the free-stream region is elevated to the first vibrational energy level, this will significantly affect the value of T_{vib}. It is important to note that this is only an issue in defining T_{vib} and the simulation is, in fact, quite accurate. For these flow conditions, the post-shock translational-rotational temperature is close to 5500 K and the post-shock vibrational temperature reaches 4500 K at its maximum. For nitrogen, these conditions would not initiate significant dissociation reactions and so the assumption to exclude chemistry for this flow is reasonable.

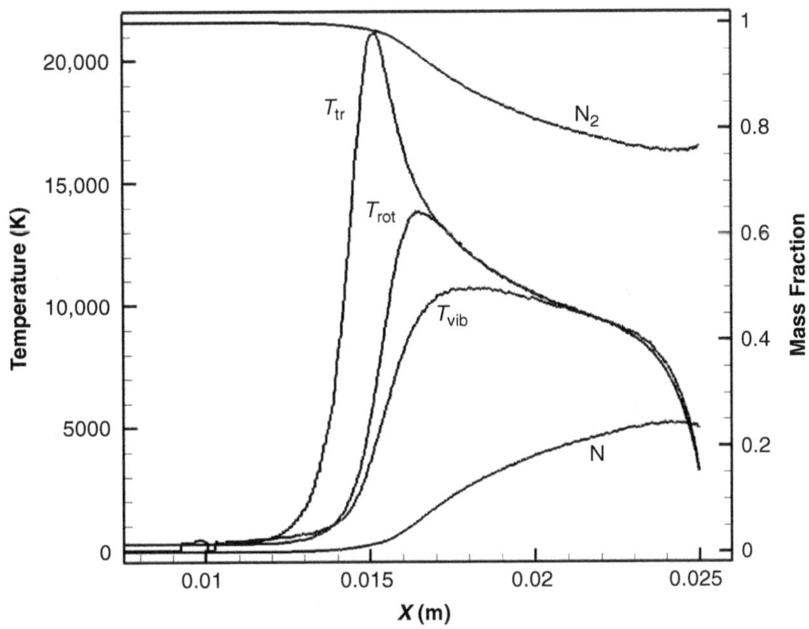

Figure 7.24 Stagnation line profiles for dissociating nitrogen flow over a cylinder.

The second simulation is representative of Mach 20 nitrogen flow over the cylinder at approximately 60 km altitude. These conditions generate significantly higher temperatures in the shock layer, which now lead to nitrogen dissociation. The freestream conditions are now $\rho = 2.88 \times 10^{-4}$ kg/m^3, $T = 245.45$ K, $V = 6281$ m/sec, and the wall temperature of the cylinder remains $T_{\text{wall}} = 1000$ K. The simulation setup is identical to that for the 70 km altitude calculation (depicted in Fig. 7.22), except the DSMC calculation is much more computationally expensive. The temperatures and mass fraction of molecular and atomic nitrogen are plotted along the stagnation streamline in Fig. 7.24. We now see the translational temperature peak above 20,000 K. The rotational temperature equilibrates with the translational temperature first at approximately 13,000 K. The vibrational temperature excites rapidly as well, reaching thermal equilibrium with translational and rotational modes at a post-shock temperature close to 10,000 K. Nitrogen dissociation is now evident. After thermal equilibrium is established, dissociation continues to remove energy from the gas and the temperature within the shock layer drops from 10,000 K to approximately 8000 K. As the gas enters the boundary layer it is then rapidly cooled to the wall temperature. Within the boundary layer the rate of dissociation clearly slows and, in the vicinity of the wall, recombination is evident by the slight increase in N_2 mass fraction and decrease in N mass fraction.

A third simulation is performed, again corresponding to Mach 20 flow at 60 km altitude. However, instead of nitrogen, the gas is now five-species air. Specifically, the freestream conditions are $\rho = 2.88 \times 10^{-4}$ kg/m^3,

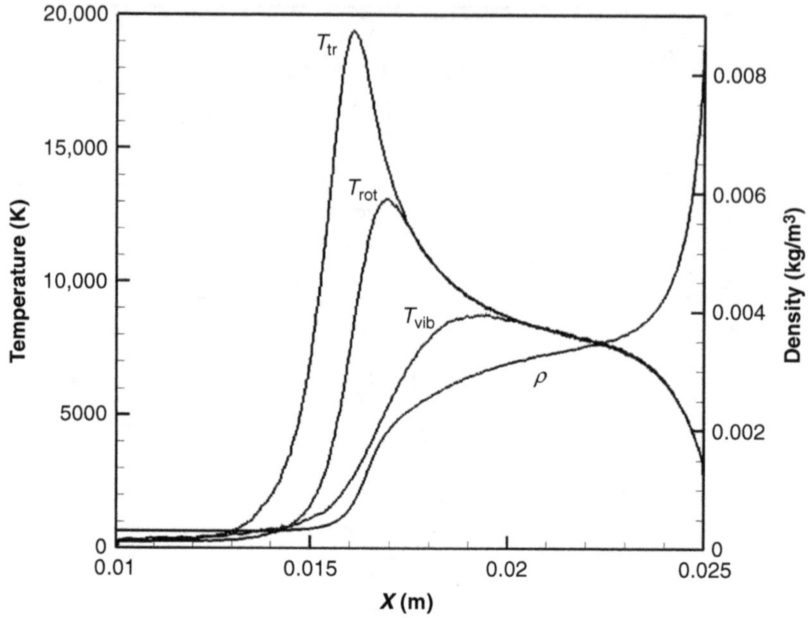

Figure 7.25 Stagnation line temperature profiles in dissociating air flow over a cylinder.

$T = 245.45$ K, $V = 6281.4$ m/sec, and the gas comprises 76.7% diatomic nitrogen and 23.3% diatomic oxygen (by mass fraction). The wall temperature of the cylinder remains $T_{wall} = 1000$ K. The solution for the mixture temperatures and mixture density along the stagnation line is plotted in Fig. 7.25. Compared to the previous simulation in nitrogen, the temperature profiles are similar but approximately 1000–2000 K lower in the shock and shock-layer.

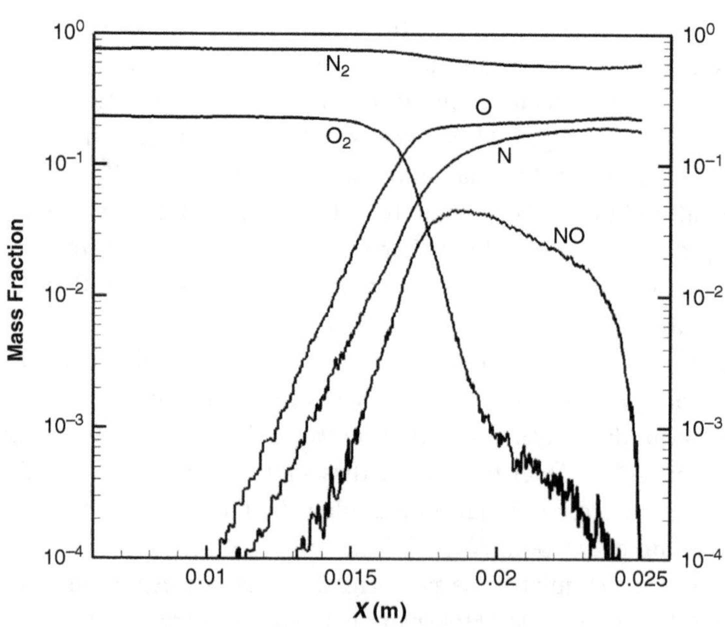

Figure 7.26 Stagnation line profiles for species mass fractions in dissociating air flow over a cylinder.

This is a result of the fact that oxygen dissociation removes energy from the flow. The mass fraction of all species along the stagnation line are plotted in Fig. 7.26. At the onset of the shock wave, oxygen dissociation is rapid and oxygen becomes fully dissociated within the shock layer and throughout the boundary layer down to the cylinder surface. The nonequilibrium trend of NO production behind the shock wave, followed by the steady dissociation of NO in the shock layer is clearly seen. One reason that quantifying the NO concentration is important is that NO is a strong radiator. Many high-energy flow experiments seek to measure NO radiative emission to infer the overall characteristics of the flow, and for this reason, understanding the concentration of NO under nonequilibrium may be important.

7.6 Summary

This chapter has described, in detail, the most common DSMC models for nonequilibrium thermochemistry and analyzed consistency with continuum models used for high-temperature reacting flows. Models for translational-rotational-vibrational energy exchange and chemical reactions were described, as well as the coupling between internal energy relaxation and chemical reaction processes. The rate of energy transfer between translation and rotation is dependent on temperature, but also on the magnitude of nonequilibrium and the direction toward the equilibrium state (compression vs. expansion). The rate of energy transfer between translation and vibration has a strong temperature dependence and is modeled based on experimental data from Millikan–White, which is available for certain species up to approximately 10,000 K. Chemical reactions are modeled based on the total collision energy of colliding particles (the TCE model), and vibrationally favored dissociation can be modeled by the VFD model. These reaction models are shown to be consistent with modified Arrhenius rate coefficient expressions used widely in continuum models of reacting flows. Finally, backward and recombination rate probabilities are best calculated using equilibrium constants (or partition functions) similar to continuum modeling approaches.

The DSMC models and algorithms presented in this chapter are well established in terms of their range of accuracy and computational efficiency. However, most models are phenomenological due to the scarcity of molecular data available for the molecular level models employed by DSMC. Although such data are currently limited, new experiments and advances in quantum chemistry predictions may provide this data in the near future, enabling improved DSMC models to be formulated for challenging, nonequilibrium, chemically reacting flow problems.

Appendix A Generating Particle Properties

In this appendix, we present algorithms for generating particle properties from prescribed distribution functions. Generating particle properties (also referred to as "sampling" particle properties) from a prescribed distribution function is important for two main reasons. First, particles must be generated and/or their properties must be specified at domain boundaries within a direct simulation Monte Carlo calculation, for example at inflow and solid wall boundaries. Second, post-collision properties of particles are often sampled from specific distribution functions, for example the equilibrium distribution functions used in the Borgnakke–Larsen (BL) model. In this appendix, we outline algorithms for generating particles based on equilibrium distributions for a volume of gas, a gas flux across a plane, a gas flux off of a solid surface, and we also outline the acceptance–rejection algorithm that is generally applicable to any distribution function.

All algorithms require the availability of successive random numbers, R, that are uniformly distributed over the range, $0 \leq R \leq 1$. In addition, a normalized distribution function (f_x) must be specified,

$$\int_a^b f_x \, dx = 1 \,, \qquad a \leq x \leq b \tag{A.1}$$

such that $f_x \, dx$ represents the probability that the value of x lies between x and $x + dx$.

A.1 Cumulative Distribution Function

In some cases, the cumulative distribution function can be used to obtain an expression for generating samples (particle properties) directly from a distribution function. The cumulative distribution function (F_x) is defined as

$$F_x \equiv \int_a^x f_x \, dx, \qquad \text{and therefore,} \;\; 0 \leq F_x \leq 1 \tag{A.2}$$

If the distribution function, f_x, is of a certain form, it may be possible to invert Eq. A.2 and solve for x. In this case, sampling the function F_x provides a means to sample values of x that follow the distribution f_x.

A.1.1 Generating Random Particle Positions

As an example, consider the trivial problem of generating a random x-position for a particle, where $a \leq x \leq b$. All x-positions in this range are equally likely and therefore $f_x = \text{Constant}$. From Eq. A.1,

$$f_x = \frac{1}{b-a} \tag{A.3}$$

and therefore,

$$F_x = \int_a^x f_x \, dx = \int_a^x \frac{1}{b-a} dx = \frac{x-a}{b-a} \tag{A.4}$$

Since $0 \leq F_x \leq 1$, we can sample a random value of F_x by drawing a random number $(0 \leq R \leq 1)$, and selecting $F_x = R$. Using Eq. A.4, the value of x that corresponds to R is then

$$x = a + R(b-a) \tag{A.5}$$

This is the expected result for the random selection of an x-position between a and b. The same result (Eq. A.5) can be used to generate random x, y, and z positions within specified cell or domain boundaries.

A.1.2 Generating Radial Positions for Axisymmetric Flow

As another example, consider generating a random radial position (r) of a particle within a uniform density axisymmetric flow, where $a \leq r \leq b$. In this case, since the cross-sectional area is proportional to r, the probability of finding a particle with a specific value of r is also proportional to r. Now, $f_r = \text{Constant} \times r$, and from Eq. A.1 we find

$$f_r = \frac{2r}{b^2 - a^2} \tag{A.6}$$

and therefore,

$$F_r = \int_a^r f_r \, dr = \frac{r^2 - a^2}{b^2 - a^2} = R \tag{A.7}$$

The final expression to generate a random value of r, for this particular case, becomes

$$r = \sqrt{a^2 + R(b^2 - a^2)} \tag{A.8}$$

A.1.3 Generating Particle Velocities from a Maxwell–Boltzmann Distribution

In a typical DSMC calculation, particle inflow boundaries are located in regions where the gas is in a known equilibrium state and the velocity distribution function is a Maxwell–Boltzmann distribution. For this reason, an efficient algorithm for sampling a Maxwell–Boltzmann distribution is desired.

Consider the Maxwell–Boltzmann distribution function for the thermal velocity of a particle in the *i*-coordinate direction ($-\infty \leq C_i' \leq +\infty$),

$$f_0(C_i') = \frac{\beta}{\sqrt{\pi}} e^{-\beta^2 C_i'^2}, \qquad \text{where } \beta = \sqrt{\frac{m}{2kT_{\text{tr}}}} \tag{A.9}$$

The cumulative distribution function is therefore

$$F(C_i') = \frac{\beta}{\sqrt{\pi}} \int_{-\infty}^{C_i'} e^{-\beta^2 C_i'^2} dC_i' = \frac{1}{2}\left[1 + \text{erf}\,(\beta C_i')\right] = R \tag{A.10}$$

Here, erf(x) represents the error function (refer to Appendix E) and there is no analytical solution for C_i' in terms of R. For such cases, where the corresponding cumulative distribution function cannot be inverted, the acceptance–rejection procedure described later in Section A.2, can be used to sample properties from any distribution function.

However, for the particular case of a Maxwell–Boltzmann velocity distribution, Eq. A.9 can be mathematically reformulated such that the resulting cumulative distribution function can be inverted. To reformulate the problem, consider the joint probability of finding *two* thermal velocity values (C_i' and Z_i') in the *i*-coordinate direction. Specifically, consider the probability of finding C_i' between C_i' and $C_i' + dC_i'$ in addition to Z_i' between Z_i' and $Z_i' + dZ_i'$. The joint probability, corresponding to a Maxwell–Boltzmann distribution is

$$\begin{aligned} f_0(C_i')\,dC_i'\, f_0(Z_i')\,dZ_i' &= \frac{\beta}{\sqrt{\pi}} e^{-\beta^2 C_i'^2}\, dC_i' \,\frac{\beta}{\sqrt{\pi}} e^{-\beta^2 Z_i'^2}\, dZ_i' \\ &= \frac{\beta^2}{\pi} e^{-\beta^2 (C_i'^2 + Z_i'^2)}\, dC_i'\, dZ_i'. \end{aligned} \tag{A.11}$$

This expression can be converted to cylindrical polar coordinates using the following transformation:

$$\begin{aligned} C_i' &= r\cos\theta \\ Z_i' &= r\sin\theta \end{aligned} \tag{A.12}$$

where

$$\frac{\partial\,(C_i', Z_i')}{\partial\,(r,\theta)} = \begin{vmatrix} (\partial C_i'/\partial r) & (\partial C_i'/\partial \theta) \\ (\partial Z_i'/\partial r) & (\partial Z_i'/\partial \theta) \end{vmatrix} = \begin{vmatrix} \cos\theta & -r\sin\theta \\ \sin\theta & r\cos\theta \end{vmatrix} = r \tag{A.13}$$

and therefore, $dC_i'\, dZ_i' = r\, dr\, d\theta$

In cylindrical polar coordinates, Eq. A.11 becomes

$$f_0(C_i')\,dC_i'\,f_0(Z_i')\,dZ_i' = \frac{\beta^2}{\pi}e^{-\beta^2r^2}r\,dr\,d\theta$$
$$= \underbrace{e^{-\beta^2r^2}d\left(\beta^2r^2\right)}_{f(\beta^2r^2)d(\beta^2r^2)} \times \underbrace{\frac{d\theta}{2\pi}}_{f(\theta)d\theta} \tag{A.14}$$

and the cumulative distribution function can now be written as

$$F(C_i', Z_i') = F(\beta^2r^2)F(\theta) \tag{A.15}$$

Each separate cumulative distribution can be inverted as follows:

$$F(\theta) = \int_0^\theta \frac{d\theta}{2\pi} = \frac{\theta}{2\pi} = R_1 \tag{A.16}$$

which leads to

$$\theta = 2\pi R_1 \tag{A.17}$$

Furthermore,

$$F(\beta^2r^2) = \int_0^{\beta^2r^2} e^{-\beta^2r^2}d(\beta^2r^2) = 1 - e^{-\beta^2r^2} = R_2 \tag{A.18}$$

and since generating a random value of $1 - R_2$ is the same as generating a random value R_2, we have

$$r = \frac{\sqrt{-\ln(R_2)}}{\beta} \tag{A.19}$$

In this manner, by drawing two independent random numbers R_1 and R_2, the simple expressions given by Eqs. A.17 and A.19 can be used to determine r and θ, and therefore determine two thermal velocities, C_i' and Z_i' using Eq. A.12.

Certain DSMC algorithms often require the generation of a relatively small number of particles, for example, particles within a single flow cell or fluxing through a single cell face. Although the use of Eq. A.12 to generate both C_i' and Z_i' values will indeed result in particle velocities following the desired Maxwell–Boltzmann distribution, the values of C_i' and Z_i' are correlated. To avoid such correlations, it is typical to generate only one velocity value and disregard the other. In this manner, particle velocity vectors (C_x, C_y, C_z) can be sampled from a specified Maxwell–Boltzmann distribution (corresponding to bulk velocity components $< C_x >, < C_y >, < C_z >$, and translational temperature T_{tr}), using the following expression:

$$C_i = \ < C_i > + C_i' = \ < C_i > + \sqrt{\frac{2kT_{tr}}{m}}\sin(2\pi R_1)\sqrt{-\ln(R_2)} \tag{A.20}$$

Here $i = x$, y, or z and $R_1 \neq R_2$. Therefore, using Eq. A.20, two independent random numbers are required to generate each component (x, y, z) of a particle velocity vector from a Maxwell–Boltzmann distribution. This simple algorithm is much more efficient than the general acceptance–rejection algorithm discussed later in Section A.2.

A.1.4 Generating Particle Internal Energies from a Boltzmann Distribution

As discussed in Chapter 6 (Section 6.4.2, Eq. 6.64), the energy distribution function for any continuous energy mode (ϵ_i) in an equilibrium gas, with corresponding ζ_i degrees of freedom, can be expressed by

$$f\left(\frac{\epsilon_i}{kT_i}, \zeta_i\right) d\left(\frac{\epsilon_i}{kT_i}\right) = \frac{1}{\Gamma\left(\zeta_i/2\right)} \left(\frac{\epsilon_i}{kT_i}\right)^{\zeta_i/2-1} e^{-\epsilon_i/kT_i}\, d\left(\frac{\epsilon_i}{kT_i}\right) \tag{A.21}$$

For the special case of $\zeta_i = 2$, corresponding to the rotational energy mode for a diatomic species and also corresponding to the vibrational mode (if vibration is fully excited), this expression reduces to

$$f\left(\frac{\epsilon_i}{kT_i}\right) d\left(\frac{\epsilon_i}{kT_i}\right) = e^{-\epsilon_i/kT_i}\, d\left(\frac{\epsilon_i}{kT_i}\right) \tag{A.22}$$

This is precisely the same form as the first term in Eq. A.14, and therefore the cumulative distribution function has the same form as Eq. A.18. This can be solved for a value of the internal energy,

$$\epsilon_i = -\ln(R)kT_i \tag{A.23}$$

which is the analogue of Eq. A.19. Here, T_i is the equilibrium temperature of the internal energy mode (i). In general, one may wish to generate particles corresponding to a state of thermal nonequilibrium where translational, rotational, and vibrational temperatures are different ($T_{\rm tr} \neq T_{\rm rot} \neq T_{\rm vib}$), yet each energy mode (i) is characterized by a Maxwell–Boltzmann distribution based on T_i.

Quantized vibrational energies may be generated in a number of ways. For example, according to the harmonic oscillator vibrational energy model, the expression to sample vibrational energy is analogous to Eq. A.23, however, using the following truncation:

$$\begin{aligned} \epsilon_{\rm vib} &= k\theta_{\rm vib} i_{\rm vib} \\ i_{\rm vib} &= {\rm floor}\left[-\ln(R) \times \left(\frac{T_{\rm vib}}{\theta_{\rm vib}}\right)\right] \end{aligned} \tag{A.24}$$

where $i_{\rm vib}$ is an integer corresponding to the quantized vibrational energy level and $\theta_{\rm vib}$ is the characteristic temperature for vibration. Although Eq. A.24 simply rounds the energy down, different binning strategies could be used. Also, the expression could be generalized to an anharmonic oscillator description, or any vibrational energy model for that matter, using the partition function corresponding to the model.

In summary, particle velocity vectors can be generated using Eq. A.20 and particle rotational and vibrational (continuous) energies can be generated using Eq. A.23, according to Maxwell–Boltzmann distribution functions. To generate quantized vibrational energies, Eq. A.24 can be used. To generate particle properties from a general distribution function, the acceptance–rejection technique is now described.

A.2 Acceptance–Rejection Sampling

To sample any continuous distribution function, $f(x)$, the acceptance–rejection algorithm can be employed. The standard inequality used in most DSMC implementations is $u \leq f(y)/M$. Here, the value y is drawn from a uniform distribution and $f(y)$ is evaluated and normalized by a constant that is greater than or equal to the maximum value of the distribution function, $M \geq f(x)|_{\max}$. A second value, u, is then drawn from a uniform distribution, and if the inequality holds, then the value y is accepted, and a sample value is determined as $x = y$. If the inequality does not hold, the value y is rejected and the process is repeated. In this manner, the resulting set of x values follow $f(x)$.

For completeness, we outline the algorithm in the text that follows. To sample a random variable X from a continuous distribution $f(X)$, and obtain a value x for X, the algorithm is as follows:

(1) Generate a value y for a random variable Y having a distribution $g(Y)$.
(2) Generate a value u for a random variable U that is independent of Y.
(3) Choose a constant M, and check if

$$u \leq f(y)/[Mg(y)]$$

 (a) If the preceding expression holds, then accept the generated value y, by setting $x = y$.
 (b) If the preceding expression does not hold, then reject the value y, go back to (1) and regenerate another value of y.

Using the aforementioned algorithm to generate a series of x values, the probability distribution function $f(x)$ is correctly sampled. When using the acceptance–rejection method within a DSMC calculation, the random variable Y is typically set using a uniform distribution, $g(Y) = 1$. Similarly, the random variable U is typically set using another uniform distribution in the interval $(0, 1)$. As mentioned previously, M is set as a value equal to or greater than the maximum value of the distribution function, i.e., $M \geq f(x)|_{\max}$. With these typical settings for DSMC calculations, the inequality used in the acceptance–rejection algorithm simplifies to $u \leq f(y)/M$.

The acceptance–rejection algorithm is also described in detail for post-collision sampling of continuous internal energy in Chapter 6, Section 6.4.4, for sampling quantized vibrational energy in Chapter 7, Section 7.3.3, and also for sampling the properties of particles fluxing across a plane, in Section A.3 in the text that follows. In some cases, it is desirable to sample from a Chapman–Enskog velocity distribution (Eqs. 5.67 and 5.68) instead of a Maxwell–Boltzmann distribution. An efficient algorithm for sampling the Chapman–Enskog distribution has been developed by Garcia and Alder (1998). Finally, algorithms for sampling the generalized Chapman–Enskog distribution, for multispecies mixtures, have been developed by Stephani, Goldstein and Varghese (2013).

A.3 Generation of Particle Flux Across a Planar Element

Often, the computational domain boundaries of a DSMC calculation are represented by a large number of planar elements, such as a subset of cell faces or sometimes a separate triangulated surface. For inflow surfaces, the flux of simulation particles through each planar element and the properties of the particles may be specified at each simulation timestep.

Consider a planar element of area A and with a unit normal vector in Cartesian coordinates $\hat{n} = (n_x\bar{x}, n_y\bar{y}, n_z\bar{z})$ directed *into* the computational domain. It is convenient to carry out all operations in face-normal coordinates $(\bar{x}_f, \bar{y}_f, \bar{z}_f)$ as depicted in Fig. A.1, where the $\bar{x}_f$ direction is aligned with $\hat{n}$, and the $\bar{y}_f$ and $\bar{z}_f$ directions are normal to $\hat{n}$.

First, the dot product between the bulk velocity vector,

$$\overline{V} = \left(< C_x > \bar{x}, < C_y > \bar{y}, < C_z > \bar{z}\right) \tag{A.25}$$

and the face unit normal vector is used to define a speed ratio (s_n) in the face-normal direction,

$$s_n = \frac{\overline{V} \cdot \hat{n}}{C_m} \tag{A.26}$$

where $C_m = \sqrt{2kT_{\text{tr}}/m} = 1/\beta$ is the mean thermal speed. Rewriting the expression for free molecular mass flux across a plane (Eq. 1.154 derived in Chapter 1 with s_3 now replaced by s_n) we can write the molecular flux as

$$F_n = \frac{nC_m}{2\sqrt{\pi}} \left\{ e^{-s_n^2} + \sqrt{\pi} s_n \left[1 + \text{erf}(s_n)\right] \right\} \tag{A.27}$$

where n is the number density. Therefore, F_n represents the local flux of molecules per square meter per second in the $\hat{n}$ direction. The flux of simulated particles is then given by

$$F_{\text{DSMC}} = \frac{F_n A \Delta t}{W_p} \tag{A.28}$$

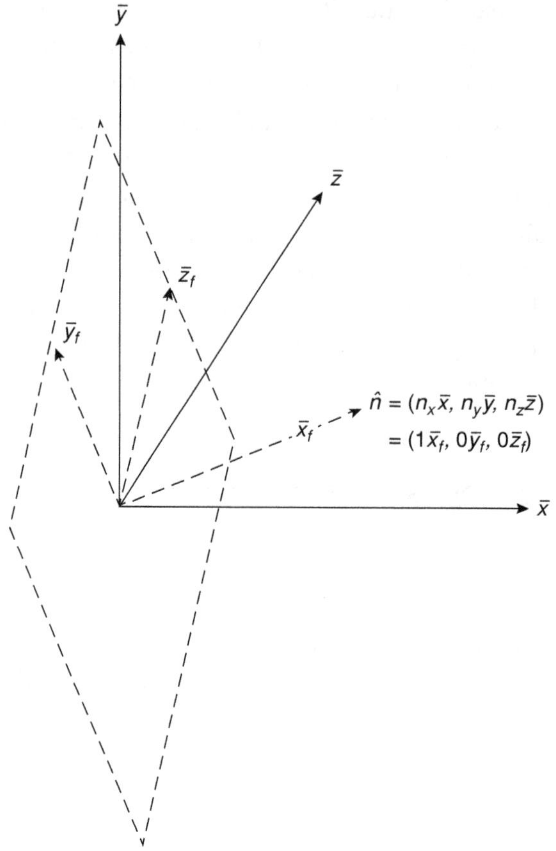

Figure A.1 Face-normal coordinate system used to determine particle flux through a planar element and particle properties.

where A is the area of the element under consideration, Δt is the DSMC timestep, and W_p is the particle weight. Finally, the number of particles generated to simulate the flux through the element is

$$N_{\text{gen}} = \text{floor}\left[F_{\text{DSMC}} + R\right], \quad \text{where } 0 \le R \le 1 \text{ is a random number,} \tag{A.29}$$

since only an integer number of particles can be generated at each timestep.

The properties of each of these N_{gen} particles are then sampled from equilibrium distribution functions. The thermal velocity components in the $\bar{y}_f$ and $\bar{z}_f$ directions (normal to $\hat{n}$) are sampled from Maxwell–Boltzmann distribution functions using Eq. A.20 derived previously in Section A.1.3. Specifically,

$$\begin{aligned} C'_{y_f} &= C_m \sin(2\pi R_1)\sqrt{-\ln(R_2)} \\ C'_{z_f} &= C_m \sin(2\pi R_3)\sqrt{-\ln(R_4)} \end{aligned} \tag{A.30}$$

Note here that $< C_{y_f} >$ and $< C_{z_f} >$ were set to zero, so that only the thermal velocity components (C'_{y_f} and C'_{z_f}) are calculated. The bulk velocity will be added to the thermal velocities in a later step, after switching back to the Cartesian coordinate system.

The particle velocity component in the $\bar{x}_f$ direction (i.e., the $\hat{n}$ direction) requires special attention. Specifically, only positive values of this velocity component should be sampled. This involves sampling from a partial, and biased, Maxwell–Boltzmann distribution. The relevant distribution function was previously derived in Chapter 1 to determine the incident flux of a particle property to a planar surface element in free molecular flow, and was given in Eq. 1.147. Referring to Eq. 1.147 and related discussion, the appropriate distribution function for the thermal velocity component in the $\hat{n}$ direction (C'_{x_f}) is identical to Eq. 1.147, where $s_3 = s_n$, $C'_3 = C'_{x_f}$, and $Q = 1$:

$$\begin{aligned}\int_{C'_{x_f}=-s_n/\beta}^{\infty} & f(C'_{x_f})\, dC'_{x_f} \\ &= \int_{C'_{x_f}=-s_n/\beta}^{\infty} \frac{1}{\sqrt{\pi}} \left(\beta C'_{x_f} + s_n\right) e^{-\beta^2 C'^2_{x_f}}\, dC'_{x_f}\end{aligned} \tag{A.31}$$

In this manner, particles with thermal velocities less than $-s_n/\beta$ in the $\hat{n}$ direction will not flux through the planar surface into the simulation domain.

To sample values of C'_{x_f} from this distribution function, the acceptance–rejection algorithm (outlined in Section A.2) is used. This procedure requires a maximum value of the distribution function, $M = f(C'_{x_f})|_{\max}$, and then determining the ratio of $f(C'_{x_f})/M$. The maximum of (C'_{x_f}) occurs for

$$\beta C'_{x_f} = \left\{\sqrt{s_n^2 + 2} - s_n\right\} /2 \tag{A.32}$$

and therefore,

$$\frac{f(y)}{M} = \frac{f(y)}{f(y)|_{\max}} = K(y + s_n)e^{-y^2} \tag{A.33}$$

where

$$\begin{aligned} K &= \frac{2}{s_n + h} \exp\left[1/2 + s_n(s_n - h)/2\right] \\ h &= \sqrt{s_n^2 + 2} \end{aligned} \tag{A.34}$$

Here, $y = \beta C'_{x_f}$ is a random variate representing a trial value for $\beta C'_{x_f}$. To ensure sampling of the distribution tails, one might calculate a random value for C'_{x_f} in the range $-3C_m \le C'_{x_f} \le +3C_m$, or equivalently, $-3 \le y \le 3$. For velocities outside of this range, the probability density would be low (i.e., $< e^{-3^2}$), and therefore such velocities would be quite rare and could be neglected. In this case, the acceptance–rejection algorithm would consist of the following steps:

(1) Calculate the value, $y = -3 + 6R_1$, where $0 \le R_1 \le 1$ is a random number.
(2) Generate a second random number, $0 \le R_2 \le 1$.

(3) Calculate $f(y)/M$ using Eq. A.33 and check if

$$R_2 \leq f(y)/M$$

(a) If the preceding expression holds, then accept the generated value y, by setting $C'_{x_f} = y/\beta$.

(b) If the preceding expression does not hold, then reject the value y, go back to (1), and generate another value of y.

In summary, for each of the N_{gen} particles (Eq. A.29), thermal velocity components C'_{y_f}, and C'_{z_f} are generated using Eq. A.30, and component C'_{x_f} is generated using Eq. A.33 within the acceptance–rejection algorithm described earlier. These values represent thermal velocity components in the $(\bar{x}_f, \bar{y}_f, \bar{z}_f)$ coordinate frame of reference. Using the face-element normal vector, $\hat{n} = (n_x\bar{x}, n_y\bar{y}, n_z\bar{z})$, these velocity components can be transformed back into the Cartesian coordinate frame of reference, yielding thermal velocities C'_x, C'_y, and C'_z. Finally, the full particle velocity vector is obtained by adding the specified bulk flow velocity as follows:

$$\begin{aligned} C_x &= < C_x > + C'_x \\ C_y &= < C_y > + C'_y \\ C_z &= < C_z > + C'_z \end{aligned} \tag{A.35}$$

At this stage, the velocity vectors for all N_{gen} particles have been assigned in Cartesian coordinates. The internal energies of each of the N_{gen} particles are obtained by sampling the appropriate Maxwell–Boltzmann energy distribution function, using Eq. A.23 for continuous energy distributions, or Eq. A.24 in the case of a quantized vibrational energy distribution.

A.4 Generating Particle Properties Resulting from Surface Collisions

When a particle collides with a solid surface boundary in a DSMC simulation, the post-collision particle properties must be updated to achieve the desired gas–surface boundary condition. The standard boundary condition used in DSMC for collisions with solid surfaces is diffuse reflection and full thermal accommodation. This means that the particles leave the surface according to a Maxwell–Boltzmann distribution corresponding to zero bulk velocity and the surface temperature (T_{surface}).

For most materials, below the scale of a surface element (which in DSMC is often greater than or equal to the local mean free path) lies complex surface structure at the micro- to nanoscale. Gas particles may undergo many surface collisions before leaving the microstructure and returning to the gas flow. For this reason, specifying diffuse reflection relative to the planar surface element, and full thermal accommodation to the surface temperature, are excellent assumptions.

When generating particle velocities and internal energies resulting from a surface collision, it is convenient to work in face-normal coordinates (as shown in Fig. A.1). In fact, the procedure is analogous to that outlined in Section A.3 above, except that the bulk velocity in the $\hat{n}$ direction is now zero (i.e., $s_n = 0$). As a result, the distribution function for the thermal velocity component in the $\hat{n}$ direction is obtained from Eq. A.31 with $s_n = 0$, and a different normalization factor resulting from restriction that $0 \leq C'_{x_f} \leq \infty$:

$$f(C'_{x_f})\,dC'_{x_f} = 2\beta^2 C'_{x_f} e^{-\beta^2 C'^2_{x_f}}\,dC'_{x_f} \tag{A.36}$$

Therefore, the distribution for $\beta^2 C'_{x_f}$ can be written as

$$f(\beta^2 C'_{x_f})\,d\left(\beta^2 C'_{x_f}\right) = e^{-\beta^2 C'^2_{x_f}}\,d\left(\beta^2 C'_{x_f}\right) \tag{A.37}$$

A distribution function of this form leads to the identical cumulative distribution function analyzed earlier in Eq. A.18, and therefore C'_{x_f} can be obtained using a single random number, $0 \leq R \leq 1$,

$$C'_{x_f} = \frac{\sqrt{-\ln(R)}}{\beta} = C_m\sqrt{-\ln(R)} \tag{A.38}$$

The other thermal velocity components, C'_{y_f} and C'_{z_f} (normal to $\hat{n}$), are calculated using Eq. A.30 as described in Section A.3. Finally, these velocity components must be transformed from the face-normal coordinate system back to the Cartesian coordinate system (using the surface normal vector, $\hat{n}$), to obtain the final thermal velocity components C'_x, C'_y, and C'_z. The internal energies of the particle are obtained by sampling the appropriate Maxwell–Boltzmann energy distribution function, using Eq. A.23 for continuous energy distributions, or Eq. A.24 in the case of a quantized vibrational energy distribution. For full thermal accommodation, all temperatures should correspond to the surface temperature ($T_{\mathrm{tr}} = T_{\mathrm{rot}} = T_{\mathrm{vib}} = T_{\mathrm{surface}}$).

It is important to note that even with these assumptions, DSMC calculations naturally predict "velocity slip" and "temperature jump" phenomena. These are nonequilibrium phenomena where the average velocity of gas molecules at the surface is nonzero, and the average thermal energy of gas molecules at the surface is not the same as that corresponding to the surface temperature. This occurs due to an insufficient number of gas–surface collisions resulting from a combination of low density and/or small length scales for the gas–surface interaction. For example, "velocity slip" and "temperature jump" may be noticeable for flow over a sharp leading edge. In this situation only a fraction of the molecules near the surface have collided with (and accommodated to) the surface, while the remainder have not. This results in local non-Maxwell–Boltzmann velocity distributions with nonzero average velocity and a temperature not equal to the surface temperature.

Since the DSMC method simulates gas–surface interactions on a per-collision basis, there is considerable flexibility to incorporate advanced physical models. Many gas–surface collision models, including for gas–surface chemical reactions, have been proposed in the literature for a wide range of applications.

A.5 Subsonic and Fluctuating Boundary Conditions

Subsonic DSMC boundary conditions involve generating particles (or updating their properties) at boundaries and, in general, involve the same algorithms outlined in this appendix. However, the precise gas state at a subsonic boundary is not known a priori, rather it is now part of the solution. There are a number of strategies available to implement subsonic boundary conditions in DSMC, some of which are analyzed in a recent study by Farbar and Boyd (2014). The most suitable implementation for subsonic boundary conditions is often problem specific. Whatever strategy is chosen, care should be taken to properly verify the physical accuracy of the subsonic flow result for the problem of interest.

When simulating nonequilibrium flow for microscale applications, not only is the flow typically subsonic, but it can also involve real fluctuations (refer to Fig. 6.1). For such flows, the algorithms presented in this section require modification. In general, if real fluctuations are present within the gas, then such fluctuations should also be prescribed at simulation boundaries. Owing to correlations between hydrodynamic variables, care must be taken when evaluating means and variances, and particle generation (boundary conditions) must also be performed in a consistent manner. Effects of fluctuations and correlations, as well as numerical strategies to model these effects in a physically accurate manner have been described in a series of articles by Tysanner and Garcia (2004, 2005) and Garcia (2007).

Appendix B

Collisional Quantities

In this appendix, we derive a number of useful expressions specific to *pairs* of molecules in an equilibrium gas. In general, the distributions of relative velocity, collision energy, and internal energy, for molecular pairs involved *in collisions* in a gas at equilibrium are not the same as the distributions for molecular pairs in the gas system as a whole. The most clear example of this is that the distribution function for relative velocity found in collision pairs is different than for all pairs within the gas, since the collision probability is biased by relative velocity. This difference between quantities in collisions compared to quantities of the gas as a whole is important for a number of DSMC algorithms. Since DSMC mainly alters the state of the gas through collisions, the collisional distribution functions are often the most relevant for DSMC relaxation models.

B.1 Distributions for Molecule Pairs

Since the collision cross section (collision probability) is a strong function of relative velocity, let us consider the average relative velocity of molecular pairs, selected at random, from a gas in equilibrium. Let us further generalize, and write an expression for the average relative velocity (g) raised to a power (j),

$$\langle g^j \rangle = \int_{-\infty}^{\infty}\int_{-\infty}^{\infty} g^j f_0(\bar{C}) f_0(\bar{Z})\, d\bar{C}\, d\bar{Z} \tag{B.1}$$

where f_0 indicates a Maxwell–Boltzmann equilibrium velocity distribution function. As discussed later in Section B.2, since the probability of collision is also proportional to the relative velocity raised to some power (i.e., VHS-type models), the quantity in Eq. B.1 is also proportional to the number of collisions in the equilibrium gas.

Similar to the derivation of the hard-sphere collision rate in Section 1.3.5 of Chapter 1, we can evaluate the above expression in terms of the gamma function. Specifically, analogous to Eqs. 1.123 through 1.133, by reformulating the velocity vectors $\overline{C}$ and $\overline{Z}$ in terms of the relative velocity,

$\overline{g} \equiv \overline{C} - \overline{Z}$, and the center-of-mass velocity, $\overline{W} \equiv (m_A\overline{C} + m_B\overline{Z})/(m_A + m_B)$, we can rewrite Eq. B.1 as,

$$\langle g^j \rangle = \frac{(m_A m_B)^{3/2}}{(2\pi kT)^3} \int_{-\infty}^{\infty}\int_{-\infty}^{\infty} g^j \exp\left\{-\frac{1}{2kT}\left[(m_A + m_B)W^2 + m_r g^2\right]\right\} d\overline{W}\, d\overline{g} \tag{B.2}$$

Writing in spherical-polar coordinates,

$$\begin{aligned} d\overline{W} &= W^2 \sin\phi_W \, d\phi_W \, d\theta_W \, dW \\ d\overline{g} &= g^2 \sin\phi_g \, d\phi_g \, d\theta_g \, dg \end{aligned} \tag{B.3}$$

where

$$\begin{aligned} W &= |\overline{W}| \\ g &= |\overline{g}| \end{aligned} \tag{B.4}$$

we can integrate over both sets of angles, $[0 < \phi < \pi]$, $[0 < \theta < 2\pi]$, yielding

$$\begin{aligned} d\overline{W} &= 4\pi W^2 dW \\ d\overline{g} &= 4\pi g^2 dg \end{aligned} \tag{B.5}$$

Using the above expressions, we can rewrite Eq. B.2 in terms of two separate integrals:

$$\langle g^j \rangle = \int_0^{\infty} f_0(W)\, dW \int_0^{\infty} g^j f_0(g)\, dg \tag{B.6}$$

where $f_0(W)$ is the distribution of center-of-mass velocity magnitude for particle pairs within the gas at equilibrium,

$$f_0(W) = \frac{4\,(m_A + m_B)^{3/2}}{\sqrt{\pi}(2kT)^{3/2}} W^2 \exp\left[-\frac{(m_A + m_B)\, W^2}{2kT}\right] \tag{B.7}$$

and $f_0(g)$ is the distribution of relative velocity magnitude for particle pairs within the gas at equilibrium,

$$f_0(g) = \frac{4m_r^{3/2}}{\sqrt{\pi}(2kT)^{3/2}} g^2 \exp\left[-\frac{m_r g^2}{2kT}\right] \tag{B.8}$$

Since $f_0(W)$ is a normalized distribution function, the first integral in Eq. B.6 is unity, and therefore, as one might expect,

$$\langle g^j \rangle = \int_0^{\infty} g^j f_0(g)\, dg \tag{B.9}$$

The expression

$$\langle g^j \rangle = \int_0^{\infty} \frac{4m_r^{3/2}}{\sqrt{\pi}(2kT)^{3/2}}\, g^{j+2} \exp\left[-\frac{m_r g^2}{2kT}\right] dg \tag{B.10}$$

can be reduced using the following transformations:

$$\begin{aligned} x &\equiv \frac{m_r g^2}{2kT} \\ g^{j+2} &= \left(\frac{2kT}{m_r}\right)^{1+j/2} x^{1+j/2} \\ dg &= \sqrt{\frac{kT}{2m_r x}} dx \end{aligned} \tag{B.11}$$

to give the following:

$$\langle g^j \rangle = \frac{2}{\sqrt{\pi}} \left(\frac{2kT}{m_r}\right)^{j/2} \int_0^\infty x^{t-1} e^{-x} dx \tag{B.12}$$

where $t = (3 + j)/2$. This integral is the gamma function, $\Gamma(t)$, and therefore

$$\langle g^j \rangle = \frac{2}{\sqrt{\pi}} \left(\frac{2kT}{m_r}\right)^{j/2} \Gamma\left(\frac{3+j}{2}\right) \tag{B.13}$$

The average relative velocity magnitude (the case where $j = 1$) of randomly selected molecular pairs within an equilibrium gas, is therefore

$$\langle g \rangle = \sqrt{\frac{8kT}{\pi m_r}} \tag{B.14}$$

For a simple gas ($m_A = m_B = m = 2m_r$), the average relative velocity of molecular pairs becomes

$$\langle g \rangle = 4\sqrt{\frac{kT}{\pi m}} = \sqrt{2}\left\langle |\overline{C}| \right\rangle \tag{B.15}$$

where $< |\overline{C}| >$ is the average speed of molecules in an equilibrium single-species gas derived in Eq. 1.119 of Chapter 1. Furthermore, it is noted that Eq. B.15 was used in Chapter 6 to derive the equilibrium collision rate resulting from the no-time-counter (NTC) DSMC algorithm given in Eq. 6.5.

B.2 Distributions for Molecule Pairs Involved in Collisions

Now that we have derived distribution functions for molecular pairs in a gas at equilibrium, we seek to derive distribution functions for only those pairs involved in collisions. Since the collision cross section, and therefore the probability of collision, is biased by relative velocity, these distributions are not generally the same.

Let us consider the average value of a quantity Q found in collisions within an equilibrium gas. Analogous to the expression derived for a hard-sphere gas in Eq. 1.133 of Chapter 1, the differential expression for the number of collisions between A and B particles (having velocities $\overline{W}$ and $\overline{g}$ in the center-of-mass frame) per unit volume, per unit time, is

$$dZ_{\mathrm{AB}} = n_{\mathrm{A}} n_{\mathrm{B}} \frac{(m_{\mathrm{A}} m_{\mathrm{B}})^{3/2}}{(2\pi kT)^3} (\sigma_T \, g) \exp\left\{ -\frac{1}{2kT} \left[(m_{\mathrm{A}} + m_{\mathrm{B}}) W^2 + m_r g^2 \right] \right\} d\overline{W} \, d\overline{g} \tag{B.16}$$

We are interested in a quantity Q, which is a function of the relative velocity of the molecule pair, $Q = Q(g)$. In this case, we can now write the following expression for the average value of Q found in collisions within an equilibrium gas,

$$\langle Q \rangle_{\mathrm{collisions}} = \frac{\int_0^\infty \int_0^\infty Q(g) \, dZ_{\mathrm{AB}}}{\int_0^\infty \int_0^\infty dZ_{\mathrm{AB}}} \tag{B.17}$$

Here, the numerator is the summation of $Q(g)$ over all collisions, and the denominator is the total number of collisions. If the collision cross section (σ_T) is a power-law function of g, we can replace $(\sigma_T \, g)$ with Dg^j, where D is a constant of proportionality. Finally, using the same transformation as used in Eqs. B.2 through B.9, and noting that the coefficients in the numerator and denominator cancel, we arrive at the following expression for the average value of $Q(g)$ found in collisions:

$$\langle Q \rangle_{\mathrm{collisions}} = \frac{\int_0^\infty Q(g) \, g^j f_0(g) \, dg}{\langle g^j \rangle} \tag{B.18}$$

where $f_0(g)$ is given in Eq. B.8 and $\langle g^j \rangle$ is given in Eq. B.13.

For the VHS cross section model (Eq. 6.17), $\sigma_T \propto g^{-2\nu}$, and therefore $(\sigma_T \, g) \propto g^{1-2\nu}$. In this case, $j = 1 - 2\nu$, or equivalently in terms of the viscosity temperature exponent parameter, $j = 2 - 2\omega$ (see Eq. 6.24). For the VHS collision model, the average value of Q found in collisions within an equilibrium gas becomes,

$$\langle Q \rangle_{\mathrm{collisions}}^{\mathrm{VHS}} = \frac{2}{\Gamma(5/2 - \omega)} \left(\frac{m_r}{2kT} \right)^{5/2-\omega} \int_0^\infty Q(g) \, g^{2(2-\omega)} \exp\left[-\frac{m_r g^2}{2kT} \right] dg \tag{B.19}$$

This can be written as

$$\langle Q \rangle_{\mathrm{collisions}}^{\mathrm{VHS}} = \int_0^\infty Q(g) f_{\mathrm{collisions}}^{\mathrm{VHS}}(g) \, dg \tag{B.20}$$

where the distribution function for relative velocity (g) found in collisions (using the VHS model) within an equilibrium gas is

$$f_{\mathrm{collisions}}^{\mathrm{VHS}}(g) = \frac{2}{\Gamma(5/2 - \omega)} \left(\frac{m_r}{2kT} \right)^{5/2-\omega} g^{2(2-\omega)} \exp\left[-\frac{m_r g^2}{2kT} \right] \tag{B.21}$$

Clearly, due to the dependence of collision cross section (collision probability) on relative velocity, $f_{\mathrm{collisions}}^{\mathrm{VHS}}(g)$ is not equal to $f_0(g)$.

As an example, the VHS cross section model (Eq. 6.17) requires a reference cross section value. For convenience, this reference cross section was defined based on the value of $\langle g^{2\nu}\rangle^{\mathrm{VHS}}_{\mathrm{collisions}}$. By setting $Q(g) = g^{2\nu} = g^{2\omega-1}$, the integral in Eq. B.20 becomes a standard integral, leading to

$$\langle g^{2\omega-1}\rangle^{\mathrm{VHS}}_{\mathrm{collisions}} = \frac{(2kT/m_r)^{\omega-1/2}}{\Gamma(5/2-\omega)} \tag{B.22}$$

which is the same result given in Chapter 6 without derivation by Eq. 6.21.

The above expressions can be used to determine the distribution function for translational energy found in collisions. Relative to the center-of-mass motion of the collision pair, the relative translation energy of the pair is

$$\epsilon_{\mathrm{tr}} \equiv \frac{1}{2} m_r g^2 \tag{B.23}$$

We can simply rewrite Eqs. B.20 and B.21 in terms of ϵ_{tr}, as

$$\langle Q\rangle^{\mathrm{VHS}}_{\mathrm{collisions}} = \int_0^\infty Q(\epsilon_{\mathrm{tr}}) f^{\mathrm{VHS}}_{\mathrm{collisions}}(\epsilon_{\mathrm{tr}})\, d\epsilon_{\mathrm{tr}} \tag{B.24}$$

where the distribution function of translation energy found in collisions (using the VHS model) in a gas at equilibrium is

$$f^{\mathrm{VHS}}_{\mathrm{collisions}}(\epsilon_{\mathrm{tr}}) = \frac{\sqrt{2/m_r}}{\Gamma(5/2-\omega)}\left(\frac{1}{kT}\right)^{5/2-\omega} \epsilon_{\mathrm{tr}}^{3/2-\omega} \exp\left[-\frac{\epsilon_{\mathrm{tr}}}{kT}\right] \tag{B.25}$$

This is a useful expression that is often used to determine post-collision properties within DSMC, for example, refer to Chapter 6 (Sections 6.4.2, and 6.4.4), where models for energy transfer between translational and internal (rotational and vibrational) modes is discussed in detail.

As discussed in Section 6.4.2 of Chapter 6, the distribution function of energy in any continuous energy mode (ϵ_i), with corresponding ζ_i degrees of freedom, in an equilibrium gas, can be expressed by

$$f\left(\frac{\epsilon_i}{kT}, \zeta_i\right) d\left(\frac{\epsilon_i}{kT}\right) = \frac{1}{\Gamma(\zeta_i/2)}\left(\frac{\epsilon_i}{kT}\right)^{\zeta_i/2-1} e^{-\epsilon_i/kT} d\left(\frac{\epsilon_i}{kT}\right) \tag{B.26}$$

This expression applies to the energy distribution for molecules in the gas, but is also applicable for the energy distribution of molecules involved *in collisions* in the gas, as long as the degrees of freedom (ζ_i) are those that participate in the collision (i.e., those that contribute toward ϵ_i). For example, ϵ_{rot} may refer to the rotational energy of only one molecule in the collision, in which case $\zeta_{\mathrm{rot}} = 2$. However, if $\epsilon_{\mathrm{rot}} = \epsilon_{\mathrm{rot},1} + \epsilon_{\mathrm{rot},2}$ is the total rotational energy of two molecules in a collision pair, then $\zeta_{\mathrm{rot}} = 4$ is the total degrees of freedom of both molecules (refer to Eq. 6.68, for example).

In this manner, comparison of Eqs. B.26 and B.25 reveals that the effective translational degrees of freedom that participate in collisions (using the VHS model), is $\zeta_{\mathrm{tr}} = 5 - 2\omega$. For the typical range of $0.5 < \omega < 1.0$, this implies that the available translational degrees of freedom for molecules *in collisions* is higher than that of the molecules in gas (i.e., $\zeta_{\mathrm{tr}} > 3$). This is due to the fact that collision probabilities are biased by the relative velocity, and therefore by the translational energy involved in the collision.

Finally, the aforementioned expressions can be used to determine the distribution function corresponding to the total collision energy. The distribution is again given by Eq. B.26, where $\epsilon_i = \epsilon_{\text{coll}}$ (refer to Eq. 7.25), corresponding to all energy modes of both molecules in the collision pair, $\zeta_i = \zeta_T$ (refer to Eq. 7.26). As discussed in Section 7.4.1 of Chapter 7, the distribution function for total collision energy is used for the TCE chemistry model. Furthermore, the fraction of collisions that contain energy greater than a specified energy barrier can be readily determined using Eq. B.26. Specifically,

$$\frac{dN\left[\epsilon_{\text{coll}} > \epsilon_a\right]}{N} = \int_{\epsilon_a/kT}^{\infty} f\left(\frac{\epsilon_{\text{coll}}}{kT}, \zeta_T\right) d\left(\frac{\epsilon_{\text{coll}}}{kT}\right) = \frac{\Gamma_{\text{I}}\left(\zeta_T, \epsilon_a/kT\right)}{\Gamma\left(\zeta_T\right)} \tag{B.27}$$

This expression includes the incomplete gamma function,

$$\Gamma_{\text{I}}(t, \alpha) \equiv \int_{\alpha}^{\infty} x^{t-1} e^{-x} dx \tag{B.28}$$

The fraction of collisions involving collision energy greater than some threshold, as given by Eq. B.27, is a useful quantity for molecular analysis and modeling.

Appendix C Determining Post-Collision Velocities

In this appendix we outline the basic procedures to determine the post-collision velocities of DSMC particles. This is typically the last step of the overall collision algorithm and is performed after chemical reactions are processed, and after post-collision internal energies are assigned. After these processes, the remaining collision energy is distributed to the translational modes of the post-collision particles, and ultimately post-collision velocity vectors are assigned to each particle. This appendix presents algorithms to assign post-collision particle velocities for hard-sphere scattering (hard-sphere [HS], variable hard-sphere [VHS], and generalized hard-sphere [GHS] models), soft-sphere scattering (variable soft-sphere [VSS] and generalized soft-sphere [GSS]), and also presents algorithms specific to the products of dissociation, exchange, and recombination reactions.

C.1 Elastic and Inelastic Collisions

Collisions between particles are most conveniently analyzed in the center-of-mass frame of reference. The corresponding translational energy associated with a collision was previously introduced in Eq. 6.62 of Chapter 6, as $\epsilon_{\text{tr}} = m_r g^2/2$. For an elastic collision, no energy is transferred between translational and internal energy modes. In this case, the relative velocity g (and therefore ϵ_{tr}) remains constant during the collision (refer to Eq. 1.68 in Chapter 1). Note that this value of g is the same as that used to determine the collision probability (see Eq. 6.13 in Chapter 6).

For inelastic collisions, as described in Section 6.4 of Chapter 6, energy is transferred between translational and internal energy modes. After the energy redistribution is complete, there will be some amount of translational energy (ϵ_{tr}) remaining (refer to Eq. 6.77 and related discussion, or Eq. 6.104 and related algorithm below). In this case, the post-collision relative velocity is

$$g = \sqrt{\frac{2\epsilon_{\text{tr}}}{m_r}} \tag{C.1}$$

where m_r is the reduced mass of the collision pair. In either case, elastic or inelastic, the post-collision relative velocity g is known, and we proceed to choose a scattering angle and calculate final velocity vectors.

C.1.1 Hard-Sphere Scattering

For hard-sphere scattering, used by the HS, VHS, and GHS collision models, a random scattering angle is chosen. Specifically, the center-of-mass velocity (W_i) of the collision pair can be calculated using Eq. 1.58, where the pre-collision velocities are denoted by $C_{i,1}^{\text{pre}}$ and $C_{i,2}^{\text{pre}}$, and the two particles are denoted with subscripts 1 and 2,

$$W_i = \frac{m_1}{m_1 + m_2} C_{i,1}^{\text{pre}} + \frac{m_2}{m_1 + m_2} C_{i,2}^{\text{pre}} \tag{C.2}$$

where $i = x, y, z$. Due to conservation of linear momentum, W_i remains unchanged during the collision. The velocity vectors relative to the center-of-mass velocity are now randomized, such that the relative velocity magnitude is g. First, the scattering angle is determined from Eq. 5.111 of Chapter 5 as

$$\cos\chi = 2\left(\frac{b}{d}\right)^2 - 1 \tag{C.3}$$

where b is the impact parameter and d is the diameter that defines the total cross section (refer to Eq. 6.16). As discussed in Section 6.2.1, impact parameters are random in dilute gas and, in fact, the quantity $(b/d)^2$ is randomly distributed in the range [0,1]. As a result, this allows us to calculate random scattering angles, χ and θ by

$$\begin{aligned} \cos\chi &= 2R_1 - 1 \\ \sin\chi &= \sqrt{1 - \cos^2\chi} \\ \theta &= 2\pi R_2 \end{aligned} \tag{C.4}$$

where R_1 and R_2 are random numbers in the range [0,1]. This enables new relative velocity components to be determined as

$$\begin{aligned} g'_x &= g\cos\chi \\ g'_y &= g\sin\chi\cos\theta \\ g'_z &= g\sin\chi\sin\theta \end{aligned} \tag{C.5}$$

Finally, the post-collision velocity vectors of the particles are assigned, analogous to Eqs. 1.63 and 1.64, as

$$\begin{aligned} C_{i,1} &= W_i + \frac{m_2}{m_1 + m_2} g'_i \\ C_{i,2} &= W_i - \frac{m_1}{m_1 + m_2} g'_i \end{aligned} \tag{C.6}$$

where $i = x, y, z$. This completes the collision, and all particle species, internal energies, and final velocity vectors have been determined. The particles are now ready for the next movement step of the DSMC algorithm.

C.1.2 Soft-Sphere Scattering

For soft-sphere scattering, used by the VSS and GSS models, the scattering angle χ is biased. Specifically, from Eq. 6.33 we can write

$$\cos\chi = 2\left(\frac{b}{d}\right)^{(2/\alpha_{1-2})} - 1 \tag{C.7}$$

where α_{1-2} is the VSS exponent parameter specific to the species pair (1–2). Here, d has the same definition as for hard-sphere scattering (Eq. 6.16), and so the quantity $(b/d)^2$ is, again, randomly distributed in the range [0,1]. Therefore, angles for soft-sphere scattering are determined using

$$\begin{aligned} \cos\chi &= 2R_1^{(1/\alpha_{1-2})} - 1 \\ \sin\chi &= \sqrt{1-\cos^2\chi} \\ \theta &= 2\pi R_2 \end{aligned} \tag{C.8}$$

For elastic collisions, since there is no energy exchange between translational and internal energy models, the original relative velocity *vector* can be maintained, and the soft-sphere scattering angles applied directly to this vector. The original relative velocity vector is known from the pre-collision velocities of the collision pair,

$$g_i = C_{i,1}^{\text{pre}} - C_{i,2}^{\text{pre}} \tag{C.9}$$

where $i = x, y, z$. In this case, it can be shown that the final relative velocity components are given by

$$\begin{aligned} g'_x &= g_x\cos\chi + \left(\sqrt{g_y^2+g_z^2}\,\sin\theta\right)\sin\chi \\ g'_y &= g_y\cos\chi + \left(\frac{g\,g_z\cos\theta - g_x g_y\sin\theta}{\sqrt{g_y^2+g_z^2}}\right)\sin\chi \\ g'_z &= g_z\cos\chi - \left(\frac{g\,g_y\cos\theta + g_x g_z\sin\theta}{\sqrt{g_y^2+g_z^2}}\right)\sin\chi \end{aligned} \tag{C.10}$$

The center-of-mass velocity is still given by Eq. C.2, and as in the hard-sphere scattering case, remains constant. The final post-collision velocity vectors of both particles are therefore also determined by Eq. C.6, using the values of g'_x, g'_y, and g'_z, resulting from the soft-sphere expressions in Eq. C.10.

It is important to the note that it may not be required to maintain the pre-collision relative velocity vector and apply the scattering angle expressions directly to this vector. Since relative velocity vectors may be sufficiently random within a dilute gas, Eq. C.9 could be replaced with a randomized expression for g_i. In this manner, the angles χ and θ would still be biased, via Eqs. C.7 and C.10; however, this bias would now be applied to a random relative velocity vector. For inelastic collisions, where energy is exchanged between the translational and internal energy modes, the magnitude of the relative velocity changes and maintaining the relative velocity vector from the pre-collision configuration may no longer conserve energy. For inelastic collisions, and also for reactive collisions discussed next, it is preferable to randomize the relative velocity vector prior to applying soft-sphere scattering angles.

Therefore, for soft-sphere scattering of inelastic collisions (including reactive collisions), the post-collision value of g is determined by the remaining translational energy (ϵ_{tr}) by Eq. C.1. Instead of maintaining the original relative velocity vector g_i from Eq. C.9, a random vector is created using the scalar value of g. This can be achieved by using Eq. C.4 together with the following expressions:

$$\begin{aligned} g_x &= g \sin\chi \cos\theta \\ g_y &= g \sin\chi \sin\theta \\ g_z &= g \cos\chi \end{aligned} \tag{C.11}$$

Therefore, for soft-sphere scattering of inelastic collisions, Eq. C.11 replaces Eq. C.9. The remaining soft-sphere expressions are the same; namely, the final relative velocity components are obtained from Eq. C.10, and these are used in Eq. C.6 to determine the final post-collision velocity vectors.

This completes the collision, and all particle species, internal energies, and final velocity vectors have been determined based on soft-sphere scattering. The particles are now ready for the next movement step of the DSMC algorithm.

C.2 Dissociation Reaction Collisions

A collision between two simulation particles that results in a dissociation reaction involves three product particles. In the schematic diagram, Fig. C.1, these are labeled as particle P, which could be an atom or molecule, and particles A1 and A2, both of which are atoms. Special consideration is required to properly assign the velocity vectors of these product particles.

As discussed in Section 7.4 of Chapter 7, the total collision energy (ϵ_{coll}) is calculated based on the pre-collision translational and internal energies of the two particles entering the collision (labeled as P and M in Fig. C.1). If a

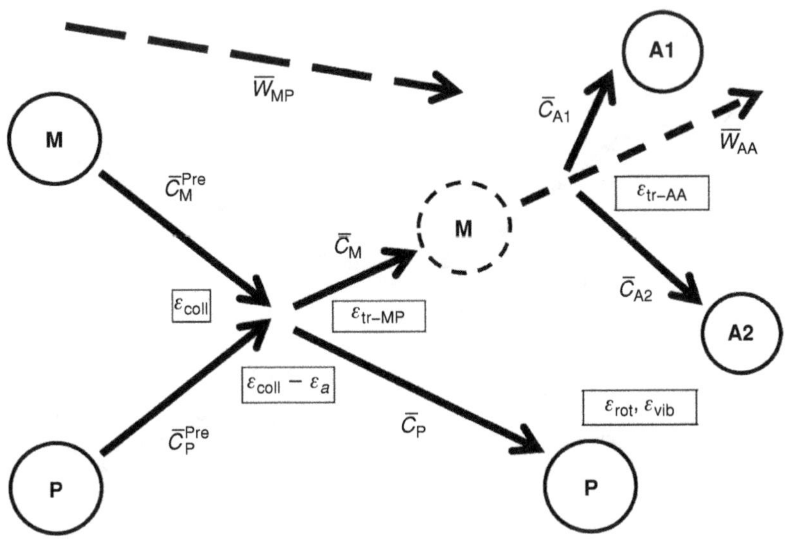

Figure C.1 Schematic showing the procedure to determine post-collision velocity vectors following a dissociation reaction.

dissociation reaction occurs, the relevant bond energy (ϵ_a) is removed from the collision energy. The remaining collision energy is then redistributed to the product species. To begin, we group the products as two particles; M is the molecule that has dissociated (composed of two atoms), and P is the other particle involved in the collision. As described in more detail in Section 7.4.2, if P is a molecule, then its internal energies (ϵ_{rot}, ϵ_{vib}) are first assigned. From the remaining collision energy, the portion assigned to the relative translational energy of M and P is referred to as ϵ_{tr-MP}, and the other portion, referred to as ϵ_{tr-AA}, is the relative translational energy of the two atoms comprising M.

In the first step, M is treated as a single particle, and the relative energy ϵ_{tr-MP} is used to determine post-collision velocities for particles P and M. The above procedures for hard-sphere or soft-sphere scattering are performed where P and M correspond to particles 1 and 2. Therefore, the expressions in Section C.1 determine velocity vectors, $C_{i,\mathrm{P}}$ and $C_{i,\mathrm{M}}$. The properties of particle P are now finalized, and it remains to properly dissociate particle M.

Particle M is properly dissociated using the remaining collision energy, ϵ_{tr-AA}, to determine the velocity vectors of the two atoms comprising M. In this case, as depicted in Fig. C.1, the center-of-mass velocity of the two atoms is set as $W_{i,\mathrm{AA}} = C_{i,\mathrm{M}}$, determined in the previous step. The above procedures for hard-sphere or soft-sphere scattering are performed, where the two atoms (A1 and A2) now correspond to particles 1 and 2. Therefore, the expressions in Section C.1 determine velocity vectors, $C_{i,\mathrm{A1}}$ and $C_{i,\mathrm{A2}}$. The properties of all three product particles, P, A1, and A2, are now finalized, and the dissociation collision is complete.

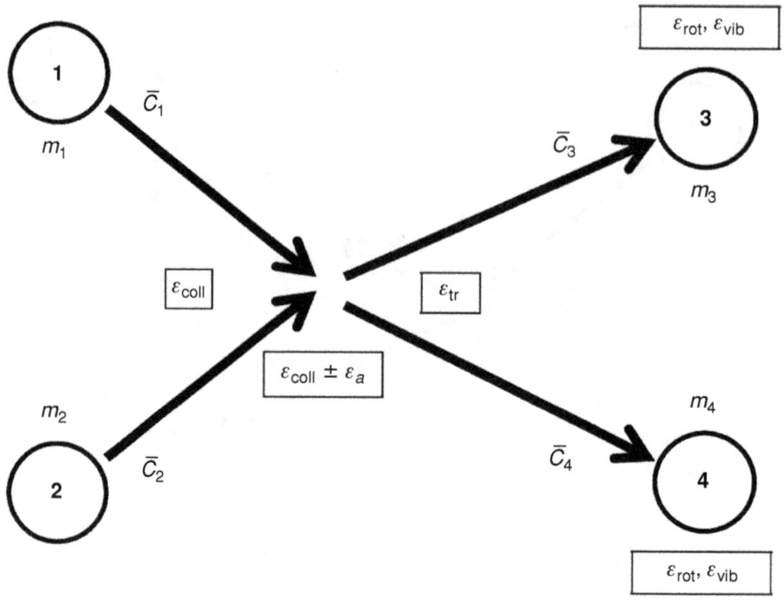

Figure C.2 Schematic showing the procedure to determine post-collision velocity vectors following an exchange reaction.

In principle, either hard-sphere or soft-sphere scattering can be used to determine velocities for M and P and also for A1 and A2. However, as discussed in Section C.1.2 for soft-sphere scattering, the pre-collision relative velocity vector can not be preserved for a reactive collision, and therefore the soft-sphere expressions specific to an inelastic collision should be used (i.e., Eq. C.11 replaces Eq. C.9). However, since soft-sphere collision models are typically parameterized based on viscosity and diffusivity data, or nonreactive cross section data, it is not clear that these models accurately describe the post-collision deflection angles of reaction products. If the accuracy of these soft-sphere models for reactive collisions is unknown, it may be prudent to simply use hard-sphere scattering laws for reactive collisions.

C.3 Exchange Reaction Collisions

A collision between two simulation particles that results in an exchange reaction involves only two product particles. However, special consideration is again required to properly assign the velocity vectors of these product particles. As discussed in Section 7.5.4 and depicted schematically in Fig. C.2, the total collision energy (ϵ_{coll}) is calculated based on the relative translational energy and internal energies of the two particles entering the collision. Depending on the direction of the reaction, it may be an exothermic or endothermic reaction, and therefore the relevant bond energy (ϵ_a) may be added to, or subtracted from, the total collision energy. The remaining collision energy is then redistributed to the product species. To begin, internal energy (ϵ_{rot}, ϵ_{vib}) is redistributed among the two product particles as outlined in Section 7.5.4 (and also in Section 7.4.2). The remaining collision energy, ϵ_{tr}, is then used to determine post-collision velocity vectors for the two product particles.

The procedure to determine post-collision velocity vectors is similar to the procedures described in Section C.1, however, the center-of-mass velocity and reduced mass require further consideration, since the masses of the product molecules may be different than the masses of the pre-collision particles. It is most convenient to label four separate particles with specific masses (m_1, m_2, m_3, and m_4) as depicted in Fig. C.2.

Similar to the procedures for inelastic nonreactive collisions (Section C.1) and dissociation collisions (Section C.2), the required information consists of the remaining translational energy, ϵ_{tr}, and a center-of-mass velocity for the collision, such that momentum is conserved. In the case of an exchange reaction, the center-of-mass velocity (W_i) is calculated using Eq. C.2 based on the reactant particle velocities ($C_{i,1}$ and $C_{i,2}$ in Fig. C.2). However, in addition, the following ratio between post-collision and pre-collision center-of-mass velocities is required,

$$W_{\text{ratio}} = \frac{m_1 + m_2}{m_3 + m_4} \tag{C.12}$$

and will be used in a later step.

The relative velocity for the product particles (g) is calculated from ϵ_{tr} using Eq. C.1, however, the reduced mass of the product particles should be used,

$$m_r = \frac{m_3 m_4}{m_3 + m_4} \tag{C.13}$$

since it is ϵ_{tr} that is conserved in the collision.

Given this value for g, the post-collision relative velocity components (g'_x, g'_y, and g'_z) are calculated using either hard-sphere or soft-sphere (inelastic) scattering expressions presented in Section C.1. The final post-collision velocity vectors of the two product particles are then calculated as

$$\begin{aligned} C_{i,3} &= W_i W_{\text{ratio}} + \frac{m_4}{m_3 + m_4} g'_i \\ C_{i,4} &= W_i W_{\text{ratio}} - \frac{m_3}{m_3 + m_4} g'_i \end{aligned} \tag{C.14}$$

The properties of both product particles, 3 and 4, are now finalized, and the exchange reaction collision is complete.

C.4 Recombination Reaction Collisions

Recombination collisions involve two primary particles colliding in the presence of a third-body particle. As shown in Fig. C.3, we consider the two

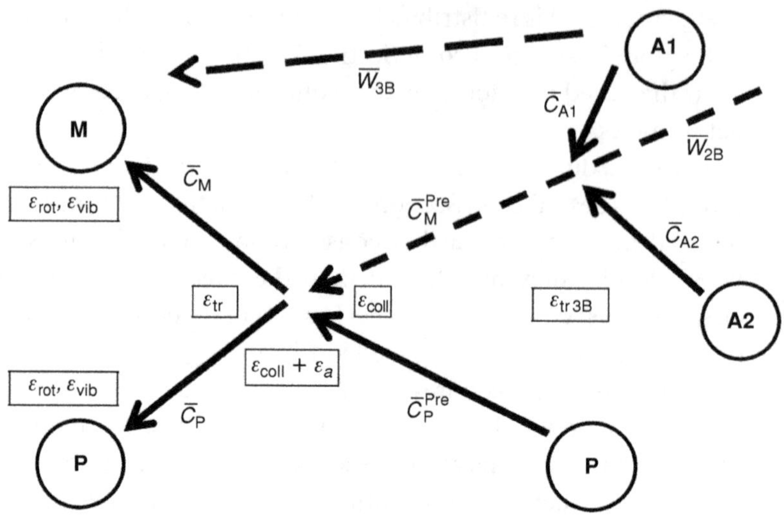

Figure C.3 Schematic showing the procedure to determine post-collision velocity vectors following a recombination reaction.

primary particles to be atoms (labeled as A1 and A2), which collide with a third-body particle (either an atom or a molecule, labeled as P). Note that the schematic for recombination (Fig. C.3) is essentially the reverse of the schematic for dissociation (Fig. C.1). Ultimately, particles A1 and A2 recombine into a new molecular species (labeled M) where energy exchange may occur with particle P. The velocity vectors of the two product particles, M and P, are to be determined.

First, as discussed in Section 7.5.4 (step 4a), calculation of the total collision energy (ϵ_{coll}) involves calculating the relative translational energy for a three-particle system (ϵ_{tr3B}). To calculate this energy, the center-of-mass velocity of the two primary particles, A1 and A2, is calculated as

$$W_{i,2\text{B}} = \frac{m_{\text{A1}}}{m_{\text{A1}} + m_{\text{A2}}} C_{i,\text{A1}} + \frac{m_{\text{A2}}}{m_{\text{A1}} + m_{\text{A2}}} C_{i,\text{A2}} \tag{C.15}$$

Also, the relative velocity squared for particles A1 and A2 is calculated as

$$g_{2\text{B}}^2 = (C_{x,\text{A1}} - C_{x,\text{A2}})^2 + (C_{y,\text{A1}} - C_{y,\text{A2}})^2 + (C_{z,\text{A1}} - C_{z,\text{A2}})^2 \tag{C.16}$$

Next, the relative velocity vector between the primary particle pair and the third body (P) is calculated as

$$g_{i,3\text{B}} = W_{i,2\text{B}} - C_{i,\text{P}} \tag{C.17}$$

and therefore we can calculate

$$g_{3\text{B}}^2 = g_{x,3\text{B}}^2 + g_{y,3\text{B}}^2 + g_{z,3\text{B}}^2 \tag{C.18}$$

Furthermore, the reduced mass corresponding to the primary particle pair is

$$m_{r2\text{B}} = \frac{m_{\text{A1}} m_{\text{A2}}}{m_{\text{A1}} + m_{\text{A2}}} \tag{C.19}$$

and the reduced mass relating to the third body is defined as

$$m_{r3B} = \frac{m_{r2B} m_P}{m_{r2B} + m_P} \tag{C.20}$$

Finally, the relative translational energy of the three-particle system is

$$\epsilon_{tr3B} = \frac{1}{2}\left(m_{r2B} g_{2B}^2 + m_{r3B} g_{3B}^2\right) \tag{C.21}$$

This translation energy (ϵ_{tr3B}) is then used to compute the total collision energy (ϵ_{coll}). This total collision energy is then redistributed among the energy modes of the recombined products as described next.

Since recombination reactions are endothermic, the relevant bond energy (ϵ_a) is added to the total collision energy. As outlined in Section 7.5.4 (and also in Section 7.4.2), the collision energy is first redistributed to the internal energy modes (ϵ_{rot}, ϵ_{vib}) of the two product particles, M and P (M is certainly a molecule and P may be an atom or a molecule). The remaining collision energy, ϵ_{tr}, is used to determine the final velocity vectors of both M and P.

In this case, the relative velocity g is determined from ϵ_{tr} using Eq. C.1, where the reduced mass m_r is that corresponding to the product particles, M and P. Recall, this is the correct choice, since it is collision energy that is conserved and not relative velocity. Next, referring to Fig. C.3, the center-of-mass velocity is calculated as

$$W_i = \frac{m_M}{m_M + m_P} C_{i,M}^{pre} + \frac{m_P}{m_M + m_P} C_{i,P}^{pre} \tag{C.22}$$

where $C_{i,M}^{pre} = W_{i,2B}$ from Eq. C.15 (i.e., the pre-collision center-of-mass velocity of the two primary atoms entering the recombination reaction).

Using these values for g and W_i, the post-collision relative velocity components (g'_x, g'_y, and g'_z) are calculated using either hard-sphere or soft-sphere (inelastic) scattering expressions presented in Section C.1. The final post-collision velocity vectors of the two product particles are then calculated using Eq. C.6. The properties of both product particles, M and P, are now finalized, and the recombination reaction collision is complete.

Appendix D Macroscopic Properties

In this appendix, we summarize some of the most common macroscopic quantities of interest and present the equations to compute them from the properties of DSMC particles. These quantities include average velocities, densities, temperatures, pressures, mean free path, and mean collision time, for individual species and also for gas mixtures. Surface properties such as surface pressure, heat flux, and shear stress, are also discussed.

D.1 Gas Properties

To calculate macroscopic quantities, averages must be taken over a sufficient number of samples (i.e., particles, and their molecular properties). For steady-state simulations, samples are accumulated within each cell over many timesteps during steady state. For unsteady simulations, samples may be accumulated only over a single timestep, or perhaps over a small number of timesteps, before the sampled information is discarded and replaced with new sampled information. In either case, samples are accumulated within a flow volume of interest over some number of timesteps. Averages are then taken over these samples to obtain macroscopic quantities. The flow volume of interest is typically the collision cell, which is approximately 0.5–1 cubic mean free paths (λ); however, sampling volumes can be larger if desired.

In each sampling volume, a number of quantities should be stored. These quantities were discussed earlier in Section 6.2.4 of Chapter 6, where it was recommended that memory be allocated for them in each collision cell. Here we focus on nine variables, stored for each species (s). These variables represent cumulative sums of a specific particle property, summed for all particles within the sampling volume (within the sampling cell), and summed over a number of sampling timesteps. The nine variables are denoted by square brackets with a subscript (s) referring to the particular species, and they include:

$$\left[\sum N_p\right]_s$$
$$\left[\sum C_x\right]_s, \left[\sum C_y\right]_s, \left[\sum C_z\right]_s$$
$$\left[\sum C_x^2\right]_s, \left[\sum C_y^2\right]_s, \left[\sum C_z^2\right]_s$$
$$\left[\sum \epsilon_{\text{rot}}\right]_s, \left[\sum \epsilon_{\text{vib}}\right]_s$$

All of the standard macroscopic gas properties of interest can be calculated from these nine variables (for each species).

D.1.1 Single-Species Quantities

In general, the gas properties of each species in the mixture are computed first, and these properties are then averaged to obtain properties of the gas mixture.

The number density of each species is determined from

$$n_s = \frac{\left[\sum N_p\right]_s W_p}{N_{\Delta t-\text{samp}} V} \tag{D.1}$$

where $\left[\sum N_p\right]_s$ is the cumulative sum of the number of particles (N_p) within the sampling cell, $N_{\Delta t-\text{samp}}$ is the number sampling timesteps over which the sums were accumulated, V is the sampling cell volume, and W_p is the particle weight. The expressions presented in this appendix assume that all particles within the same cell have the same particle weight. If this is not the case, for example if different species have different weights, then the correct particle weight should be used and the expressions modified accordingly.

The mass density is then

$$\rho_s = n_s m_s \tag{D.2}$$

where m_s is the mass of one particle of species s.

The bulk velocity components for each species are calculated from

$$\begin{aligned} < C_x >_s &= \left[\sum C_x\right]_s / \left[\sum N_p\right]_s \\ < C_y >_s &= \left[\sum C_y\right]_s / \left[\sum N_p\right]_s \\ < C_z >_s &= \left[\sum C_z\right]_s / \left[\sum N_p\right]_s \end{aligned} \tag{D.3}$$

where $[\sum C_i]_s$ is the cumulative sum of the particle velocity components within the sampling cell.

A translational temperature can be defined for each coordinate direction as

$$
\begin{aligned}
T_{x,s} &= \frac{m_s}{k}\left\{\left[\sum C_x^2\right]_s / \left[\sum N_p\right]_s - < C_x >_s < C_x >_s\right\} \\
T_{y,s} &= \frac{m_s}{k}\left\{\left[\sum C_y^2\right]_s / \left[\sum N_p\right]_s - < C_y >_s < C_y >_s\right\} \\
T_{z,s} &= \frac{m_s}{k}\left\{\left[\sum C_z^2\right]_s / \left[\sum N_p\right]_s - < C_z >_s < C_z >_s\right\}
\end{aligned}
\tag{D.4}
$$

where $\left[\sum C_i^2\right]_s$ is the cumulative sum of the square of the particle velocity components within the sampling cell. These temperatures can be averaged to obtain the overall translational temperature for each species in the gas,

$$T_{\text{tr},s} = (T_{x,s} + T_{y,s} + T_{z,s})/3 \tag{D.5}$$

The partial pressure of each species is then

$$p_s = n_s k T_{\text{tr},s} \tag{D.6}$$

A rotational temperature for each species is calculated from

$$T_{\text{rot},s} = \frac{2}{\zeta_{\text{rot},s} k}\left\{\left[\sum \epsilon_{\text{rot}}\right]_s / \left[\sum N_p\right]_s\right\} \tag{D.7}$$

where $[\sum \epsilon_{\text{rot}}]_s$ is the cumulative sum of the rotational energy of particles within the sampling cell, and $\zeta_{\text{rot},s}$ is the rotational degrees of freedom corresponding to species (s).

A vibrational temperature for each species can be calculated using a similar expression,

$$T_{\text{vib},s} = \frac{2}{\zeta_{\text{vib},s} k}\left\{\left[\sum \epsilon_{\text{vib}}\right]_s / \left[\sum N_p\right]_s\right\} \tag{D.8}$$

where $[\sum \epsilon_{\text{vib}}]_s$ is the cumulative sum of the vibrational energy of particles within the sampling cell, and $\zeta_{\text{vib},s}$ is the vibrational degrees of freedom corresponding to species (s). However, it is important to be consistent with the vibrational energy model used in the DSMC collision algorithms and boundary condition algorithms. As an example for the simple harmonic oscillator (SHO) model, used frequently throughout Chapters 6 and 7, the vibrational temperature is determined by

$$T_{\text{vib},s} = \theta_{\text{vib},s} / \ln\left\{1 + \frac{k\theta_{\text{vib},s}}{[\sum \epsilon_{\text{vib}}]_s / [\sum N_p]_s}\right\} \tag{D.9}$$

where θ_{vib} is the characteristic temperature of vibration, a required input parameter for the SHO model.

An average temperature corresponding to all energy modes can be calculated for each species as

$$T_{\text{avg},s} = \frac{\zeta_{\text{tr},s} T_{\text{tr},s} + \zeta_{\text{rot},s} T_{\text{rot},s} + \zeta_{\text{vib},s} T_{\text{vib},s}}{\zeta_{\text{tr},s} + \zeta_{\text{rot},s} + \zeta_{\text{vib},s}} \tag{D.10}$$

Here, $\zeta_{\mathrm{tr},s} = 3$ for all species and $\zeta_{\mathrm{rot},s} = \zeta_{\mathrm{vib},s} = 0$ for atomic species. For diatomic species, $\zeta_{\mathrm{rot},s} = 2$ and the effective vibrational degrees of freedom ($\zeta_{\mathrm{vib},s}$) should be calculated consistent with the vibrational energy model used. For example, for the SHO model

$$\zeta_{\mathrm{vib},s} = \frac{2\theta_{\mathrm{vib},s}/T_{\mathrm{vib}}}{e^{\theta_{\mathrm{vib},s}/T_{\mathrm{vib}}} - 1} \tag{D.11}$$

As seen in Eq. D.11, the available vibrational degrees of freedom are best calculated using the vibrational temperature of the gas mixture (T_{vib}). This quantity is defined in Eq. D.19, which therefore may need evaluation before the average species temperatures (Eq. D.10) are calculated. For flows with little vibrational energy content, the calculation of T_{vib} may be problematic. Specifically, it may be determined that $T_{\mathrm{vib}} = 0$ K due to the lack of any particles with energies above the ground vibrational energy level. In this case, one could use the translational temperature of the mixture, T_{tr} (defined in Eq. D.15) in Eq. D.11 for the effective vibrational degrees of freedom. For polyatomic species, the degrees of freedom in these equations require appropriate modification.

D.1.2 Gas Mixture Quantities

Now that the macroscopic gas properties have been calculated for each species (s), the overall properties of the gas mixture can be evaluated by averaging over the total number of species (N_s).

The mixture number density and mixture mass density are given by

$$n = \sum_{s=1}^{N_s} n_s \tag{D.12}$$

and

$$\rho = \sum_{s=1}^{N_s} \rho_s \tag{D.13}$$

The mixture mass velocity (weighted by mass fraction as defined in Eq. 5.72 of Chapter 5) is evaluated as

$$\begin{aligned} C_{0_x} &= \left\{ \sum_{s=1}^{N_s} (\rho_s < C_x >_s) \right\} / \rho \\ C_{0_y} &= \left\{ \sum_{s=1}^{N_s} (\rho_s < C_y >_s) \right\} / \rho \\ C_{0_z} &= \left\{ \sum_{s=1}^{N_s} (\rho_s < C_z >_s) \right\} / \rho \end{aligned} \tag{D.14}$$

Mixture translational temperatures in each coordinate direction, in addition to the overall translational temperature of the mixture, are also weighted by mass fraction (refer to Eq. 5.80 in Chapter 5),

$$
\begin{aligned}
T_x &= \left\{ \sum_{s=1}^{N_s} (\rho_s T_{x,s}) \right\} / \rho \\
T_y &= \left\{ \sum_{s=1}^{N_s} (\rho_s T_{y,s}) \right\} / \rho \\
T_z &= \left\{ \sum_{s=1}^{N_s} (\rho_s T_{z,s}) \right\} / \rho \\
T_{\text{tr}} &= \left\{ \sum_{s=1}^{N_s} (\rho_s T_{\text{tr},s}) \right\} / \rho
\end{aligned}
\tag{D.15}
$$

The pressure of the mixture (refer to Eq. 5.81 in Chapter 5) is

$$
p = \sum_{s=1}^{N_s} p_s = nkT_{\text{tr}} \tag{D.16}
$$

Rotational and vibrational temperatures of the mixture are simply averages of the temperatures determined for each individual species. Specifically, the rotational temperature of the mixture is given by

$$
T_{\text{rot}} = \left\{ \sum_{s=1}^{N_s} (\rho_s T_{\text{rot},s}) \right\} / \rho_{\text{polyatomic}} \tag{D.17}
$$

where the mass-fraction weighting only includes polyatomic species (i.e., that contain internal energy),

$$
\rho_{\text{polyatomic}} = \sum_{s \neq s_{\text{atomic}}}^{N_s} \rho_s \tag{D.18}
$$

In the same manner, the vibrational temperature of the mixture is calculated by

$$
T_{\text{vib}} = \left\{ \sum_{s=1}^{N_s} (\rho_s T_{\text{vib},s}) \right\} / \rho_{\text{polyatomic}} \tag{D.19}
$$

Finally, the overall average temperature of the mixture can be evaluated by

$$
T_{\text{avg}} = \left\{ \sum_{s=1}^{N_s} (\rho_s T_{\text{avg},s}) \right\} / \rho \tag{D.20}
$$

It is often useful to calculate the local mean free path (λ) and mean collision time (τ_{coll}) at all points in the flow. These values may be used to assess whether the DSMC simulation is adequately resolved. That is, these values can be used to determine if the local cells size is smaller than λ and if the simulation timestep is smaller than τ_{coll}. Also, the local value of λ may be required to perform adaptive mesh refinement (AMR) and the local value of τ_{coll} may be used to set the local timestep size, if a variable time-stepping algorithm is used.

For a VHS-type gas, the mean free path for particles of species (s) before colliding with any other particle in the mixture is

$$\frac{1}{\lambda_s} = \sum_{q=1}^{N_s} \left\{ \left(\frac{T_{\text{ref},s-q}}{T_{\text{tr}}} \right)^{\nu_{s-q}} n_q \pi d_{\text{ref},s-q}^2 \sqrt{1 + \frac{m_s}{m_q}} \right\} \tag{D.21}$$

and the mean collision time for particles of species (s) before colliding with any other particle in the mixture is

$$\frac{1}{\tau_{\text{coll},s}} = \sum_{q=1}^{N_s} \left\{ \left(\frac{T_{\text{tr}}}{T_{\text{ref},s-q}} \right)^{1/2-\nu_{s-q}} 2 n_q d_{\text{ref},s-q}^2 \sqrt{\frac{2\pi k T_{\text{ref},s-q}}{m_{r,s-q}}} \right\} \tag{D.22}$$

Here, $T_{\text{ref},s-q}$, $d_{\text{ref},s-q}$, and ν_{s-q}, are the VHS model parameters specific to species pair $s - q$.

Finally, the average mean free path for all particles in the mixture is evaluated as

$$\lambda = \sum_{s=1}^{N_s} \left(\lambda_s \frac{n_s}{n} \right) \tag{D.23}$$

and the average mean collision time as

$$\tau_{\text{coll}} = \sum_{s=1}^{N_s} \left(\tau_{\text{coll},s} \frac{n_s}{n} \right) \tag{D.24}$$

The collision rate can also be evaluated directly from a DSMC simulation simply by counting the number of collisions performed in a given cell (refer to Example 6.1 in Chapter 6). Using the mean thermal speed calculated within the cell, this collision rate can be used to calculate a value for the mean collision time and mean free path for the gas.

D.2 Surface Properties

In contrast to continuum simulations, surface properties in DSMC are not calculated from gradients of macroscopic properties, rather they are calculated directly from the momentum and energy transferred to/from the surface during each particle-surface collision. Specifically, the following sampled quantities should also be stored for each surface element,

$$\left[\sum W_p N_{ps}\right]$$

$$\left[\sum mW_p \left(C_x^{\text{post}} - C_x^{\text{pre}}\right)\right]$$

$$\left[\sum mW_p \left(C_y^{\text{post}} - C_y^{\text{pre}}\right)\right]$$

$$\left[\sum mW_p \left(C_z^{\text{post}} - C_z^{\text{pre}}\right)\right]$$

$$\left[\sum \frac{mW_p}{2} \left(C_x^{2\text{ post}} + C_y^{2\text{ post}} + C_z^{2\text{ post}} - C_x^{2\text{ pre}} + C_y^{2\text{ pre}} + C_z^{2\text{ pre}}\right)\right]$$

$$\left[\sum W_p \left(\epsilon_{\text{rot}}^{\text{post}} - \epsilon_{\text{rot}}^{\text{pre}}\right)\right]$$

$$\left[\sum W_p \left(\epsilon_{\text{vib}}^{\text{post}} - \epsilon_{\text{vib}}^{\text{pre}}\right)\right]$$

These quantities could be stored for each species if desired. However, often one is only interested in the overall heat flux and drag due to the entire gas mixture. As listed previously, each term in square brackets represents the cumulative sum taken over all particles, of all species, that have collided with a particular surface element over a number of sampling timesteps. Specifically, the *difference* in particle properties (momentum and translational energy, as well as internal energy) before and after each surface collision is recorded and summed for all collisions. These seven cumulative sum variables, stored for each surface element, can then be used to calculate the local macroscopic surface properties for each element.

The number flux, defined as the number of particles impacting the surface per unit time, per unit area, is calculated as

$$n_f = \frac{\left[\sum W_p N_{ps}\right]}{N_{\Delta t-\text{samp}} \Delta t A} \tag{D.25}$$

where $N_{\Delta t-\text{samp}}$ is the number sampling timesteps over which the sums were accumulated, Δt is the simulation timestep, and A is the area of the surface element.

The net flux of momentum, in the x, y, and z coordinate directions, per unit area, to the surface element is calculated as

$$\begin{aligned}
F_{M_x} &= \frac{-\left[\sum mW_p \left(C_x^{\text{post}} - C_x^{\text{pre}}\right)\right]}{N_{\Delta t-\text{samp}} \Delta t A} \\
F_{M_y} &= \frac{-\left[\sum mW_p \left(C_y^{\text{post}} - C_y^{\text{pre}}\right)\right]}{N_{\Delta t-\text{samp}} \Delta t A} \\
F_{M_z} &= \frac{-\left[\sum mW_p \left(C_z^{\text{post}} - C_z^{\text{pre}}\right)\right]}{N_{\Delta t-\text{samp}} \Delta t A}
\end{aligned} \tag{D.26}$$

These expressions can be summed over all surface elements (N_e) to readily determine the net force, in each coordinate direction, acting on the surface as a whole,

$$F_x^{\text{total}} = \sum_{e=1}^{N_e} (F_{M_x} A)_e$$
$$F_y^{\text{total}} = \sum_{e=1}^{N_e} (F_{M_y} A)_e \quad \text{(D.27)}$$
$$F_z^{\text{total}} = \sum_{e=1}^{N_e} (F_{M_z} A)_e$$

These net force expressions can be transformed into surface normal coordinates (as shown in Fig. A.1) to obtain expressions for the surface pressure and shear-stress tensor. For example, the surface pressure is calculated as

$$p = -\left(F_{M_x} n_x + F_{M_y} n_y + F_{M_z} n_z\right) \quad \text{(D.28)}$$

where $\hat{n} = \left(n_x \bar{x}, n_y \bar{y}, n_z \bar{z}\right)$ is the surface unit normal vector in Cartesian coordinates. The transformations to obtain the shear stress tensor are left to the reader.

Finally, the net flux of energy per unit area (the surface heat flux) is given by

$$q = \frac{E_{\text{net}}}{N_{\Delta t-\text{samp}} \Delta t A} \quad \text{(D.29)}$$

where

$$E_{\text{net}} = \left[\sum W_p \left(\epsilon_{\text{rot}}^{\text{post}} - \epsilon_{\text{rot}}^{\text{pre}}\right)\right] + \left[\sum W_p \left(\epsilon_{\text{vib}}^{\text{post}} - \epsilon_{\text{vib}}^{\text{pre}}\right)\right] + \left[\sum \frac{m W_p}{2} \left(C_x^{2\ \text{post}} + C_y^{2\ \text{post}} + C_z^{2\ \text{post}} - C_x^{2\ \text{pre}} + C_y^{2\ \text{pre}} + C_z^{2\ \text{pre}}\right)\right] \quad \text{(D.30)}$$

The total heat flux to the surface as a whole is then

$$q^{\text{total}} = \sum_{e=1}^{N_e} (qA)_e \quad \text{(D.31)}$$

These expressions determine the most common surface properties of interest, however, the molecular nature of DSMC enables the study of many interesting aspects of the gas–surface interaction. For example, if surface chemistry is simulated, where species are formed or destroyed, then the preceding expressions can be readily modified to include such effects.

Appendix E

Common Integrals

E.1 Standard Integrals

A definite integral that appears frequently in kinetic theory and statistical mechanics is the following:

$$I_n(a) \equiv \int_0^\infty x^n e^{-ax^2}\, dx \tag{E.1}$$

where $a > 0$ and n is a nonnegative integer. Values of this integral for specific values of n are

$$\begin{aligned} I_0(a) &= \frac{1}{2}\left(\frac{\pi}{a}\right)^{1/2} \\ I_1(a) &= \frac{1}{2a} \\ I_2(a) &= \frac{1}{4}\left(\frac{\pi}{a^3}\right)^{1/2} \\ I_3(a) &= \frac{1}{2a^2} \\ I_4(a) &= \frac{3}{8}\left(\frac{\pi}{a^5}\right)^{1/2} \\ I_5(a) &= \frac{1}{a^3} \end{aligned} \tag{E.2}$$

Starting with I_0 and I_1, which are established in Eq. E.2, each integral can be found from a lower integral in the series by means of the relation

$$I_{n+2} = -\frac{dI_n}{da} \tag{E.3}$$

which follows obviously from the defining formula (Eq. E.1). The integral from $-\infty$ to $+\infty$ is twice the value given above if n is even, and the integral from $-\infty$ to $+\infty$ is zero if n is odd.

E.2 The Error Function

Another definite integral that appears frequently in kinetic theory and statistical mechanics is the error function,

$$\mathrm{erf}(a) \equiv \frac{2}{\sqrt{\pi}} \int_0^a e^{-x^2} dx \tag{E.4}$$

and the complimentary error function is

$$\mathrm{erfc}(a) = 1 - \mathrm{erf}(a) \tag{E.5}$$

Note that

$$\mathrm{erf}(-a) = -\mathrm{erf}(a) \tag{E.6}$$

$$\mathrm{erf}(0) = 0 \tag{E.7}$$

and

$$\mathrm{erf}(\infty) = 1 \tag{E.8}$$

For general values of the argument a, various series expansions, tabulations, and curve fits are available in the mathematical literature.

E.3 The Gamma Function

Another definite integral that appears frequently in kinetic theory and statistical mechanics is the gamma function,

$$\Gamma(t) \equiv \int_0^\infty x^{t-1} e^{-x} dx \tag{E.9}$$

The following reduction formula holds for the gamma function

$$\Gamma(t+1) = t\,\Gamma(t) \tag{E.10}$$

and for t equal to zero or a positive integer,

$$\Gamma(t+1) = t! \tag{E.11}$$

The incomplete gamma function also appears frequently,

$$\Gamma_{\mathrm{I}}(t,\alpha) \equiv \int_\alpha^\infty x^{t-1} e^{-x} dx \tag{E.12}$$

and the reduction formula is

$$\Gamma_{\mathrm{I}}(t,\alpha) = (t-1)\Gamma_{\mathrm{I}}(t-1,\alpha) + \alpha^{t-1} e^{-\alpha} \tag{E.13}$$

Finally, note that

$$\Gamma_{\mathrm{I}}(1/2, \alpha) = \sqrt{\pi}\, \mathrm{erfc}(\sqrt{\alpha}) \tag{E.14}$$

where erfc() is the complimentary error function given in Eq. E.5.

References

Abe, T. (1994). "Direct simulation Monte Carlo method for internal-translational energy exchange in nonequilibrium flow." *Rarefied Gas Dynamics: Theory and Simulations*. In *Proceedings of the 18th International Symposium on Rarified Gas Dynamics*, University of British Columbia, Vancouver, BC, Canada, 1994, pp. 103–113.

Alsmeyer, H. (1976). "Density profiles in argon and nitrogen shock waves measured by the absorption of an electron beam." *Journal of Fluid Mechanics* **74**(3), 497–513.

Annis, B., and Malinauskas, A. (1971). "Temperature dependence of rotational collision numbers from thermal transpiration." *The Journal of Chemical Physics* **54**(11), 4763–4768.

Appleton, J., Steinberg, M., and Liquornik, D. (1968). "Shock-tube study of nitrogen dissociation using vacuum-ultraviolet light absorption." *The Journal of Chemical Physics* **48**(2), 599–608.

Baker, L. L., and Hadjiconstantinou, N. G. (2005). "Variance reduction for Monte Carlo solutions of the Boltzmann equation." *Physics of Fluids* **17**, 051703.

Bender, J. D., Valentini, P., Nompelis, I., Paukku, Y., Varga, Z., Truhlar, D. G., Schwartzentruber, T., and Candler, G. V. (2015). "An improved potential energy surface and multi-temperature quasiclassical trajectory calculations of $N_2 + N_2$ dissociation reactions." *The Journal of Chemical Physics* **143**(5), 054304.

Bergemann, F., and Boyd, I. D. (1994). "New discrete vibrational energy model for the direct simulation Monte Carlo Method." *Progress in Astronautics and Aeronautics* **158**, 174–183.

Bertin, J. (1994). *Hypersonic Aerothermodynamics*. AIAA, Washington.

Bhathnagor, P., Gross, E. G., and Krook, M. (1954). "A model for collision processes in gases." I. Small amplitude processes in charged and neutral one-component systems. *Physical Review* **94**(3), 511–525.

Billing, G. D. and Wang, L. (1992). "Semiclassical calculations of transport coefficients and rotational relaxation of nitrogen at high temperatures." *The Journal of Physical Chemistry* **96**(6), 2572–2575.

Bird, G. (1963). "Approach to translational equilibrium in a rigid sphere gas." *Physics of Fluids (1958–1988)* **6**(10), 1518–1519.

Bird, G. (1994). *Molecular Gas Dynamics and the Direct Simulation of Gas Flows.* Oxford University Press, New York.

Bird, G. (2013). *The DSMC method.* CreateSpace Independent Publishing Platform.

Borgnakke, C., and Larsen, P. S. (1975). "Statistical collision model for monte carlo simulation of polyatomic gas mixture." *Journal of Computational Physics* **18**(4), 405–420.

Bose, D., and Candler, G. V. (1996). "Thermal rate constants of the $N_2 + O \rightarrow NO + N$ reaction using ab initio 3A and 3A potential energy surfaces." *The Journal of Chemical Physics* **104**(8), 2825–2833.

Bourgat, J.-F., Desvillettes, L., Le Tallec, P., and Perthame, B. (1994). "Microreversible collisions for polyatomic gases and Boltzmann's theorem." *European Journal of Mechanics B: Fluids* **13**(2), 237–254.

Boyd, I. D. (1990*a*). "Analysis of rotational nonequilibrium in standing shock waves of nitrogen." *AIAA Journal* **28**(11), 1997–1999.

Boyd, I. D. (1990*b*). "Rotational–translational energy transfer in rarefied nonequilibrium flows." *Physics of Fluids A: Fluid Dynamics (1989–1993)* **2**(3), 447–452.

Boyd, I. D. (2007). "Modeling backward chemical rate processes in the direct simulation Monte Carlo Method." *Physics of Fluids (1994–present)* **19**(12), 126103.

Boyd, I. D., Chen, G., and Candler, G. V. (1995). "Predicting failure of the continuum fluid equations in transitional hypersonic flows." *Physics of Fluids (1994–present)* **7**(1), 210–219.

Boyd, I. D., and Gokcen, T. (1994). "Computation of axisymmetric and ionized hypersonic flows using particle and continuum methods." *AIAA Journal* **32**(9), 1828–1835.

Burt, J. M., and Josyula, E. (2014). "Efficient direct simulation Monte Carlo modeling of very low Knudsen number gas flows." In *52nd Aerospace Sciences Meeting*.

Burt, J. M., Josyula, E., and Boyd, I. D. (2011). "Techniques for reducing collision separation in direct simulation Monte Carlo calculations." In *42nd AOAA Thermophysics Conference*, Honolulu.

Burt, J. M., Josyula, E., and Boyd, I. D. (2012). "Novel Cartesian implementation of the direct simulation Monte Carlo method." *Journal of Thermophysics and Heat Transfer* **262**, 258–270.

Carnevale, E., Carey, C., and Larson, G. (1967). "Ultrasonic determination of rotational collision numbers and vibrational relaxation times of polyatomic gases at high temperatures." *The Journal of Chemical Physics* **47**(8), 2829–2835.

Chapman, S., and Cowling, T. G. (1952). *The Mathematical Theory of Non-uniform Gases: An Account of the Kinetic Theory of Viscosity, Thermal Conduction and Diffusion of Gases.* Cambridge University Press, Cambridge.

Choquet, I. (1994). "Thermal nonequilibrium modeling using the direct simulation monte carlo method: Application to rotational energy." *Physics of Fluids (1994–present)* **6**(12), 4042–4053.

Deschenes, T. R., and Boyd, I. D. (2011). "Extension of a modular particle-continuum method to vibrationally excited, hypersonic flows." *AIAA Journal* **49**(9), 1951–1959.

Dietrich, S., and Boyd, I. D. (1996). "Scalar and parallel optimized implementation of the direct simulation Monte Carlo Method." *Journal of Computational Physics* **126**(2), 328–342.

Donev, A., Bell, J. B., Garcia, A. L., and Alder, B. J. (2010). "A hybrid particle-continuum method for hydrodynamics of complex fluids." *Multiscale Modeling and Simulation* **8**(3), 871–911.

Donev, A., Garcia, A. L., and Alder, B. J. (2008). "Stochastic event-driven molecular dynamics." *Journal of Computational Physics* **227**(4), 2644–2665.

Fan, J. (2002). "A generalized soft-sphere model for Monte Carlo Simulation." *Physics of Fluids (1994–present)* **14**(12), 4399–4405.

Farbar, E., and Boyd, I. D. (2014). "Subsonic flow boundary conditions for the direct simulation Monte Carlo method." *Computers and Fluids* **102**, 99–110.

Galitzine, C., and Boyd, I. D. (2015). "An adaptive procedure for the numerical parameters of a particle simulation." *Journal of Computational Physics* **281**, 449–472.

Ganzi, G., and Sandler, S. I. (1971). "Determination of thermal transport properties from thermal transpiration measurements." *The Journal of Chemical Physics* **55**(1), 132–140.

Gao, D., Zhang, C., and Schwartzentruber, T. E. (2011). "Particle simulations of planetary probe flows employing automated mesh refinement." *Journal of Spacecraft and Rockets* **48**(3), 397–405.

Garcia, A. L. (2000). *Numerical Methods for Physics*. Prentice Hall, Englewood Cliffs, NJ.

Garcia, A. (2007). "Estimating hydrodynamic quantities in the presence of microscopic fluctuations." *Communications in Applied Mathematics and Computational Science* **1**(1), 53–78.

Garcia, A. L., and Alder, B. J. (1998). "Generation of the Chapman–Enskog distribution." *Journal of Computational Physics* **140**(1), 66–70.

Gimelshein, N., Gimelshein, S., and Levin, D. (2002). "Vibrational relaxation rates in the direct simulation Monte Carlo method." *Physics of Fluids* **14**, 4452–4455.

Gmurczyk, A. S., Tarczynski, M., and Walenta, Z. A. (1979). "Shock Wave Structure in the Binary Mixtures of Gases with Disparate Molecular Masses." 11th International symposium on rarefied gas dynamics, edited by Campargue, R., Commissariat a l'Energie Atomique, Paris, Vol. 1, pp. 333–341.

Gombosi, T. (1994). *Gaskinetic Theory*. Cambridge University Press, New York.

Grad, H. (1963). "Asymptotic theory of the Boltzmann equation." *Physics of Fluids (1958–1988)* **6**(2), 147–181.

Gross, E. P., and Jackson, E. A. (1959). "Kinetic models and the linearized boltzmann equation." *Physics of Fluids (1958–1988)* **2**(4), 432–441.

Haas, B. L. and Boyd, I. D. (1993). "Models for direct Monte Carlo simulation of coupled vibration–dissociation." *Physics of Fluids A: Fluid Dynamics (1989–1993)* **5**(2), 478–489.

Haas, B. L., Hash, D. B., Bird, G. A., Lumpkin III, F. E., and Hassan, H. (1994). "Rates of thermal relaxation in direct simulation Monte Carlo methods." *Physics of Fluids (1994–present)* **6**(6), 2191–2201.

Hanson, R., and Baganoff, D. (1972). "Shock tube study of nitrogen dissociation rates using pressure measurements.' *AIAA Journal* **10**, 211–215.

Hassan, H., and Hash, D. B. (1993). "A generalized hard-sphere model for monte carlo simulation." *Physics of Fluids A: Fluid Dynamics (1989–1993)* **5**(3), 738–744.

Healy, R., and Storvick, T. (1969). "Rotational collision number and eucken factors from thermal transpiration measurements." *The Journal of Chemical Physics* **50**(3), 1419–1427.

Hirschfelder, J. O., Curtiss, C. F., Bird, R. B., and Mayer, M. G. (1954), *Molecular Theory of Gases and Liquids*. Wiley New York.

Homolle, T. M. M., and Hadjiconstantinou, N. G. (2007). "A low-variance deviational simulation Monte Carlo for the Boltzmann equation." *Journal of Computational Physics* **226**, 2341–2358.

Hornung, H. (1972). "Induction time for nitrogen dissociation', *The Journal of Chemical Physics* **56**(6), 3172–3173.

Kannenberg, K. C., and Boyd, I. D. (2000). "Strategies for efficient particle resolution in the direct simulation Monte Carlo method." *Journal of Computational Physics* **157**(2), 727–745.

Kennard, E. (1938). *Kinetic Theory of Gases*. McGraw-Hill, New York.

Kewley, D. and Hornung, H. (1974). "Free-piston shock-tube study of nitrogen dissociation." *Chemical Physics Letters* **25**(4), 531–536.

Kim, J. G. and Boyd, I. D. (2013). "State-resolved master equation analysis of thermochemical nonequilibrium of nitrogen." *Chemical Physics* **415**, 237–246.

Kistemaker, P., Tom, A. and De Vries, A. (1970). "Rotational relaxation numbers for the isotopic molecules of N_2 and CO." *Physica* **48**(3), 414–424.

Koshi, M., B. S. S. M. and Asaba, T. (1978). "Dissociation of nitric oxide in shock waves." In *Proceedings of the 17th Symposium (International) on Combustion* Leeds, UK, **1**, 553–562.

Koura, K. (1997). "Monte Carlo direct simulation of rotational relaxation of diatomic molecules using classical trajectory calculations: Nitrogen shock wave." *Physics of Fluids (1994–present)* **9**(11), 3543–3549.

Koura, K. (1998). "Monte carlo direct simulation of rotational relaxation of nitrogen through high total temperature shock waves using classical trajectory calculations." *Physics of Fluids (1994–present)* **10**(10), 2689–2691.

Koura, K., and Matsumoto, H. (1991). "Variable soft sphere molecular model for inverse-power-law or Lennard-Jones potential." *Physics of Fluids A: Fluid Dynamics (1989–1993)* **3**(10), 2459–2465.

Koura, K., and Matsumoto, H. (1992). "Variable soft sphere molecular model for air species." *Physics of Fluids A: Fluid Dynamics (1989–1993)* **4**(5), 1083–1085.

Kunc, J., Hash, D., and Hassan, H. (1995). "The ghs interaction model for strong attractive potentials." *Physics of Fluids (1994–present)* **7**(5), 1173–1175.

Landau, L., and Teller, E. (1936). "Theory of sound dispersion." *Physikalische Zeitschrift der Sowjet-Union* **10**, 34.

Lin, W., Varga, Z., Song, G., Paukku, Y., and Truhlar, D. G. (2016). "Global triplet potential energy surfaces for the N_2 (x1σ)+ O (3p) no (x2π)+ N (4s) reaction." *The Journal of Chemical Physics* **144**(2), 024309.

Lordi, J. A., and Mates, R. E. (1970). "Rotational relaxation in nonpolar diatomic gases." *Physics of Fluids (1958–1988)* **13**(2), 291–308.

Lumpkin III, F. E., Haas, B. L., and Boyd, I. D. (1991). "Resolution of differences between collision number definitions in particle and continuum simulations." *Physics of Fluids A: Fluid Dynamics (1989–1993)* **3**(9), 2282–2284.

Magin, T. E., and Degrez, G. (2004*a*). "Transport algorithms for partially ionized and unmagnetized plasmas." *Journal of Computational Physics* **198**(2), 424–449.

Magin, T. E., and Degrez, G. (2004*b*). "Transport properties of partially ionized and unmagnetized plasmas." *Physical Review E* **70**(4), 046412.

Matsumoto, H., and Koura, K. (1991). "Comparison of velocity distribution functions in an argon shock wave between experiments and Monte Carlo calculations for Lennard-Jones Potential." *Physics of Fluids A: Fluid Dynamics (1989–1993)* **3**(12), 3038–3045.

Millikan, R., and White, D. (1963). "Systematics of vibrational relaxation." *Journal of Chemical Physics* **39**, 3209–3213.

Nompelis, I., and Schwartzentruber, T. (2013). "Strategies for parallelization of the dsmc method. In '51st AIAA Aerospace Sciences Meeting." Vol. AIAA, Paper 2013 1204, Grapevine, TX.

Norman, P., Valentini, P., and Schwartzentruber, T. (2013). "Gpu-accelerated classical trajectory calculation direct simulation Monte Carlo applied to shock waves." *Journal of Computational Physics* **247**, 153–167.

Nyeland, C., and Billing, G. D. (1988). "Transport coefficients of diatomic gases: Internal-state analysis for rotational and vibrational degrees of freedom." *The Journal of Physical Chemistry* **92**(7), 1752–1755.

Panesi, M., Jaffe, R. L., Schwenke, D. W., and Magin, T. E. (2013). "Rovibrational internal energy transfer and dissociation of N_2 (1σg+)- N (4su) system in hypersonic flows." *The Journal of Chemical Physics* **138**(4), 044312.

Park, C. (1990). *Nonequilibrium Hypersonic Aerothermodynamics*. Wiley, New York.

Park, C. (1993). "Review of chemical-kinetic problems of future NASA missions. I. Earth entries." *Journal of Thermophysics and Heat Transfer* **7**(3), 385–398.

Parker, J. (1959). "Rotational and vibrational relaxation in diatomic gases." *Physics of Fluids* **2**, 449–462.

Paukku, Y., Yang, K. R., Varga, Z., and Truhlar, D. G. (2013). "Global ab initio ground-state potential energy surface of N_4." *The Journal of Chemical Physics* **139**(4), 044309.

Pfeiffer, M., Nizenkov P., Mirza, A., and Fasoulas, S. (2016). "Direct simulation Monte Carlo modeling of relaxation processes in polyatomic gases." *Physics of Fluids (1994–present)*, **28**(2), 027103.

Poovathingal, S., Schwartzentruber T. E., Murray, V., and Minton, T. K. (2015). Molecular simulations of surface ablation using reaction probabilities from molecular beam experiments and realistic microstructure. AIAA Paper 1449.

Poovathingal, S., Schwartzentruber, T. E., Murray, V. J., and Minton, T. K. (2016). "Molecular simulation of carbon ablation using beam experiments and resolved microstructure." *AIAA Journal* **54**(1), 1–12.

Present, R. (1958). *Kinetic Theory of Gases*. McGraw-Hill, New York.

Ramshaw, J. D., and Chang, C. (1996). "Friction-weighted self-consistent effective binary diffusion approximation." *Journal of Non-Equilibrium Thermodynamics* **21**(3), 223–232.

Schwartzentruber, T. E., and Boyd, I. D. (2006). "A hybrid particle-continuum method applied to shock waves." *Journal of Computational Physics* **215**(2), 402–416.

Schwartzentruber, T. E., Scalabrin, L. C., and Boyd, I. D. (2007). "A modular particle–continuum numerical method for hypersonic non-equilibrium gas flows." *Journal of Computational Physics* **225**(1), 1159–1174.

Schwartzentruber, T. E., Scalabrin, L. C., and Boyd, I. D. (2008*a*). "Hybrid particle-continuum simulations of hypersonic flow over a hollow-cylinder-flare geometry." *AIAA Journal* **46**(8), 2086–2095.

Schwartzentruber, T. E., Scalabrin, L. C., and Boyd, I. D. (2008*b*). "Hybrid particle-continuum simulations of nonequilibrium hypersonic blunt-body flowfields." *Journal of Thermophysics and Heat Transfer* **22**(1), 29–37.

Schwartzentruber, T. E., Scalabrin, L. C., and Boyd, I. D. (2008*c*). "Multiscale particle-continuum simulations of hypersonic flow over a planetary probe." *Journal of Spacecraft and Rockets* **45**(6), 1196–1206.

Stephani, K., Goldstein, D., and Varghese, P. (2013). "A non-equilibrium surface reservoir approach for hybrid dsmc/Navier–Stokes particle generation." *Journal of Computational Physics* **232**(1), 468–481.

Tysanner, M. W., and Garcia, A. L. (2004). "Measurement bias of fluid velocity in molecular simulations." *Journal of Computational Physics* **196**(1), 173–183.

Tysanner, M. W., and Garcia, A. L. (2005). "Non-equilibrium behaviour of equilibrium reservoirs in molecular simulations." *International Journal for Numerical Methods in Fluids* **48**(12), 1337–1349.

Valentini, P., and Schwartzentruber, T. E. (2009*a*). "A combined event-driven/time-driven molecular dynamics algorithm for the simulation of shock waves in rarefied gases." *Journal of Computational Physics* **228**(23), 8766–8778.

Valentini, P., and Schwartzentruber, T. E. (2009*b*). "Large-scale molecular dynamics simulations of normal shock waves in dilute argon." *Physics of Fluids (1994–present)* **21**(6), 066101.

Valentini, P., Schwartzentruber, T. E., Bender, J. D., and Candler, G. V. (2016). "Dynamics of nitrogen dissociation from direct molecular simulation", *Physical Review Fluids* **1**, 043402.

Valentini, P., Schwartzentruber, T. E., Bender, J. D., Nompelis, I., and Candler, G. V. (2015). "Direct molecular simulation of nitrogen dissociation based on an ab initio potential energy surface'." *Physics of Fluids (1994–present)* **27**(8), 086102.

Valentini, P., Tump, P. A., Zhang, C., and Schwartzentruber, T. E. (2013). "Molecular dynamics simulations of shock waves in mixtures of noble gases." *Journal of Thermophysics and Heat Transfer* **27**(2), 226–234.

Valentini, P., Zhang, C., and Schwartzentruber, T. E. (2012). "Molecular dynamics simulation of rotational relaxation in nitrogen: Implications for rotational collision number models." *Physics of Fluids (1994–present)* **24**(10), 106101.

Varga, Z., Meana-Pañeda, R., Song, G., Paukku, Y., and Truhlar, D. G. (2016). "Potential energy surface of triplet N_2O_2." *The Journal of Chemical Physics* **144**(2), 024310.

Venkattraman, A., and Alexeenko, A. A. (2012). "Binary scattering model for Lennard-Jones potential: Transport coefficients and collision integrals for non-equilibrium gas flow simulations." *Physics of Fluids (1994–present)* **24**(2), 027101.

Vincenti, W., and Kruger, C. (1967). *Introduction to Physical Gas Dynamics*. Wiley, New York.

Wadsworth, D. C., and Wysong, I. J. (1997). "Vibrational favoring effect in dsmc dissociation models." *Physics of Fluids (1994–present)* **9**(12), 3873–3884.

Wagner, W. (1992). "A convergence proof for bird's direct simulation Monte Carlo method for the Boltzmann equation." *Journal of Statistical Physics* **66**(3–4), 1011–1044.

Wang, W.-L., and Boyd, I. D. (2003). "Predicting continuum breakdown in hypersonic viscous flows." *Physics of Fluids (1994–present)* **15**(1), 91–100.

Wray, K. L. (1962). "Shock-tube study of the coupling of the O_2–Ar rates of dissociation and vibrational relaxation." *The Journal of Chemical Physics* **37**(6), 1254–1263.

Wright, M. J., Bose, D., Palmer, G. E., and Levin, E. (2005). "Recommended collision integrals for transport property computations. Part 1: Air species." *AIAA Journal* **43**(12), 2558–2564.

Wright, M. J., Hwang, H. H., and Schwenke, D. W. (2007). "Recommended collision integrals for transport property computations, Part ii: Mars and Venus entries." *AIAA Journal* **45**(1), 281–288.

Wysong, I. J., and Wadsworth, D. C. (1998). "Assessment of direct simulation Monte Carlo phenomenological rotational relaxation models." *Physics of Fluids (1994–present)* **10**(11), 2983–2994.

Zhang, C., and Schwartzentruber, T. E. (2012). "Robust cut-cell algorithms for dsmc implementations employing multi-level Cartesian grids." *Computers and Fluids* **69**, 122–135.

Zhang, C., and Schwartzentruber, T. E. (2013). "Inelastic collision selection procedures for direct simulation Monte Carlo calculations of gas mixtures." *Physics of Fluids (1994–present)* **25**(10), 106105.

Zhang, C., Valentini, P., and Schwartzentruber, T. E. (2014). "Nonequilibrium-direction-dependent rotational energy model for use in continuum and stochastic molecular simulation." *AIAA Journal* **52**(3), 604–617.

Index